MW00836761

Defect Assessment for Integrity Management of Pipelines

Defect Assessment for Integrity Management of Pipelines

Y. Frank Cheng
University of Calgary
Calgary, Canada

Published by John Wiley & Sons, Inc., Hoboken, New Jersey.
Published simultaneously in Canada.

For general information on our other products and services or for technical support, please contact our Customer Care Department within the United States at (800) 762-2974, outside the United States at (317) 572-3993 or fax (317) 572-4002.

Wiley also publishes its books in a variety of electronic formats. Some content that appears in print may not be available in electronic formats. For more information about Wiley products, visit our web site at www.wiley.com.

Library of Congress Cataloging-in-Publication Data applied for:

Hardback ISBN: 9781119815396

Cover Design: Wiley
Cover Image: © Frank Cheng

Set in 9.5/12.5pt STIXTwoText by Straive, Pondicherry, India

To Jianshu and Winston

Contents

Preface

Defect assessment is an essential component which is integral to the integrity management program of pipelines. By processing and analyzing the data collected by in-line inspection (ILI) tools and from other sources such as historical records, operating control and monitoring system, and offline/aboveground inspections with various models, formulas and numerical algorisms, the defect assessment provides information about performance condition of the pipelines, including prediction of failure pressure, determination of fitness-for-service (FFS), and further, estimation of the remaining service life. The defect assessment also contributes to failure risk and reliability evaluation, and recommendations of proper measures and actions for pipeline failure mitigation and control.

The pipeline defect assessment technique has evolved in the past several decades, experiencing development of three levels of technical progress, i.e., Levels I, II, and III methods. Targeting determination of the stress and strain distributions at the defects and evaluation of pipeline FFS and failure pressure, the three levels of methods distinguish themselves mainly by improved accuracy of the defect sizing, inclusion of the interaction of multiple defects, and solving highly nonlinear problems in defect assessment on pipelines, respectively. Nowadays, Levels I and II methods have been extensively used in industry for improved integrity management, while the Level III method, which relies on finite element (FE) modeling and analysis for solving nonlinearity at pipeline defects, has found its applications mainly in engineering research community due to background knowledge requirement and computational complexity.

In the last decade, my research group has been focusing on Level III defect assessment on pipelines, developing various FE-based models and methods to determine the stress and strain distributions at corrosion defects based on accurate definition of the defect dimension under pipeline operating conditions, evaluating their effect on FFS of the pipelines and predicting the failure pressure. Moreover, the assessment targets not only a single corrosion defect on pipelines, but also multiple defects between which a mutual interaction may exist to further degrade the pipeline integrity. For various orientations the corrosion defects are aligned with each other, critical spacings between them are defined to determine if an interaction exists so that they should be assessed either together or separately.

The major contribution of my group to development of the Level III defect assessment technique is, based on the mechano-electrochemical interaction theoretical concept I proposed in 2013, to integrate the mechanical force with electrochemical force, developing a multi-physics field coupling model for defect assessment while considering the dynamic nature of corrosion defects in actual service environments. Prior to that, the corrosion defects have been usually treated as metal-loss features, while ignoring the dynamic process of defect growth due to corrosion reactions. This is regarded as "revolutionary" to pipeline defect assessment techniques. The novel Level III defect assessment method, at the first time of its kind, enables prediction of the rate of corrosion defect growth on pipelines under the synergism of mechanical and electrochemical forces, reproducing the reality and thus providing more accurate and reliable results.

In addition to corrosion defects, the mechano-electrochemical interaction integrated Level III assessment method has also expanded its use to other types of surface anomalies such as dents, buckles and wrinkles, as well as combinations of different types of defects. Moreover, the defect assessment applies on both straight pipes and pipeline elbows where the defects experience different mechanical and corrosion conditions. Criteria and methods are developed to evaluate the pipeline performance and predict burst failure.

The book starts with an overview of pipeline integrity management program in Chapter 1, where the basic principle, main components and methods, and design pathway of integrity management of pipelines are introduced. Various threats to degrade the pipeline integrity in the field are reviewed, and common ILI tools for detecting surface defects are summarized. Chapter 2 introduces the historical development of defect assessment techniques, while focusing on the principles, criteria, and applications of Levels I and II methods. Commentary remarks are given to analyze the limitations of the two levels of assessment method. In Chapter 3, the FE-based Level III defect assessment method is detailed in terms of the principles, criteria, and applications for pipeline FFS determination and failure pressure prediction. The assessment applies for both single and multiple corrosion defects, straight pipes and pipeline elbows, internal and external defects, and the defects on pipelines under vibration induced by running of ILI tools. Chapters 4 and 5 contain the important innovation of

Level III defect assessment method by integrating mechanical and electrochemical forces at corrosion defects, considering the synergism of stress/strain and electrochemical corrosion and its effect on pipeline performance and failure during service. The fundamentals of mechano-electrochemical interaction for pipeline corrosion are imparted in Chapter 4, followed by development of a multi-physics field coupling model for defect assessment. The defects are either regularly shaped or with complex shapes encountered in the field, where a definitive method is proposed to accurately size the defects. Particularly, when a corrosion defect is present on a pipe in suspension under soil-erosive conditions, additional mechanical factors such as surface loading and a non-uniform stress distribution in the suspended pipe segment are considered and modeled. Moreover, the defect growth rate on pipelines is modeled and predicted under both mechanical stress and electrochemical corrosion effects, and the results help estimate the remaining life of corroded pipelines in the field. In addition to single corrosion defect, multiple corrosion defects where a mutual interaction exists are modeled with the novel Level III assessment method. The adjacent corrosion defects are oriented either longitudinally, circumferentially, or overlapped with each other. Critical spacings between them are defined to determine if an interaction exists to degrade the pipeline integrity. Furthermore, a new criterion based on anodic current density, i.e., corrosion rate, at the adjacent area between the corrosion defects is proposed and validated to evaluate the defect interaction. In Chapter 6, dent assessment on pipelines is included, where the dent assessment principle, uniqueness, challenge, and failure criteria are reviewed. A new method to define the critical strain at a dent is introduced. In addition to dent assessment, the combinations between a dent and a gouge, corrosion, and a crack are modeled and assessed on pipelines. Finally, assessment of buckles on pipelines and buckling failure analysis by FE-based models are included in Chapter 7. Buckling failure of pipelines usually occurs under pipe–soil interactions, where an axial compressive load or bending moment is generated on the pipelines. The critical compressive force or the critical bending moment is defined for pipelines containing a dent or corrosion defect where buckling failure potentially occurs, while considering the parametric effects such as pipe dimension, defect size, internal pressure, and steel properties. A new method for prediction of burst capacity of corroded pipelines under a combined bending moment and axial compressive load is proposed.

I acknowledge numerous fruitful discussions I have had with many industry partners and academic colleagues. I am indebted to the dedicated and unfailing assistance and contributions provided by the students and postdoctoral fellows that I have the pleasure to supervise to study defect assessment on pipelines in my research group. They are Drs. Luyao Xu, Jialin Sun, Jian Zhao, Zhuwu Zhang, Yi Shuai, and Guojin Qin. Thank you very much for your hard work and research accomplishments!

Research grants from the Canada Research Chairs Program, Natural Science and Engineering Research Council of Canada (NSERC), Mitacs, and many industrial organizations have created the favorable conditions that helped to support an active research environment that has both contributed to and enabled the writing of this book. I am grateful and indebted to the assistance provided by these programs, agencies, and organizations, as well as the University of Calgary's Schulich School of Engineering and the Department of Mechanical & Manufacturing Engineering.

Finally, I thank my wife, Jianshu, and my son, Winston, who have provided encouragements and have supported the creation of this book.

Y. Frank Cheng
Calgary, Alberta, Canada

List of Abbreviations and Symbols

2D	2-dimensional
3D	3-dimensional
AC	Alternating current
ACVG	Alternating current voltage gradient
API	American Petroleum Institute
ASME	American Society of Mechanical Engineering
BS	British Standard
BS&W	Basic sediments and water
CEPA	Canadian Energy Pipeline Association
CFR	Code of federal regulations
CIS	Close interval survey
CO_2	Carbon dioxide
CP	Cathodic protection
CSA	Canadian Standardization Association
CSE	Copper sulfate electrode
CTOD	Crack tip opening displacement
DC	Direct current
DCVG	Direct current voltage gradient
DFDI	Ductile fracture damage index
DNV	Det Norske Veritas
DSAW	Double submerged arc-welded
EAC	Environmentally assisted cracking
ECA	Engineering critical assessment
ECDA	External corrosion direct assessment
EIS	Electrochemical impedance spectroscopy
EMAT	Electromagnetic acoustic transducer
EPRG	European Pipeline Research Group
ERW	Electric resistance-welded

FAD	Failure assessment diagram
FE	Finite element
FERC	Federal Energy Regulatory Commission
FFS	Fitness-for-service
H	Hydrogen atom
H_2	Hydrogen molecule
H_2S	Hydrogen sulfide
HAZ	Heat-affected zone
HE	Hydrogen embrittlement
HEDE	Hydrogen-enhanced decohesion
HELP	Hydrogen-enhanced local plasticity
HIB	Hydrogen-induced blistering
HIC	Hydrogen-induced cracking
HVAC	High voltage alternating current
HVDC	High voltage direct current
ICCP	Impressed current cathodic protection
ICDA	Internal corrosion direct assessment
ILI	In-line inspection
LOF	Lack of fusion
LOP	Lack of penetration
MAOP	Maximum allowable operating pressure
M-C	Mechanical–chemical
M-E	Mechano-electrochemical
MFL	Magnetic flux leakage
MIC	Microbiologically influenced corrosion
MnS	Manganese sulfide
NACE	National Association of Corrosion Engineers
NDT	Non-destructive testing
NEB	National Energy Board
NSC	Net Section Collapse
PDCA	Plan-Do-Check-Act
PE	Polyethylene
PHMSA	Pipeline and Hazardous Materials Safety Administration
ROW	Right-of-way
RP	Recommended practice
RPA	Rectangular parabola area
R-O	Ramberg-Osgood
ROW	Right-of-way
SBD	Strain-based design
SCADA	Supervisory control and data acquisition

SCC Stress corrosion cracking
SCCDA SCC direct assessment
SCE Saturated calomel electrode
SCF Stress concentration factor
SF Safety factor
SHE Standard hydrogen electrode
SL Suspension length
SLD Strain limit damage
SME Subject matter expert
SMYS Specified minimum yield strength
S-N Stress–Number of cycles
SP Shape parameter of a dent
SRB Sulfate-reducing bacteria
SSC Sulfide stress cracking
UKOPA UK Onshore Pipeline Association
UT Ultrasonic tool
XFEM Extended finite element method
a Activity
$\tilde{a}$ M-C activity
$\bar{a}$ Electrochemical activity
$\bar{\bar{a}}$ M-E activity
$2a$ Length of the secondary axis of a semi-ellipsoidal corrosion defect
b_a Anodic Tafel slope
b_c Cathodic Tafel slope
c_1 Length of the primary semi-axis of the bigger semi-ellipsoidal corrosion defect
c_2 Length of the primary semi-axis of the smaller semi-ellipsoidal corrosion defect
$2c$ Length of the primary axis of a semi-ellipsoidal corrosion defect
C_1 A constant obtained through burst test on a non-indented pipe
C_2 Elongation rate of pipe steel measured in uniaxial tensile testing
A_0 Cross-sectional area of a pipe before corrosion occurs
A Area
A_{eff} Effective area
A_{P} A coefficient depending on dent geometry
B_{P} A coefficient depending on pipe dimension
c Curvature coefficient of a pipe elbow
C_{P} A coefficient depending on steel properties
d Depth

d_1	Depth of the top defect for two overlapped corrosion defects
d_2	Depth of the bottom defect for two overlapped corrosion defects
d_{ave}	Average defect depth
d_{clus}	Depth of the defect cluster
d_e	Equivalent depth of multiple defects
d_g	Maximum depth of a gouge
d_i	Maximum depth of the composite defect
d_{max}	Maximum depth of an irregularly shaped corrosion defect
D	Pipe outer diameter
D_{e0}	Simplified DFDI value before spring-back
D_{eform}	Damage resulted from deforming
$D_{e,k}$	Damage during the *k*th load increment
D_{em}	Maximum DFDI at a dent
D_{et}	An indicator of the limit state for a pipeline to carry no further load
D_{max}	Maximum pipe outer diameter
D_{min}	Minimum pipe outer diameter
E	Young's modulus
F	Faraday's constant
F_c	Critical buckling load
F_{comp}	Compressive force
F_{ref}	Reference buckling load
f	Frequency
f_1	A factor representing the difference of strains after and before spring-back of an unconstrained dent
h	Final depth of the dent after removal of the indenter
h_o	Initial displacement of the indenter applied on a pipe
i_a	Anodic reaction current density
$i_{0,\ a}$	Anodic exchange current density
$i_a^{\ e}$	Anodic current density of an elastically stressed steel in a corrosive environment
$i_a^{\ f}$	Anodic current density of a plastically stressed steel in a corrosive environment
$i^a_{far-defects}$	Anodic current density of the steel pipe far away from the corrosion defects
$i^a_{mid-defects}$	Anodic current density at the middle of two adjacent corrosion defects
i_c	Cathodic reaction current density
$i_{0,c}$	Cathodic exchange current density

I	An integral value used as the damage indicator
k	An index for either liquid or solid
K	R-O material parameter, a constant
K_1, K_2, ...	Curvature of each node in a pipe during buckling modeling
$K_{Buckling}$	Pipe curvature at a local buckling position
K_d	Stress concentration factor at a dent
K_F	Fatigue stress concentration factor
K_r	Toughness ratio
L	Length
L_1	A half of the length of the top defect for two overlapped defects
L_{clus}	Length of the defect cluster
L_e	Equivalent length of multiple defects
L_{eff}	Effective length
L_g	Length of a gouge
L_i	Total length of the composite defect
L_p	Length of a pipe segment
L_r^p	Load ratio
M	Folias factor
M_c	Critical buckling moment
M_o	Critical elastic buckling moment
n	R-O material parameter, a constant
N	Fatigue life in cycles
N_0	Initial density of dislocations prior to plastic deformation
N_1, N_2, ...	Nodes in modeling of curvature of a pipe during buckling
P	Internal pressure
P_1	Pressure at initial stage
P_2	Pressure at end stage
P_0	Tresca strength solution
$P(0)$	Initial pressure capacity
P_b	Burst pressure
P_e	Minimum external hydrostatic pressure
P_F	Failure pressure
$P_{F,add}$	Failure pressure of a pipe when additional internal defects are included
P_{FE}	Burst pressures of a defect-containing elbow predicted by FE model
P_{GM}	Burst pressure of a defect-free elbow
P_i	Maximum design internal pressure
P_{max}	Upper limit burst pressure

P_{min}	Lower limit burst pressure
$P_{multiple}$	Failure pressure of a pipeline containing multiple corrosion defects
$P_{overlapped}$	Failure pressure of a pipeline containing overlapped corrosion defects
P_{single}	Failure pressure of a pipeline containing a single corrosion defect
P_y	Critical internal pressure when pipe steel yields
Q	Length correction factor
Q_k	A general source term
r_{cc}	Ratio of the lengths of primary axis of the smaller semi-ellipsoidal defect to that of the bigger semi-ellipsoidal corrosion defect in a double ellipsoidal defect
R	Ideal gas constant
R_0	Initial pipe surface radius
R_1	External surface radius of curvature in the transverse plane through a dent
R_2	External surface radius of curvature in the longitudinal plane through a dent
R_b	Bending radius of elbow
R_c	Stress ratio during cyclic loading
R_d	Surface radius of the curvature at a dent
R_p	Pipe outer radius
R_r	Outer radius of sealing cup of the ILI tool
S_C	Circumferential spacing between two adjacent corrosion defects
S_C^{Lim}	Limiting circumferential spacing between two adjacent corrosion defects
S_L	Longitudinal spacing between two adjacent corrosion defects
S_L^{Lim}	Limiting longitudinal spacing between two adjacent corrosion defects
$S_{L,ext}^{Lim}$	Limiting longitudinal spacing between external defects
$S_{L,int}^{Lim}$	Limiting longitudinal spacing in the presence of both external and internal defects
S_{Li}	Longitudinal spacing between adjacent defect projections
t	Pipe wall thickness
T	Temperature
u	Profile functions in the longitudinal direction of a pipe
v	Profile functions in the circumferential direction of a pipe
V	Volume
V_o	Initial volume
V_m	Molar volume

w	Pipe wall deflection in the radial direction of a pipe
W	Width
W_{clus}	Width of the defect cluster
z	Chemical valence or charge number
α	A coefficient
β	Width angle of a defect
β_{e}	Equivalent width angle of multiple defects
σ	Stress
$\sigma_1, \sigma_2, \sigma_3$	Principal stresses of a pipeline
σ_{a}	Alternating stress
σ_{e}	Effective stress
σ_{eq}	Equivalent stress
σ_{exp}	Experimental stress function
σ_{F}	Failure stress
σ_{FS}	Fatigue strength
σ_{flow}	Flow stress
σ_{k}	Conductivity
σ_{m}	Mean stress
σ_{max}	Maximum stress
σ_{min}	Minimum stress
σ_{Mises}	von Mises stress
σ_{Tresca}	Tresca yield stress
σ_{u}	Ultimate tensile strength
σ_{y}	Yield strength
σ_{yhard}	Stress enhancement hardening factor during plastic deformation
σ_{θ}	Hoop stress
σ_{z}	Axial stress
ε	Strain
$\varepsilon_{\mathrm{as}}$	Strain at the dent apex after spring-back
$\varepsilon_{\mathrm{ini}}$	Strain at the dent apex before spring-back
ε_0	True strain to failure
ε_1	Bending strain in the circumferential direction
ε_2	Bending strain in the longitudinal direction
ε_3	Membrane strain in the longitudinal direction
$\varepsilon_{\mathrm{apex}}$	Equivalent strain at the dent apex
$\varepsilon_{\mathrm{crit}}$	Critical strain to initiate cracks
ε_i	Equivalent strain on the inside surface of a pipe
ε_o	Equivalent strain on the outside surface of a pipe
$\varepsilon_{\mathrm{eff}}$	Effective strain
$\varepsilon_{\mathrm{eq}}$	Equivalent strain

$\varepsilon_{\mathrm{lim}}$ Strain limit
$\varepsilon_{\mathrm{max}}$ Maximum equivalent strain
ε_{p} Plastic strain
ε_{x} Strain in the axial direction of a pipe
ε_{y} Strain in the circumferential direction of a pipe
ε_{z} Strain in the radial direction of a pipe
ε^{b} Bending strain
ε^{m} Membrane strain
γ_{xy} Shear strain
Δ Pipe ovality
$\Delta\varepsilon$ Cyclic strain range
θ Angular position of a corrosion defect on pipe elbow
θ_{b} Orientation of bending load
θ_{incl} Inclination angle of corrosion defect relative to the axial direction of a pipeline
φ Electrical potential
φ_{eq} Equilibrium electrode potential
$\varphi_{a,eq}$ Equilibrium potential of anodic reaction
$\varphi^{0}_{a,eq}$ Standard equilibrium potential of anodic reaction
$\varphi_{c,eq}$ Equilibrium potential of cathodic reaction
$\varphi^{0}_{c,eq}$ Standard equilibrium potential of cathodic reaction
Φ Axial routing angle of the pipe
μ Chemical potential
μ_{0} Chemical potential of solid in a standard state
μ_{0}' Standard chemical potential of solid considering the M-C interaction
$\Delta\mu$ Chemical potential difference
ΔP Pressure difference
$\Delta\varphi^{e}_{a,eq}$ Change of electrochemical anodic equilibrium potential under an elastic stress
$\Delta\varphi^{p}_{a,eq}$ Change of electrochemical anodic equilibrium potential under a plastic stress
χ Compressibility coefficient of solid
υ An orientation-dependent factor
η_{a} Anodic activation overpotential
η_{c} Cathodic activation overpotential
ν Poisson's ratio

1

Pipeline Integrity Management

1.1 Introduction

Pipelines provide an effective and efficient means to transport oil, natural gas, and petrochemical products across provinces, countries, and even continents, meeting continuously increasing energy demands. The oil and gas transmission pipelines around the world are up to 3,500,000 km, with about 32,000 km of new pipelines constructed each year [Hopkins, 2007]. The total length can be multiplied many times if gathering and distribution pipelines are included. The world's energy consumption is predicted to increase by 71% from 2003 to 2030, with fossil fuels continuing to supply much of the energy used worldwide [Department of Energy, 2006]. It is thus expected that pipeline construction and operation activities will continue growing. In recent years, with great efforts made to combat climate change and achieve the net-zero emission target globally, pipelines have been used for safe, economical, and highly efficient transportation of "green" energies and fuels such as hydrogen gas, hydrogen/natural gas blends, biofuels, and supercritical carbon dioxide (CO_2) [Ogden et al., 2018; Reuß et al., 2019; Cerniauskas et al., 2020]. The new energy pipelines are expected to experience rapid development in the next decade.

Energy transportation by pipelines is safe. Statistics showed that, in the United States, 1.7 fatalities to operators, personnel, and the public per year were caused by oil and gas pipeline accidents. As a comparison, transportation of oil and gas by rail and truck resulted in 2.4 and 10.2 fatalities per year, respectively [Hansen and Dursteler, 2017]. Pipeline transportation of hydrocarbon products was 4.5 times safer than rail on a like-for-like basis from analysis of the North American data [Green and Jackson, 2015].

Defect Assessment for Integrity Management of Pipelines, First Edition. Y. Frank Cheng.

The integrity of pipelines can be adversely affected by many factors in the field, such as corrosion, stress corrosion cracking (SCC), fatigue, mechanical damage, stray current, materials and manufacturing faults, equipment and component failures, geotechnical factors, incorrect operation, and external interference such as excavation [Godin, 2014; Canadian Energy Pipeline Association, 2015]. Although occurring occasionally, pipeline failures can result in energy loss, environmental and ecological impact, and, sometimes, death [Cheng, 2016]. Thus, pipeline incidents usually attract wide attention from news media and the public. One of the most widely reported pipeline incidents is the rupture and release of Enbridge's oil pipeline in Marshall, Michigan, on July 25, 2010, which resulted in the largest inland oil spill and one of the costliest spills in US history [National Transportation Safety Board, 2012]. Following the spill, the volatile hydrocarbon diluents evaporated, leaving the heavier bitumen to sink in the water column. Thirty-five miles of the Kalamazoo River were closed for clean-up until June 2012.

Safety is the top priority for pipeline operators. The concept of Integrity First has been accepted by pipeline companies and become integral to corporation culture [Canadian Energy Pipeline Association, 2013]. In today's pipeline industry, an integrity management program has been developed and implemented to ensure the safety, reliability, and longevity of the pipeline system by mitigating and preventing pipeline failure, achieving the goal of zero pipeline incidents. Particularly, defect assessment is a critical component of a well-developed pipeline integrity management program. Development of models and methods for accurate and reliable assessment of various defects, such as corrosion, cracks, dents, and other anomalies, detected on pipelines is critical to determination of the pipeline fitness-for-service (FFS), prediction of failure pressure and estimation of the remaining service life of the pipelines [Qin and Cheng, 2021].

1.2 Overview of Threats to Pipeline Integrity

During long-term service of pipelines in the field, the integrity of the pipeline system can be compromised by multiple types of threats or their combinations. According to Canadian Energy Pipeline Association (CEPA), metal loss including corrosion, cracking, and external inference remains the leading cause of incidents occurring on CEPA member operators' oil/gas transmission pipelines [Canadian Energy Pipeline Association, 2021]. Collectively, these accounted for 82% of the total incidents over the period from 2016 to 2020, as seen in Figure 1.1. Other factors affecting the pipeline

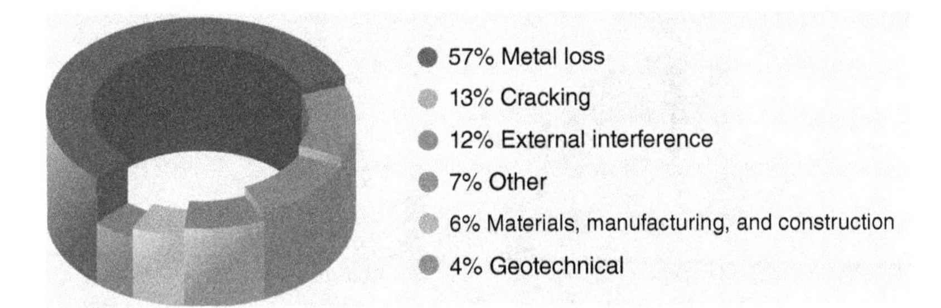

Figure 1.1 Causes of rights-of-way incidents 2016–2020 occurring on CEPA member operators' pipelines. *Source:* From Canadian Energy Pipeline Association [2021].

integrity included geohazards, external interference, and some unidentified reasons.

In the United States, the leading cause of accidents impacting people or the environment on liquid pipeline systems is corrosion according to the statistics of the Pipeline and Hazardous Materials Safety Administration (PHMSA). The second and third leading causes are equipment failure and material failure of pipe or weld, respectively. These three leading causes accounted for 65% of accidents since 2010 [Pipeline and Hazardous Materials Safety Administration, 2020]. Other factors included excavation damage, incorrect operation, natural force, and others. Similarly, the main causes resulting in onshore gas pipeline failures in the period of 2005–2020 included corrosion, equipment failure, material failure of pipe or weld, excavation damage, natural force, and others [Pipeline and Hazardous Materials Safety Administration, 2021]. Figures 1.2 and 1.3 show the statistical analysis of total number of accidents and their causes for PHMSA-regulated liquid and gas pipelines, respectively, in the United States [Pipeline and Hazardous Materials Safety Administration, 2020; 2021].

1.2.1 Corrosion

Corrosion has been recognized as one of the primary mechanisms causing pipeline failures in North America. As stated, corrosion, as the most important reason causing failures of transmission pipelines in Canada, was responsible for 46% of all reported failure incidents from 2015 to 2019 [Canadian Energy Pipeline Association, 2021]. In a comparative analysis of pipeline performance issued by the National Energy Board (NEB) in Canada, the primary cause of ruptures on NEB-regulated pipelines between 1991 and 2009 was corrosion-related cracking (38%) and metal loss (27%) [National

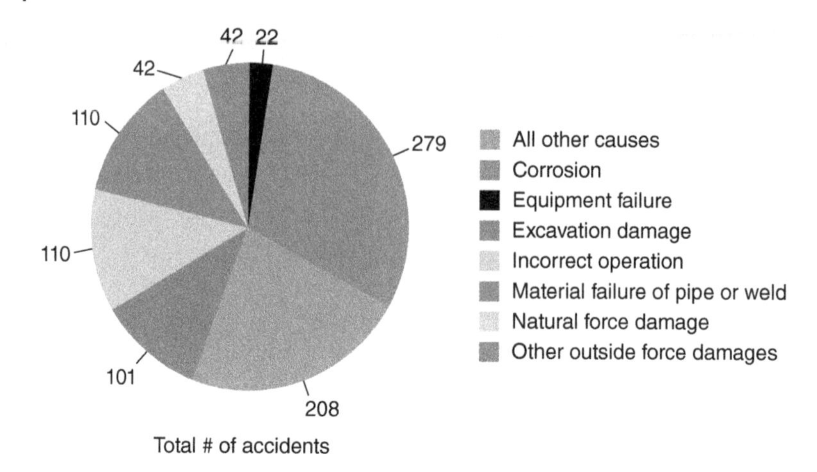

Figure 1.2 Statistical analysis of total number of accidents and their causes for PHMSA-regulated liquid pipelines in the United States from 2010 to 2019. *Source:* From Pipeline and Hazardous Materials Safety Administration [2020].

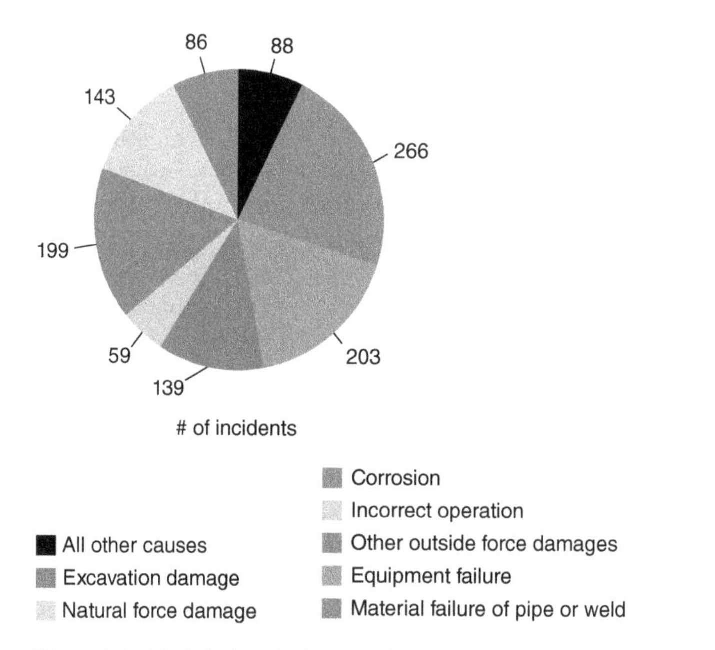

Figure 1.3 Statistical analysis of total number of accidents and their causes for PHMSA-regulated gas pipelines in the United States from 2005 to 2020. *Source:* From Pipeline and Hazardous Materials Safety Administration [2021].

Energy Board, 2011], where the cracking included hydrogen-induced cracking (HIC) and mechanical damage delayed cracking, SCC, and corrosion fatigue, and the metal loss included both internal and external corrosion, as well as scratches. For both liquid and gas pipelines operated in the United States, data from PHMSA showed that corrosion was the first leading cause of reported accidents, representing 30.5% (2010–2019) and 22.5% (2005–2020), respectively, of cases [Pipeline and Hazardous Materials Safety Administration, 2020; 2021].

For underground pipelines, corrosion can occur both externally and internally. The strategy for external corrosion control is application of high-performance coatings combined with cathodic protection (CP), an electrochemical technique mitigating and even stopping steel corrosion by cathodically polarizing the steel structure from its corrosion potential in the corrosive environment to a more negative potential. In an impressed current CP (ICCP) system, the cathodic polarization of the steel structure is realized by provision of electrons from an external power supply, i.e., a rectifier, to the structure. Generally, the coatings provide the first line of defense to protect the pipelines from environmental attack such as corrosion. When the coatings degrade or fail, the CP can provide further protection at the coating failures. The demand for CP current, i.e., electrons flowing from the rectifier to the steel structure, depends on coating performance. When the coatings are intact and provide excellent protection to the substrate steel, the CP current demand is zero. As the coatings degrade, the CP current demand increases. After the coatings completely lose the protective ability, e.g., the coatings are missed over an extensive area from the pipeline, the CP demand reaches a maximum level, and the pipeline corrosion is controlled by CP only [Cheng and Norsworthy, 2017].

The combination of coatings and CP technique does not always work properly for corrosion control on pipelines. The CP current can become shielding from reaching pipe steel in corrosive environments and lose the protective ability when the coatings fail. Three typical scenarios have been identified where the CP current is shielded while the coatings degrade and fail, as shown in Figure 1.4 [Cheng and Norsworthy, 2017]. First, a coating, primarily a polymeric coating such as epoxy or a polyethylene (PE)-based product, contains pinholes, defects, or breakages, which are usually introduced during coating manufacturing and pipeline construction, while the coating adhesion to the steel substrate is still maintained. These local features can be penetrated by water, chemical species, and gases during service in the underground environment. As a result, a corrosive environment is generated at the bottom of the features to cause corrosion of the pipe steel. Owing to geometric limitation (i.e., a small aspect ratio of the width to the

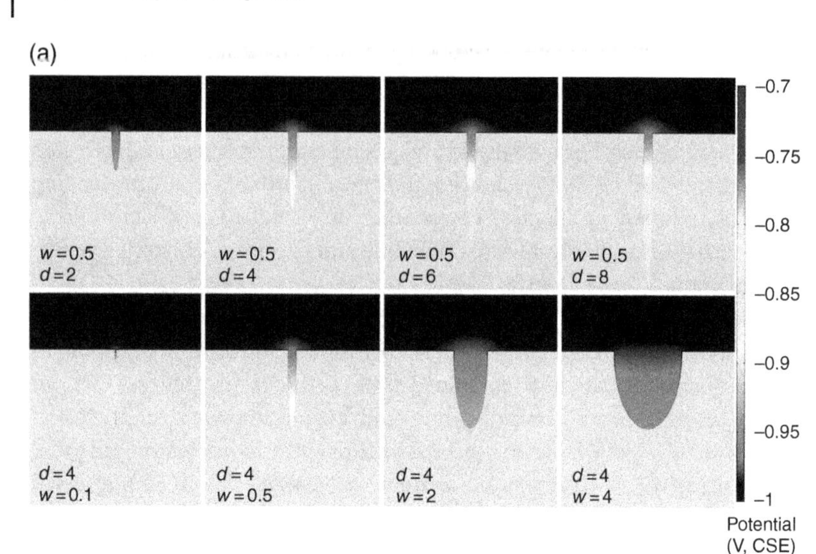

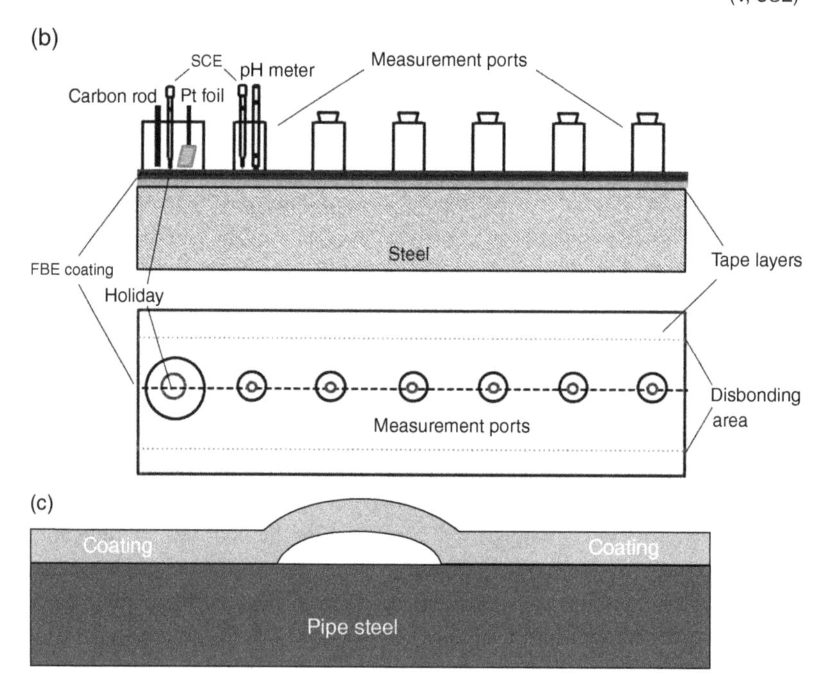

Figure 1.4 Scenarios identified on pipelines where the CP current is shielded from reaching pipe steel while the coatings degrade and fail (a) A pinhole (*d*: depth, *w*: width) contained in the coating while the adhesion to the steel substrate is still maintained. (b) A coating disbonded at a holiday. (c) A defect-free, impermeable coating membrane, disbonded from the pipe steel. *Source:* (a) From Xu and Cheng [2014] / with permission of Elsevier; (b) from Kuang and Cheng [2015b].

depth of the features) or deposit of corrosion products inside the features, the CP current is shielded from reaching the bottom of the features for corrosion protection, as shown schematically in Figure 1.4a [Xu and Cheng, 2014]. Second, the coating is disbonded from the pipe steel at a holiday (or a defect). Although CP is applied, the disbonding crevice under the coating is shielded from the CP current owing to geometric limitation, i.e., a long, narrow disbonding crevice. Separated anode and cathode are generated where anodic dissolution (i.e., corrosion of the steel) occurs under the disbondment, especially at the disbonding bottom, and cathodic reaction (e.g., electrochemical reduction of dissolved oxygen or water) occurs at the holiday, which is under CP, as seen in Figure 1.4b [Kuang and Cheng, 2015a; 2015b]. Finally, the CP current is shielded from a defect-free coating membrane, which is usually composed of PE components, such as PE tape, when the coating is disbonded from the pipe steel due to a poor surface treatment of the steel or improper coating application (Figure 1.4c). Generally, the PE is highly resistant to permeation of water, chemicals, and gases under pipeline operating conditions. The intrinsic properties make the coating impermeable to the CP current [Fu and Cheng, 2011].

In addition to the factors related to coating properties and performance as mentioned above, dry soil environments with an extremely high resistivity can break the current flow circuit of the CP system, making the CP nonfunctional [Cheng and Norsworthy, 2017]. When water is trapped under disbonded coating, corrosion would occur in the absence of CP. Furthermore, microbiologically influenced corrosion (MIC) can happen on pipelines externally. In soil environments, many types of microorganisms that cause (usually accelerate) corrosion of metals, such as the anerobic sulfate-reducing bacteria (SRB), exist. It was found that, compared to the corrosion rate of 0.0473 mm/y in abiotic soils, the corrosion rate of pipeline steels could be up to 0.282 mm/y when SRB were contained in the soil [Liu and Cheng, 2017a]. An increased moisture content in the soils favored the growth of SRB, accelerating the steel MIC. In addition, the CP could facilitate bacterial attachment to pipe steels. When a layer of the so-called "biofilm" is formed on the steel surface, the effectiveness of CP for corrosion protection decreases [Liu and Cheng, 2017b]. The CP potential of −0.85 V (copper sulfate electrode, CSE) is not sufficient to fully protect the steels from MIC induced by SRB. While a negative shift of the CP potential to −1 V (CSE) can prevent uniform MIC, pitting corrosion still occurs under the biofilm. The external corrosion of pipelines can also be accelerated by alternating current (AC) or direct current (DC) interference, where the interference sources such as high voltage AC (HVAC) or high voltage DC (HVDC) power lines and AC-powered rail transit systems are collocated with buried pipelines [Cheng, 2021]. The AC and DC inferences can accelerate steel corrosion and, sometimes, initiate pitting corrosion. Moreover, the coating properties and performance become degraded in the presence of AC interference. The applied CP potential can be shifted from the design value toward either positive or negative direction, reducing corrosion protectiveness of the CP. With growing numbers of HVAC transmission lines collocated with the pipelines, nowadays, AC corrosion has been recognized as a serious threat to the integrity of pipelines [Canadian Energy Pipeline Association, 2014].

For long-distance transmission pipelines, internal corrosion is not regarded as a big threat to pipeline integrity although pipeline failure incidents attributable to internal corrosion have been reported. The quality of the fluids (i.e., oil and natural gas) carried in the transmission pipelines has been ensured through processing before they are transported by the pipelines. For example, according to US Federal Energy Regulatory Commission (FERC) regulations, the transported oils are subject to a limitation on basic sediments & water (BS&W) of less than 0.5 vol% [Federal Energy Regulatory Commission, 2011]. Corrosion of steels does not happen in oil unless water and wetted sands entrained in the oils are separated from the oily phase and settle down on the pipe floor, making the steel water wetted over an extended period [National Academy of Sciences, 2013]. Moreover, trace amounts of salts dissolving in water can increase its conductivity. A more aggressive environment is that a biofilm forms on the pipe wall surface, separating the underneath environment from the carried fluid. A microbiological community develops under the biofilm, which are usually composed of petroleum sludge, solid sands, water, chemicals, corrosion products, and microorganisms, to cause internal MIC [Lenhart et al., 2014; Liu and Cheng, 2018]. It is not uncommon that pipelines leak due to MIC-induced perforation [Cole and Marney, 2012]. Generally, pipeline operators conduct periodic pigging to clean out various deposits including biofilms. The pigging removes the environment that might otherwise cause corrosion. In-line inspection (ILI) tools allow the operators to collect information about the geometry and location of pipeline flaws and, further, to analyze the FFS of the corroded pipelines [Canadian Energy Pipeline Association, 2015]. The industry also depends on appropriate chemical treatment programs to add inhibitors and biocides to mitigate and control internal corrosion and MIC [Banff Pipeline Workshop, 2015]. Owing to both the strictly enforced BS&W limit in the pipelined fluids and periodic pigging, along with the appropriate chemical treatment program, internal corrosion does not occur as frequently as external corrosion on transmission pipelines, although there have been reports of pipeline-leaking incidents resulting from internal corrosion.

1.2.2 Environmentally Assisted Cracking

Environmentally assisted cracking (EAC) is a general term for fracture that occurs when a susceptible metal is under tensile stress in a corrosive environment. The EAC is different from a mechanical fracture in that corrosion reaction participates in (usually accelerates) the fracture process. In other words, the EAC occurs due to both mechanical stress and corrosion reaction that act simultaneously and usually synergistically. Generally, the EAC is categorized into three types of fracture, i.e., SCC, corrosion fatigue, and HIC.

1.2.2.1 Stress Corrosion Cracking

The SCC is defined as a cracking process of metals due to the combined and synergistic interaction of mechanical tensile stress and corrosion reactions [Jones, 1992]. Pipeline SCC occurs under a combination of factors from environmental (e.g., coatings and coating failure modes, CP shielding effect, soil chemistry and resistivity, temperature, and aeration), stress (i.e., internal pressure and its fluctuations, ground movement, hoop and longitudinal stresses, and bending stress), and materials (e.g., steel grade, mechanical properties, chemical composition, microstructure, metallurgical defects, and welding metallurgy) aspects [Cheng, 2013]. SCC used to be a major threat to integrity and safety of pipelines. According to NEB of Canada, approximately 38% of the primary causes for pipeline failures were due to cracking from 1991 to 2009 [National Energy Board, 2011]. Particularly, the SCC caused pipeline failures by 10–13%. Pipeline SCC has been categorized into two types, i.e., high-pH SCC and near neutral-pH SCC, based on the pH of electrolytes contacting the pipe steel. Thus, the pH refers to the pH value of the aqueous environment at the crack location rather than the soil pH [Cheng, 2013].

The environments to cause high pH SCC on pipelines are aerobic, concentrated carbonate–bicarbonate electrolytes with a pH range of 8–10 [National Energy Board, 1996]. The environments result from CP current penetrating the so-called permeable coatings, such as asphalt, to reach the pipe steel surface, cathodically polarizing the steels. Hydroxyl ions are produced by cathodic reduction of dissolved oxygen to elevate solution pH. The CO_2, which is from the soils by decay of organic matter, is dissolved in the electrolytes. When non-dissolvable deposits such as $Mg(OH)_2$ and $Ca(OH)_2$ are formed in the alkaline electrolytes and present on the coating, the CP current flow is blocked. As a result, the cathodic polarization applied on the pipe steels is stopped. During relaxation of the steel potential from the initial CP value, the steel may pass a potential window which is susceptible to SCC. It has been accepted that the high pH SCC of pipelines follows a dissolution-based mechanism, where corrosion at the crack tip dominates the crack propagation process [Parkins, 2000]. Stress corrosion cracks tend to propagate along grain boundaries due to preferential dissolution occurring locally, which is attributed to an enhanced electrochemical corrosion activity at the grain boundaries by factors such as impurity accumulation, dislocation pinning and lattice distortion [Cheng, 2013]. Thus, the high pH SCC follows an intergranular growth mode, often with small branches [National Energy Board, 1996]. The fracture surface normally exhibits a dark, discolored layer of oxides, primarily magnetite. The last portion of the pipe wall to fracture (i.e., the rapidly fractured region) remains a shiny silver color. The high pH SCC of pipelines is temperature sensitive.

The environments to cause near-neutral pH SCC on pipelines are anerobic, diluted bicarbonate electrolytes with a pH range of 5.5–7 [National Energy Board, 1996]. The environments are generated under disbonded, impermeable coatings, such as PE tape, which can shield CP current from reaching pipe steel. Thus, the steel is at its corrosion potential in the trapped electrolyte, where CO_2 coming from the soils is dissolved. It has been accepted that hydrogen (H) atoms participate in the near-neutral pH SCC process on pipelines, accompanying anodic dissolution of the steel inside the crack [Parkins, 2000]. Stress corrosion cracks are usually wide due to corrosion, and corrosion products accumulate inside the crack. Due to the hydrogen effect on the crack propagation process, the crack growth rate is usually hundreds of times greater than the cracking process dominated by corrosion reaction only. The near-neutral pH SCC follows a transgranular growth mode [National Energy Board, 1996]. The near-neutral pH SCC of pipelines is independent of temperature under pipeline operating conditions.

Both high pH and near-neutral pH SCC of pipelines occur as colonies of multiple parallel cracks that are generally perpendicular to the direction of the highest stress, i.e., primarily hoop stress resulting from the internal pressure, on the pipelines. These cracks can vary in depth and length and grow in two directions, i.e., the axial and transverse directions. They tend to coalesce or link together to form longer cracks. At some point, these cracks may reach a critical depth and length combination that can result in rupture. A leak will occur if a crack grows through the pipe wall thickness before it reaches a critical length for rupture.

A buried pipeline is subject to stress of several types from various sources, all of which contribute to SCC. The operating pressure of the pipeline is normally the greatest source of stress, i.e., hoop stress, on the pipe wall. The soil that surrounds the pipeline can apply longitudinal stress due to ground movement, generating the so-called pipe-soil interaction. Pipe manufacturing processes, such as welding and bending, can create specific residual stresses even after post-treatment for stress-relieving.

Susceptible pipe materials are another condition to develop SCC. The SCC occurrence on pipelines is not selective of pipe steels. Experiences showed that SCC occurred on a wide variety of grades of steel from X25 to X65 [National Energy Board, 1996]. There is a strong effect of nonmetallic inclusions on the susceptibility of pipeline steels to SCC. For example, in X70 steel, the primary inclusions are enriched in aluminum oxides, which are brittle and incoherent in the steel matrix [Liu et al., 2009]. Microcracks and interstices can form at the boundary between the inclusions and the steel matrix. Moreover, both electric resistance-welded (ERW) and double submerged arc-welded (DSAW) pipes were found to be involved in SCC-related failures [National Energy Board, 1996], which was attributed to low fracture toughness of the welds or higher-than-normal residual stress developing locally [TransCanada Pipelines, 1996]. It was also found that the coarse-grained heat-affected zone (HAZ) adjacent to the DSAW was significantly more susceptible to cracking than base steel in near-neutral pH environments. The average cracking velocities were about 30% higher in the HAZ than in the base steel [Canadian Energy Pipeline Association, 1996a].

1.2.2.2 Corrosion Fatigue

Corrosion fatigue is the conjoint action of cyclic stress and corrosion reaction to cause material fracture in corrosive environments. The basic role of corrosion is to decrease the number of cycles of stress resulting in material failure when compared to mechanical fatigue failure. Fatigue failure typically includes three stages. In stage 1, cracks are initiated on the external surface of a smooth metal surface under cyclic stress conditions. The crack initiation, which usually takes a long time of period, results from mobility of dislocations, which are generated inside the metal due to cyclic stressing, toward the metal surface, producing substantial dislocation slip bands. For corrosion fatigue, the cracks can be initiated by additional mechanisms such as preferential dissolution of the dislocation slip bands and emergence points, pitting corrosion, and localized corrosion at grain boundaries [Zhao et al., 2012]. Thus, the time required to initiate corrosion fatigue cracks is much shorter than the incubation time for mechanical fatigue cracks under an identical stress condition. In stage 2, the initiated cracks grow steadily on the steel under cyclic stress. The crack growth direction is perpendicular to the stressing direction, usually leaving a regular striation pattern on the fracture surface. The pattern can thus be used to identify if the fracture is caused by fatigue failure. This stage usually takes about 50–75% of the total fatigue time. When the crack growth is accompanied by a corrosion reaction, the cracking process is accelerated due to a synergistic effect of cyclic stress and corrosion on crack propagation. Thus, the main role of corrosion in stage 2 is to increase the crack growth rate. Stage 3 occurs when the remaining cross-section of the metal is too small to withstand the applied stress. A mechanical fracture occurs, forming a final shear lip at 45° relative to the metal surface. Since the fracture is so rapid at this stage, corrosion is not involved.

Crack propagation on pipelines depends not only on the amplitude of internal pressure, which results in the primary stress, i.e., hoop stress but also on pressure fluctuations [Canadian Energy Pipeline Association, 1996b]. Internal pressure on pipelines fluctuates frequently, and the pressure fluctuation is critical for crack growth. Experimental testing established [Canadian Energy Pipeline Association, 1996b] that precracks would not grow under constant-load or constant-displacement conditions on pipe steels. Cracking was observed where dynamic loading conditions were applied. For full-scale testing, no crack growth was found under static load conditions, even at a stress level as high as 80% of the yield strength of the steel [Canadian Energy Pipeline Association, 1996c]. Furthermore, a direct relationship between the maximum stress and the crack growth rate was found for X65 pipeline steel in a near-neutral pH solution at low R ratios (i.e., the ratio of the minimum to the maximum stresses in circumferential direction of the pipe), e.g., 0.5 (typical of oil pipeline operation) and 0.7 [Canadian Energy Pipeline Association, 1996a]. When the R ratio was up to 0.85 that is typical of gas pipeline operation, the stress level did not affect the crack growth rate. Thus, pressure fluctuations on pipelines, which are like cyclic stress conditions, contribute to the cracking process.

1.2.2.3 Hydrogen-Induced Cracking

HIC can be a form of SCC in many metal–environment combinations. However, the HIC may occur under constant stress, but the corrosion process is absent (e.g., HIC of carbon steel pipes in high-pressure hydrogen gas environments). These are considered a different cracking mechanism from SCC. H atoms permeating in steels can also induce cracking in the absence of external stress [Jin and Cheng, 2010]. The H atoms can be generated in pipeline systems by many mechanisms such as corrosion and welding. Once entering pipe steels, the hydrogen effect depends on distribution of H atoms in the steels. If the H atoms disperse uniformly in the crystalline lattice with an average content below the H solubility under given temperature and pressure conditions, the steels can tolerate the H atoms without apparent side effects [Hirth, 1980]. However, H atoms tend to accumulate at metallurgical features contained in the steels, such as nonmetallic inclusions, dislocations, grain boundaries, precipitates, voids, and microcracks, due to a higher energy level at these features than the steel matrix [Hirth, 1984]. Even the average content of H atoms in steels is below the H solubility, the local H concentration may exceed a threshold value to induce hydrogen degradation.

The modes of hydrogen degradation of pipelines can be varied, depending on the steel grade, hydrogen content, metallurgical microstructures, defect features, etc. Basically, as H atoms enter, the brittleness of the steels increases and the toughness and ductility decrease, especially at high H contents, resulting in the so-called hydrogen embrittleness (HE). For low-strength pipeline steels, hydrogen bubbling (blistering) probably occurs when H atoms accumulate at internal voids to form hydrogen (H_2) molecules, elevating the local pressure. Although blisters are often found in low-strength steels exposed to corrosive environments, hydrogen-induced blistering and blister cracking were also observed on high-strength pipeline steels such as X100 steel, as shown in Figure 1.5 [Jin and Cheng, 2010]. As the strength grade increases, the steels become more susceptible to HIC. The HIC refers to the brittle fracture of a ductile alloy under applied stress in the presence of hydrogen. Particularly, fracture occurs at stresses below yield strength of the material. Figure 1.6 shows a hydrogen-induced crack in X100 steel [Jin and Cheng, 2010]. For many steels, there exists a threshold stress below which the HIC does not occur. Generally, the threshold stress level to induced HIC decreases as the strength of the material increases. However, it was found [Xue and Cheng, 2011] that, even in the absence of an external stress, hydrogen, once exceeding a threshold content, can result in cracking of X80 pipeline steel. HIC usually occurs with a delayed time to failure after H is trapped at regions with a high triaxial stress. The HIC is featured with a sharp, singular crack, rather than extensive branching or secondary cracks. Furthermore, cracking of pipeline steels, primarily downhole and upstream gathering lines, can occur in the presence of hydrogen sulfide (H_2S) under applied stress, which is called sulfide stress cracking (SSC) [Leyer et al., 2005]. For SSC, the environmental severity depends primarily on solution pH and the H_2S partial pressure [National Association of Corrosion Engineers (NACE), 2003].

Many models and mechanisms have been proposed to explain the HIC phenomenon [Hirth, 1984]. The commonly accepted mechanisms include hydrogen-enhanced decohesion (HEDE) and hydrogen-enhanced local

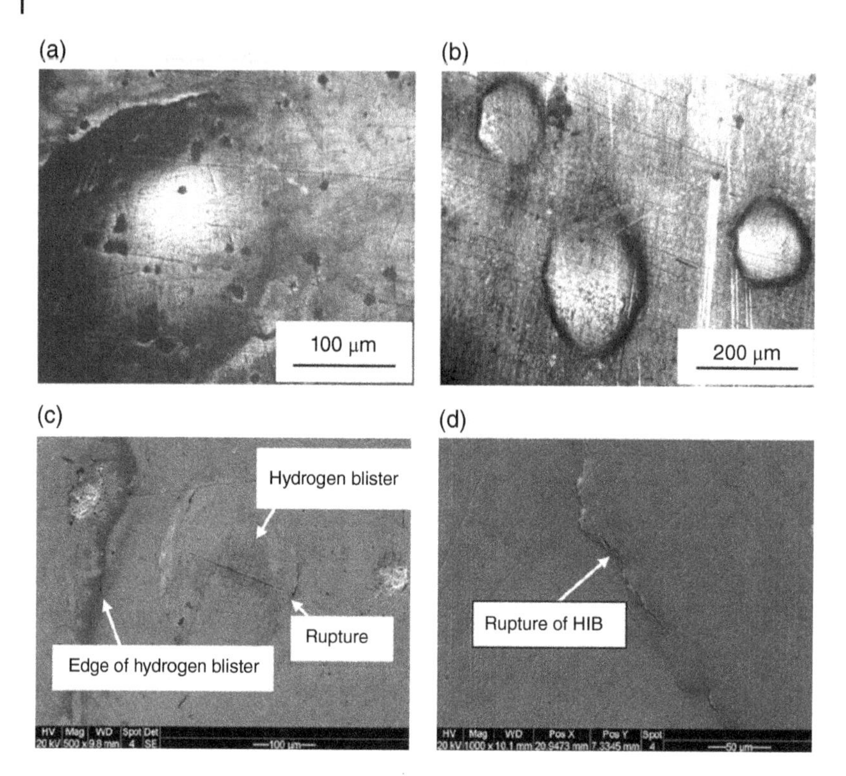

Figure 1.5 Surface morphology of hydrogen blisters on an X100 pipeline steel after 20 hours of H-charging at cathodic current densities of (a) 50 mA/cm^2 and (b) 100 mA/cm^2, as well as ruptured blisters (c and d), where HIB refers to hydrogen-induced blistering. *Source:* From Jin and Cheng [2010] / reproduced with permission from Elsevier.

plasticity (HELP) [Djukic et al., 2019]. According to the HEDE mechanism, H atoms can weaken the bonds between iron atoms due to H segregation at grain boundaries or dislocations, resulting in brittle and smooth fractures with limited plasticity. Initiation or propagation of cracks occurs if the local stress concentration exceeds the cohesive stress [Katzarov and Paxton, 2017]. The main limitation of the mechanism is that a direct measurement of cohesive force is difficult or unavailable. The HELP proposes that H atoms accumulating at crack-tip facilitate mobility of dislocation due to a local drop of yield stress and the activation energy for dislocation motion. As a result, plastic deformation of the steel is enhanced [Birnbaum, 1994]. Dislocations pile up at metallurgical features such as grain boundaries and finally failure occurs. The fracture surface usually shows plastic deformation characters and slip

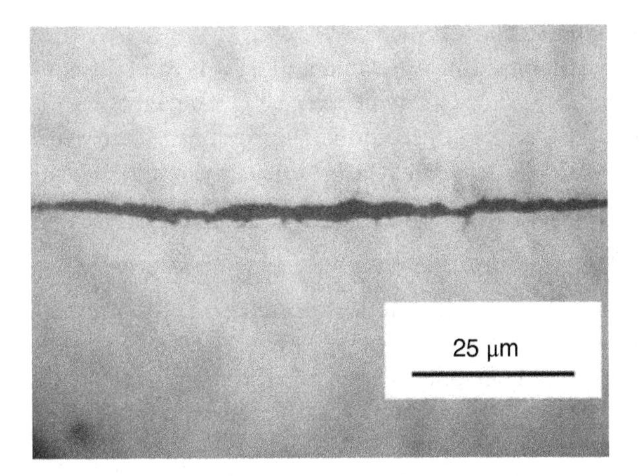

Figure 1.6 Optical view of a hydrogen-induced crack in X100 pipeline steel upon H-charging in a soil solution. *Source:* From Jin and Cheng [2010] / reproduced with permission from Elsevier.

bands at the crack tip. The HELP mechanism has been verified in a wide variety of structured metals [Lynch, 2012].

The HIC problem is particularly important in pipeline welds. It was recognized that most failures occurring at weld joints were related to HIC [Bauman, 2009]. Both welding process and preferential corrosion at the welds generate more H atoms at the welds, as compared with the base steel [Zhang and Cheng, 2010; Biro et al., 2012]. Moreover, due to microstructural and compositional heterogeneity, as well as various types of inclusions as metallurgical defects, H atoms possess a high permeability and preferentially accumulate at the welding region, especially at the HAZ [Zhang and Cheng, 2009; Xue and Cheng, 2013; Gan et al., 2018]. In addition to the local H accumulation, the welds usually suffer from a high stress concentration [Sun and Cheng, 2021a]. Moreover, the fracture toughness of the weld region is often inferior to that of base steel. All of them make the pipeline welds prone to HIC occurrence.

1.2.3 Manufacturing Defects

Defects can be generated during the manufacturing process for rolling steel plates and transforming steel billets into pipes. Generally, the manufacturing defects on a pipe are classified into two groups, i.e., manufacturing defects on pipe body and manufacturing defects on weldment.

1.2.3.1 Manufacturing Defects on Pipe Body

From pipe billets to steel pipes, a heat treatment process is carefully controlled to produce the designed pipes while removing residual stress. If the heating temperature is too low, the deformation resistance of the steels will be increased, while the plasticity is reduced. Particularly, when the heating temperature cannot make a complete conversion of the steel microstructure of austenite grains, the possibility of generating cracks will be remarkably increased. On the contrary, when the heating temperature is too high, steel oxidation will happen, resulting in decarburization and excessive austenite grain-coarsening. Metallurgically, an inappropriate heat treatment could produce a metallographic phase called Widmanstatten. The Widmanstatten is an overheating structure and can reduce the strength of steels at room temperature and increase brittleness.

In addition to property degradation and microstructural problems, various manufacturing defects, such as laminations, nonmetallic inclusions, and gouges, can be produced on pipe bodies [Cosham and Hopkins, 2004]. Lamination defects are commonly produced in the manufacturing process of seamless pipes, and they are unwanted discontinuities lying parallel to the pipe surface. The laminations are usually marked by a concentration of nonmetallic materials when nonmetallic inclusions are rolled-out during manufacturing. For example, it was found that the lamination defect was caused owing to the existence of sulfides in the steel plate [Li et al., 2007]. The stripe and plate sulfides were distributed in the ferrite along the rolling direction during cold rolling, resulting in the lamination problem. The larger the size of the sulfides, the higher the possibility of generation of lamination defects. Lamination defects can significantly reduce the bearing load, and impact toughness and fatigue strength of pipe steels [Li et al., 2019a]. Laminations can become initiation sites for fatigue cracks and HIC on the pipe body.

Nonmetallic inclusions are inevitably contained in pipe steels during metallurgical processing. The adverse effect of the inclusions on properties and performance of the steels depends on the type and characteristics of the inclusions if they are not well controlled during the manufacturing process. The volume fraction and mutual spacing of the inclusions are the contributing factors of steel toughness. Generally, the toughness of steels decreases with increased volume fraction or decreased spacing of the inclusions such as manganese sulfide (MnS) [Garrison Jr and Wojcieszynski, 2007]. The nonmetallic inclusions are usually the sites to initiate cracks due to preferentially localized corrosion and the transition from corrosion pits to cracks [Jin and Cheng, 2011; Peng et al., 2013]. Figure 1.7 shows that a crack initiates at combined inclusions of aluminum oxide and titanium oxide in an

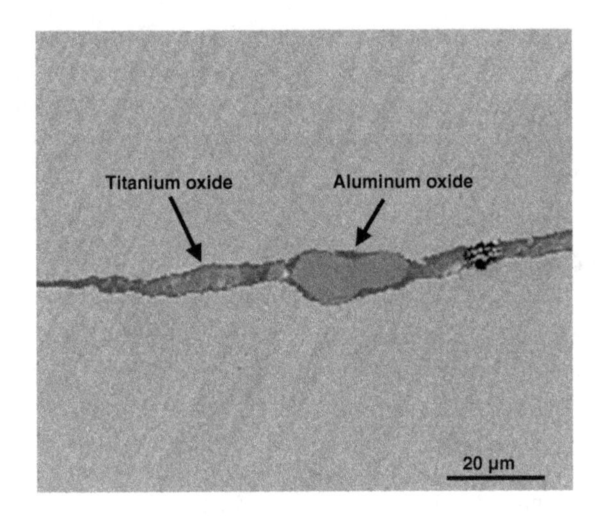

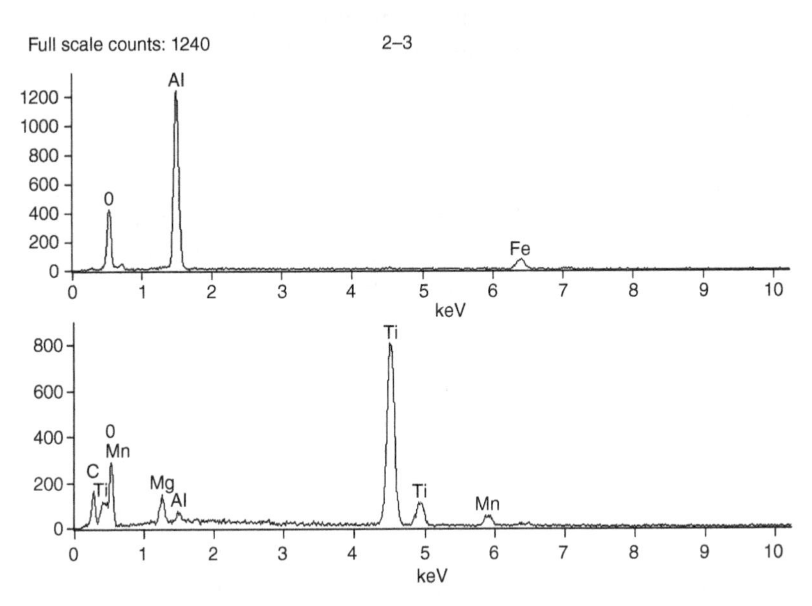

Figure 1.7 A crack initiated at combined inclusions of aluminum oxide and titanium oxide in an X100 pipeline steel. *Source:* Dong et al. [2009] / reproduced from permission from Elsevier.

X100 pipeline steel [Dong et al., 2009]. Moreover, the inclusions tend to trap H atoms, decreasing the resistance of pipe steels to HIC [Peng et al., 2018].

Gouges can be produced on pipe body during both manufacturing and construction stages. A gouge is created when steel material is removed from the pipe wall mechanically. This usually happens when the teeth of a backhoe scrape across the pipe. The gouge, especially the longitudinally oriented gouge, can reduce the burst strength of pipelines [MacDonald and Cosham, 2005]. The sharp edges of the gouge act as stress concentrators and potentially cause initiation of cracks.

1.2.3.2 Manufacturing Defects at Welds

The other type of manufacturing defect on pipelines is weld defects. Welds are recognized as the primary weak sites where pipelines fail. In addition to some common failure phenomena occurring at the pipeline welds during service, such as preferential corrosion, delayed cold cracking, hydrogen degradation, and fatigue fracture, weld defects can be generated during manufacturing. Generally, manufacturing defects at pipeline welds can be classified into planar defects, volumetric defects, and irregular shape defects [Feng et al., 2017]. Specifically, planar defects include incomplete penetration, lack of fusion and undercuts, etc. Volumetric defects include porosity, inclusions, and bevel grinding. Common irregular shape defects are irregular weld outline and misalignment of weld joints, etc. Table 1.1 lists some main manufacturing defects at pipeline welds and their illustrative images.

All types of weld defects can result in degradation of pipeline integrity. It was observed that the welds were an area prone to initiation of fatigue cracks [Jakubczak and Glinka, 1986]. Generally, misalignment of welded parts significantly reduces the fatigue life of pipes. In the absence of misalignment, a lack of penetration also reduces the resistance of welds to fatigue cracking [Deshmukh et al., 2014]. Undercut also decreases the resistance to cracking. It was found that the fatigue life reduced significantly with increased undercut depth and width [Mashiri et al., 2001]. Porosity is a cavity-type discontinuity formed by entrapment of gas in the weld metal during solidification. It can be present as isolated pores, multiple pores in a cluster, or elongated cavities also known as wormholes.

Cracks at welds produced during welding are one of the most serious threats to pipeline integrity. In the weld metal of fusion welds (e.g., seam welds and multiple process girth welds), cracks can be generated by solidification mechanisms in both the longitudinal and transverse directions. The cracks are usually formed in the pipe mill for seam welds or during pipeline construction for girth welds. Post-welding inspections by techniques such as ultrasonic testing must be performed to detect the welding cracks. However, due to the detection limit of the techniques and some other reasons, not all cracks can be identified. These cracks will continue to grow at welds under the synergism of external stress, local metallurgical features, heterogeneous mechanical properties, and/or corrosion reaction.

Table 1.1 Some main manufacturing defects at pipeline welds and the illustrative images.

Weld defect	Image
Linear porosity	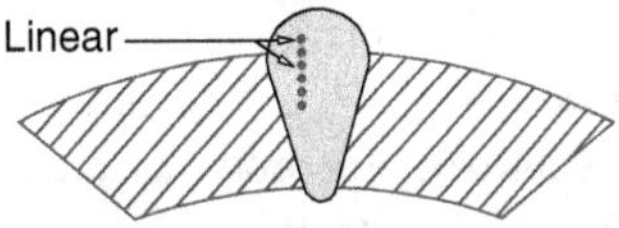
Cluster porosity	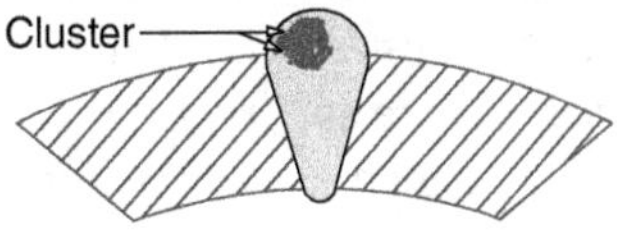
Slag	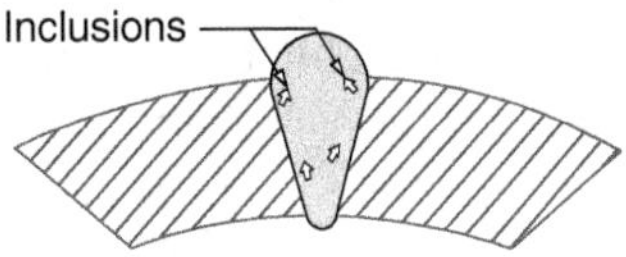
Lack of fusion (LOF)	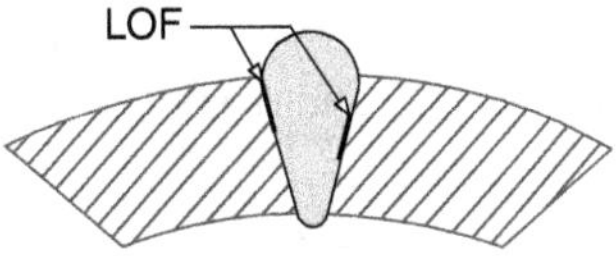
Lack of penetration (LOP)	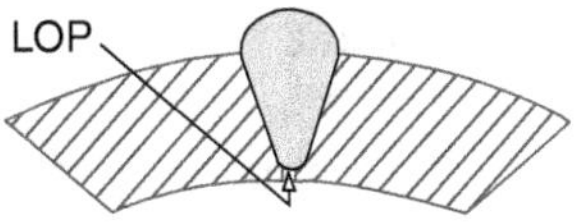
Cracks	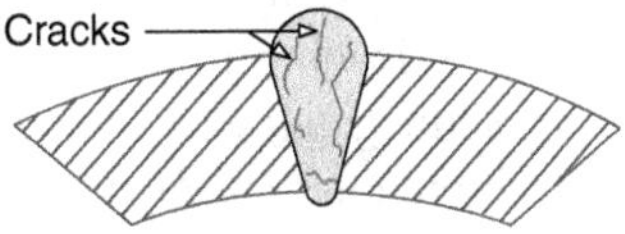
Undercut	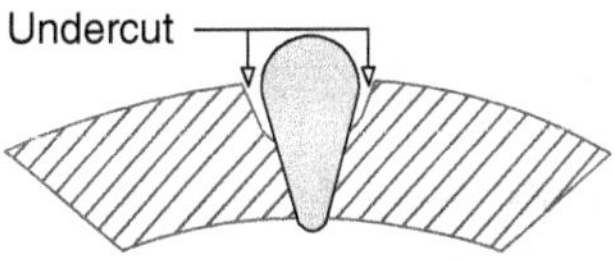
Burn through	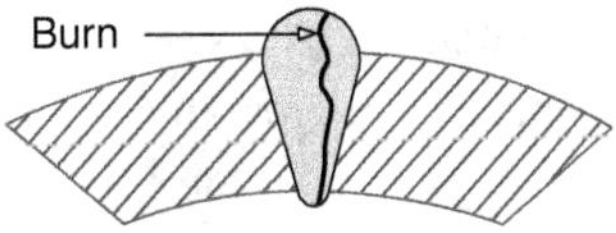

1.2.4 Construction Damage

Damage that occurs during pipeline construction is another source of danger that can threaten integrity and safety of the pipelines. It was reported that mechanical damage introduced during the construction stage of pipelines, along with corrosion, was the main cause of pipeline failures [MacDonald et al., 2007]. Statistics showed that at least one of every three incidents in gas and liquid products pipelines was attributed to mechanical damage [Rosenfeld, 2002]. There are two forms of construction damage. The first and the most serious damage is called encroachment damage that occurs when a pipe is struck by heavy equipment such as excavators, graders and ditchers during excavation or backfilling, producing a shallow residual dent plus a gouge. The other form of damage is dents which are usually located on the bottom half of a pipeline. Mechanical damages can be insignificant if they are shallow scrapes or plain dents on the pipe surface. However, formation of sharp, deep dents with gouges is dangerous [Zarea et al., 2012]. Generally, encroachment damage presents an immediate hazard to pipeline integrity, while dent damage requires a long-term maintenance and safety evaluation.

1.2.4.1 Encroachment Damage

Excavation is the most common source of encroachment damage to pipelines. In addition, roadwork, farming, and offshore anchoring can cause incidents by hitting the pipelines. Encroachment damage by heavy equipment on a pipe can create gouges or scrapes. A gouge is surface damage on a pipe, leading to removal of material (i.e., pipe steel) [Allouti et al., 2014]. Some gouges are shallow and may look insignificant to compromise the pipeline integrity. However, plastic deformation, friction and even melting could occur at the contact point. For the gouge with a sufficient depth, the steel can become heavily strain-hardened, decreasing its ductility and toughness. During formation of gouges, a high tensile strain develops at the bottom of the gouged area, resulting in initiation of cracks in the steel. About 80% of encroachment damage causes pipeline failures right away [Rosenfeld, 2002]. Even if the cracks do not fail the pipeline immediately, the cracks can grow in service and may cause catastrophic consequences.

1.2.4.2 Dents

The most common form of construction damage is a dent. A dent is a permanent inward indentation or plastic deformation of the pipe wall caused by an external force, but without causing metal loss. Dents include two types, plain dent and kink dent. The plain dent has a smooth profile and is usually caused by rocks. The kink dent presents an abrupt change in the curvature of the pipe surface [Naghipour et al., 2018], and is usually caused by heavy equipment [Rosenfeld et al., 2002]. A dent can be constrained or unconstrained based on its rebounding ability under the influence of internal pressure. Typically, a dent caused by excavating equipment is constrained and confined on top of the pipe surface, while a rock-induced dent is usually unconstrained and most likely happens on the bottom of a pipe [Rosenfeld et al., 2002]. Plain dents are usually not regarded as a threat to pipeline integrity as they do not reduce the burst strength of pipelines [Zarea et al., 2012]. However, plain dents can be dangerous if they are on or near pipe welds by facilitating crack initiation and propagation [Alexander, 1999], and reducing fatigue life of the pipeline [Fowler et al., 1994]. When a dent is combined with a gouge, the combined defect presents a hazardous defect to pipeline integrity.

A local stress and strain concentration is developed at the gouge or the dent, making the pipeline more sensitive to stress-related failures [Alexander, 1999]. Moreover, a strain hardening occurs at the base of the damaged area, making the steel more brittle and thus more susceptible to cracking [Zheng et al., 2005]. Furthermore, the damage caused by dent, gouge, or dent and gouge combination to pipelines could be delayed. The delayed damage is often related to fatigue failure and/or reduced burst resistance. According to statistics in the United States, delayed damage accounted for 20% of mechanical damage [Rosenfeld et al., 2002]. Due to the difficulty to identify promptly and thus to control effectively, the delayed damage can cause catastrophic accidents on pipelines.

1.2.5 Geotechnical Hazards

Over 98% of oil/gas transmission pipelines are buried and are under complex stress conditions. In addition to the hoop stress resulting from the internal operating pressure, the pipelines also experience external forces transmitted through the soil and surface loads, especially in unstable geotechnical areas [Trifunac and Todorovska, 2004; Kenny and Jukes, 2015; Sun and Cheng, 2021a]. Natural geohazards such as landslides, long-term slope movement, land subsidence, frost heave and thaw settlement, seismic fault movement and offshore ice gouging may cause pipelines to experience large deformations, elevating the stress and strain levels on the pipe. The pipeline integrity is remarkably affected by various geohazards.

1.2.5.1 Landslides

Landslide is a common geological disaster in mountainous areas. Landslides are usually accompanied by significant soil loss, resulting in considerable deformation on the affected pipelines. It was reported that a 40 m wide landslide caused movement of a pipe by 7.5 and 4.0 m in the horizontal and vertical directions, respectively [Vasseghi et al., 2021]. Almost every pipeline that traverses through areas in mountainous terrain has some vulnerability to landslide hazards [Harp, 2008].

The hazard induced by landslides depends on burial depth of the pipeline and the depth of affected soil by the landslides. Most pipelines are buried to depths of about one and a half meters. Only when the landslides penetrate to the pipe burial depth, or generate sufficient stresses at the depth, will a hazard be present on the pipe. Thus, the commonly occurring landslides, such as falling and sliding of soil and rocks, which are usually caused by small-scale seismic or extreme precipitation events, are negligible threats to pipelines. The exceptions include (1) penetration and transmission of impact stress generated by falling or sliding of large rocks to pipe burial depths, (2) erosion of soil due to rock-falling and rock-sliding to the pipe burial depths, or (3) debris flowing deep enough to affect the pipeline integrity. The landslides, when affecting the buried pipelines, can generate either axial tensile stress or normal impact stresses on the pipe. The resulting maximum stress may be much greater than the yield stress of the pipe steel.

1.2.5.2 Land Subsidence

Land subsidence is a gradual settling or sudden sinking of ground surface owing to subsurface movement of earth materials [Galloway et al., 2008]. The principal causes of land subsidence include subsurface fluid withdrawal, drainage of soils, sinkholes, mining activities, thawing permafrost, and natural consolidation. While causing downward motion of land surface, the subsidence also accompanies lateral ground movements.

Vertical displacement at the level of several meters may occur locally during land subsidence [Galloway et al., 2008], causing displacements of pipelines and additional longitudinal tensile and compressive loads. The loading level depends on the soil-pipe friction coefficient, pipeline burial depth, and pipeline section length subject to horizontal strains. The load conditions also change in the transverse direction of the pipelines. It was demonstrated [Kalisz, 2019] that the land subsidence caused increased failures of buried natural gas pipelines. The failures occurred most frequently on "old" pipelines, resulting in breakage of pipe walls, mostly near welded regions, and pipeline buckling.

1.2.5.3 Frost Heave and Thaw Settlement

Frost heave and thaw settlement are common geohazards occurring on pipelines buried in permafrost and semipermafrost areas such as Arctic and sub-Arctic regions where the soils are constantly freezing and thawing. The frost heave is an upward ground movement due to the growth of ice lenses around a pipeline, and the thaw settlement is a download ground movement due to melting of the ice-enriched soil near the pipeline [Yoosef-Ghodst, 2015].

The density and volume of the earth change with seasons. The temperature difference between the inside and the outside of a pipe also results in heat transfer, affecting the soil structure. The mechanical properties of frozen soils are affected by unfrozen water, mineral composition, ice content, loading rate, stress path, temperature, and stress–strain history [Li et al., 2019b]. When the permafrost beneath the pipeline thaws, the pipeline foundation settles, and the cohesive force between the pipe and the surrounding soil is reduced. Periodic stresses will be generated as the permafrost around the pipeline freezes, thaws, and refreezes [Cherniavsky, 2018]. If not released effectively, the stress can become sufficiently great to cause the pipe to move upward. As the pipe gets close to the ground surface, the temperature difference between the "warm" pipe and the "cold" soil increases, which further accelerates the upward movement of the pipe. Thus, the area where the frost heave occurs is always accompanied by fluctuated earth's surface. Experiences showed that, as the air temperature is below −30 °C, the temperature of the soil at the pipeline burial depth (about 1.5 m) can be −5 °C, causing freezing of the soil. It was reported [Naumov et al., 2019] that during freezing and thawing of soils, the maximum stress generated on a pipeline was much higher than the maximum yield stress of the pipe steel, and thus caused a permanent deformation of the pipe and eventually cracking. Moreover, high strain levels generally arise on the pipelines buried in the areas experiencing frost heave and thaw settlement. The strain demand by the surrounding environment requires the pipeline to possess a sufficient strain capacity. Considerations of both strain demand and strain capacity generate the so-called strain-based design (SBD) of pipelines, effectively minimizing tensile rupture and compressive local buckling in geologically unstable regions.

1.2.5.4 Earthquakes

Earthquakes can cause significant displacements in soils. Pipelines are bent upward on the ground under compressive forces and even exposed from the ground. The bending stiffness of pipe steels and the buried depth affect the degree of upward bending of the pipelines. Generally, a small deformation does not burst the pipelines [Psyrras and Sextos, 2018]. However, when the pipelines contain defects such as gouges, dents, and inclusions introduced during manufacturing or construction, the stability of the pipelines can be undermined by earthquakes, resulting in wrinkling on pipe wall. The wrinkles further increase and become localized as the loading increases. When the local deformation reaches a certain degree, the pipe wall will be torn, resulting in buckling [Psyrras and Sextos, 2018]. Another type of failure caused by earthquakes on buried pipelines is tensile fracture. The displacement of soils applies an axial tension on the pipelines, causing longitudinal straining and even rupture [Psyrras and Sextos, 2018]. The rupture usually occurs at pipe welds, where metallurgical defects and local compositional and stress heterogeneity often exist.

1.2.6 Threat Interaction

Field experiences showed that various factors or threats affecting pipeline integrity are usually present on the pipelines at the same time with a spatial adjacency or at the identical location. Interaction existing among them will further compromise structural integrity of the pipelines. Extensive studies have demonstrated that when multiple corrosion defects are present in adjacency or overlapped on pipelines, the failure pressure can be decreased remarkably [Chen and Shu, 2001; Chiodo and Ruggieri, 2009; Sun and Cheng, 2018]. Moreover, the effect depends heavily on the spacing, either longitudinally or circumferentially, between the defects [Sun and Cheng, 2019a; 2021b]. A stress concentration can occur in the middle of the adjacency or at the defects. Investigations on the mutual interaction between multiple corrosion defects have resulted in development of various interaction rules [Qin and Cheng, 2021], such as CW rule [Coulson and Worthingham, 1990], DNV-RP-F101 code [Veritas Det Norske, 2004], 6WT rule [Lamontagne, 2002], 3WT rule [Hopkins and Jones, 1992], and Sun-Cheng rule [Sun and Cheng, 2018; 2019a; 2019b]. These rules and codes were used to assess if an interaction exists between corrosion defects based on determination of the strength of pipeline steels at the corrosion defects with a specific geometry. Generally, the interaction between corrosion defects decreases the failure pressure of corroded pipelines and affects the applicability of existing defect assessment and FFS determination methods.

When a corrosion defect is adjacent to a dent in pipelines, an interaction can exist between the two features. As the interaction increases, the failure pressure of the pipelines decreases [Sun et al., 2021]. It was found that with a decreased spacing between the two features, the interaction increased, causing a decrease in the pressure-bearing capability of the pipeline. A critical spacing existed above which the interaction did not exist, and the dent and the corrosion feature could be assessed independently. The critical spacing that is critical to the interaction between the dent and the corrosion defect mainly depends on corrosion depth, corrosion length, and the dent depth.

Another typical scenario of threat interaction encountered on pipelines is accelerated localized corrosion at manufacturing and construction defects, where a high stress concentration is usually achieved. There has been substantial evidence proving that corrosion is enhanced by stress concentration due to stress-increased electrochemical activity at the defects in corrosive environments [Li and Cheng, 2008; Tang and Cheng, 2009; Cheng, 2013]. As a result, preferential corrosion occurs and grows at the defects, causing local perforation. Moreover, an increased stress can facilitate permeation of H atoms in pipeline steels. The H atoms are accumulated at various traps such as metallurgical defects combined with the stress concentration, resulting in HIC, which is particularly applicable at pipeline welds [Gan et al., 2018; Sun and Cheng, 2021a]. Furthermore, EAC events such as SCC can occur at a lower threshold stress due to pipe-soil interactions in geologically active regions than the SCC events occurring in the absence of the pipe-soil interactions. Generally, the ground movement implements additional axial tensile stress or strain on the pipelines [Cheng, 2013]. Similarly, axial compressive stress generated due to ground movement can contribute to buckling, which is relatively easy to occur when the affected pipe segment contains corrosion or mechanical defects [Shuai et al., 2020; 2021]. Thus, the pipelines operating in mountainous areas or Arctic/sub-Arctic regions should include the pipe-soil interaction in the integrity management program.

In summary, various types of threats usually co-exist on pipelines so that mutual interaction between the threats is induced to further increase the risk of pipeline failure. Therefore, a pipeline integrity management program should not only include the threats in consideration in pipeline design and operation stages but also determine the interaction among the threats.

1.3 Elements of Pipeline Integrity Management

The pipeline integrity management program is a process to develop, implement, measure, and manage the integrity of a pipeline through assessment, mitigation, and prevention of risks to ensure a safe, environmentally responsible, and reliable service [Nelson, 2002]. A typical pipeline integrity management program, as shown in Figure 1.8, consists of four essential steps,

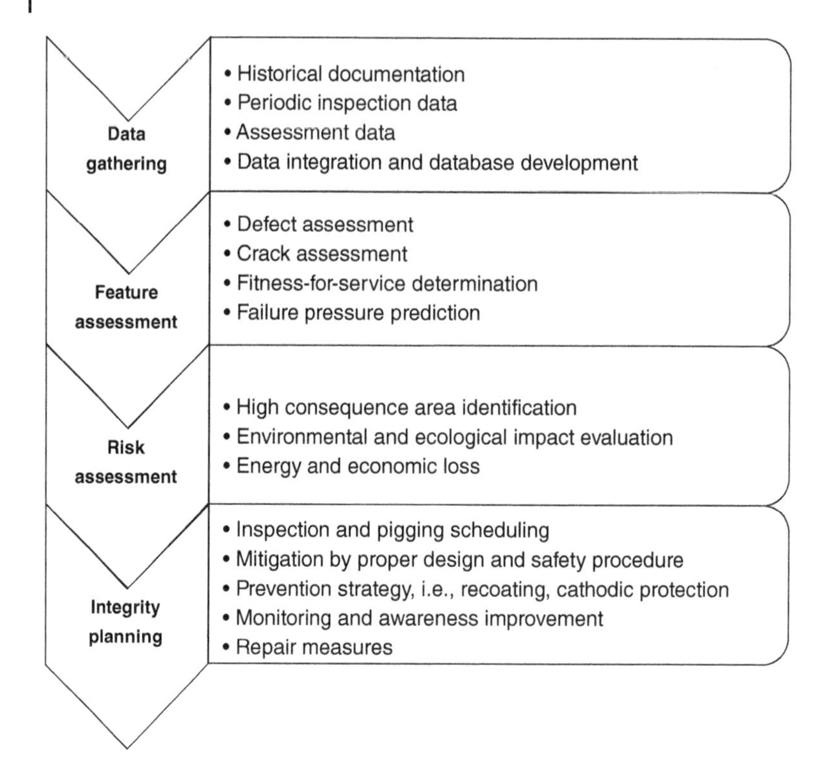

Figure 1.8 Framework of a pipeline integrity management program.

i.e., data gathering, feature assessment, risk assessment, and integrity planning [American Society of Mechanical Engineering, 2014]. In step one, substantial data are collected from literature review, sampling and testing in the field, relevant databases, and periodic inspections. Particularly, the ILI is essential to accurately identify anormal features, such as corrosion, mechanical damages (dents and scratches), and cracks, which can adversely affect pipeline integrity. The inspections are required to be conducted on the same pipe segment over a certain period so that a database can be developed for comparison and assessment purposes. In step two, the collected data, especially the data recorded by ILI tools, are analyzed by either comparison with historical records or developed models or both to assess the condition of the pipeline. Emphases are usually put on the critical features identified on the pipeline to assess and determine FFS of the pipe and, if necessary, schedule rehabilitation measures. The assessment will give critical information about the pipeline, including stress and strain distributions, pipeline failure pressure, defect growth rate, etc. In the third step, potential conseque"ces

’nd impacts resulting from pipeline failures are evaluated. These include not only energy loss and the direct financial effect on pipeline operators, but also impact on the environment, ecology, and public safety. High-consequence areas right-of-way (ROW) of the pipeline will be identified. The last step involves consideration, design, and implementation of appropriate measures to address any issues identified on the pipeline. Proper measures and actions are taken to mitigate and prevent pipeline failures. This is the goal of the entire process [Focke, 2015]. Implementing the pipeline integrity management program is the core task of pipeline operators.

In addition to the ILI and data analysis, a pipeline integrity management program also includes hydrotesting and direct assessment. Hydrotesting is conducted on pipelines either before they are placed in service or after a time of service. During hydrotesting, the pipelines to be tested are removed from service. Water or air is inserted, and a series of pressurized tests are performed, including spike testing at a pressure of 139% of maximum allowable operating pressure (MAOP) for half an hour, pressurized testing at 125% of MAOP for four hours, and leak testing at 110% of MAOP for four hours [Jacobs Consultancy Inc, 2013]. After all tests, the pipes are dewatered and dried. Hydrotesting assures that critically sized defects are identified and removed. The disadvantage is that it is a destructive test and a pass/fail test. If the testing pressure level is not sufficiently high, subcritical small cracks will remain.

The direct assessment is an indirect integrity assessment and management method based on a structured process, where knowledge of various characteristics of the pipelines and their operating history are integrated with the periodic results of inspection, examination, and evaluation to achieve a reliable and accurate assessment of the pipeline performance condition. Nowadays, the pipeline internal corrosion direct assessment (ICDA), external corrosion direct assessment (ECDA), and SCC direct assessment (SCCDA) have been developed, enabling pipeline operators to evaluate the pipeline integrity that could be affected by a wide variety of potential threats during pipeline operation. Generally, the direct assessment is complementary to various inspection methods, and cannot completely replace pipeline inspections.

1.3.1 Identification

Various tools and techniques have been developed and used to identify the threats to pipeline integrity at sufficient accuracy. There have been many methods such as close interval survey (CIS) of CP potentials, DC voltage gradient (DCVG), AC voltage gradient (ACVG), corrosion coupons, eddy current, and hydrotesting that can provide information about the threats and pipeline performance condition. The intelligent ILI tools including ultrasonic tools (UTs), magnetic flux leakage (MFL), and electromagnetic acoustic transducer (EMAT) can identify and characterize certain defect features such as corrosion, dents, and cracks at a sufficiently accurate resolution, providing direct and critical data useful for determination of the FFS of the pipelines. Compared with metal loss such as corrosion defects, cracks are more difficult to detect and size, especially for deep cracks. Due to the complex nature and continual growth of cracks on pipelines, development of high-performance ILI tools to provide the highest resolution and most accurate data is crucial. Generally, to obtain a complete assessment of the integrity of a pipeline, several ILI tools should be run in succession and the inspection records are then compared to specify the type, location, and severity of anomalies detected. It is noted that running of ILI tools is usually scheduled at a certain time interval based on individual pipeline operators' management program.

1.3.1.1 ILI Tools and the Applications

The ILI tools used for inspecting a pipeline are based on nondestructive testing (NDT) principle. The ILI tools can identify and measure metal loss features caused by corrosion or mechanical scratches, and anormal geometrical features such as dents, cracks, and crack-like flaws located in either the axial or the circumferential direction. The tools are moved with flow of the fluid carried in the pipeline, which is not necessarily removed from service. The ILI usually requires the inspected pipelines to operate in a controlled condition, where the fluid flow may be reduced [Eiber, 2003]. Depending on the types of detected features present on the pipeline, the ILI tools can be classified as metal loss tools, crack tools, geometry detection tools, etc. Some tools can detect and size both metal loss and cracks [Xie and Tian, 2018]. While other NDT techniques such as visual inspection, onsite metallography, liquid penetration, magnetic particle inspection, and eddy current can also detect and assess anomalies, the main smart ILI tools, i.e., MFL, UT, and EMAT, used in pipelines are reviewed and discussed below.

Magnetic Flux Leakage The MFL is one of the most widely used techniques for inspection of metal loss such as corrosion and scratches on pipelines. The principle of MFL inspection is based on an axially oriented magnetic field applied by huge magnets into the pipe wall thickness, where the magnetic flux leaves the pipe wall at a metal loss area present on either internal or external pipe surface. The leaked flux is picked up to deduce the area of the metal loss. Conventionally, the

MFL tools are sensitive to the circumferential dimension of a metal loss, along with its depth. Long, narrow corrosion along the axial direction cannot be detected [Eiber, 2003]. Another limitation of MFL is that long and rough defects can be confused by a series of pits. The MFL cannot directly measure actual pipe wall thickness. The MFL tools detect metal loss with an accuracy of ±10% of the wall thickness with a confidence limit of 80%. The length and depth of the metal loss area are used to predict the severity of an anomaly.

Nowadays, circumferential MFL tools are developed to produce a magnetic flux in the circumferential direction so that they are more accurate at sizing axially oriented features, but with a lowered accuracy than conventional MFL [Vanaei et al., 2017]. Three-axis high-resolution MFL tools have been developed to utilize three-directional sensors to detect the size of the magnetic field, enabling measurements of the axial, circumferential and radial data for metal loss area characterization [Feng et al., 2017].

Efforts have also been made to use MFL in crack detection on pipelines owing to the benefit that the MFL tools do not require a liquid medium to operate. Generally, the MFL has some limitations for detection of crack defects, especially at girth welds. It was found that the inclination of cracks affected the amplitude of MFL tools. When the distance between two adjacent cracks is less than 5 mm, their magnetic leakage fields may overlap, reducing the detecting accuracy. The three-axis high-resolution MFL can detect cracks with an opening exceeding 0.5 mm only [Wang and Chen, 2015]. It is thus suggested [Feng et al., 2017] that the MFL can be used for basic screening of defects and cracks, without considering an accurate sizing, especially for gas pipelines where a liquid medium is not available.

Ultrasonic Tools Nowadays, the UT has been the most accurate ILI technique for metal loss measurements on pipelines. The basic principle of UT is to send an ultrasound wave that is perpendicular to the pipe wall thickness and measure the time for the wave signal to be reflected from the inside and outside surfaces of the pipe. Metal losses on both internal and external pipe surfaces are detected by changes in the time of transmission of the signal, which are then converted into the actual pipe wall thickness. The UT must be run in a liquid medium to couple the ultrasonic signal to the pipe wall thickness [Eiber, 2003]. Thus, the UT is usually used in liquid pipelines, while the MFL tools work in both liquid and natural gas pipelines. The UT is poor at detecting and sizing short or narrow features such as corrosion pits, pinholes, and axial and circumferential slotting [Vanaei et al., 2017]. The measurement accuracy is affected by fluid flow speed during the ILI run and depends on the presence of a clean surface condition of the pipe interior. The corrosion depth detection accuracy of UT tools is about ±0.3 to ±0.6 mm [Caleyo et al., 2007], with a confidence level of 95%. The UT is much more accurate, with a greater resolution, to detect metal loss than the MFL, especially for heavy-wall pipes.

UTs for crack detection use an ultrasound wave that is oriented at approximately 45° to the pipe surface. Based on pulse echo time technology, the reflected signals from cracks are used to detect the crack's presence and depth. When the ultrasonic signal transmission direction is parallel to the fracture surface of a crack, the crack cannot be detected. Generally, the UT is used to inspect cracks either perpendicular or parallel to the axial direction of the pipe, i.e., circumferential cracks (usually using multiple sensors) or axial cracks [Zhou et al., 2019]. The UT employs either liquid-filled wheels or operates in a liquid bath to detect SCC and fatigue cracks [Maxey et al., 1998]. While the common UT tools cannot inspect cracks with a depth above 4 mm (0.16 in), newly developed UT tools overcome uncertainty beyond 4 mm, with a high accuracy sizing axial cracks as deep as 100% of the pipe wall thickness [Hennig, 2017]. The obtained information for inspected cracks or crack colonies includes the length, depth, and orientation of the pipeline axis. To date, the UT crack detection tools have been successfully used in liquid pipelines for mainly axial crack characterization. However, there have been a few works using UT tools to detect cracks at grith welds. There was one report describing the UT technique used to detect crack defects inside a girth weld. The detection limits for the cracks include a length of 30 mm and above, and a depth smaller than 1–2 mm [Xiang and Jia, 2015].

Electromagnetic Acoustic Transducer Compared with MFL and UT techniques, the EMAT is new and can detect all kinds of cracks, weld characteristics, and wall thickness variations. The EMAT, which is also known as an eddy current-acoustic inspection technology, generates ultrasound through Lorentz forces without requiring a coupling liquid agent, which is a big advantage, making it applicable in gas pipelines [Xie and Tian, 2018]. The EMAT has been successfully used to detect and size stress corrosion cracks [Kania et al., 2012], crack-like defects [Hilvert and Beuker, 2015], and corrosion defects [Hirao and Ogi, 1999]. Attempt was made to use the EMAT to detect girth weld cracks, especially the circumferential cracks, where pairs of EMAT converters were distributed to have full circumferential coverage [Canni et al., 2015]. According to statistics of EMAT-inspected 13 pipelines from 2008 to 2011, it was shown that the EMAT detection rate of axial cracks exceeded 90%, and the success rate of EMAT inspection for the depth of an axial crack ranged 86–100%. However, the inspected length was the total of the crack subject to a crack cluster or interaction of multiple cracks. It was more difficult to determine the length of a crack than its depth. The success rate of inspection for the length of axial SCC was 0–33% only [Kania et al., 2014]. Generally, the EMAT becomes more accurate as the defect depth increases, but the inspection results tend to be overestimated. Cracks of 2 mm in depth and 40 mm in length can be highly identified, but any blunt crack with a depth of 1–2 mm is unlikely to be identified. The inspection results are not reliable if the cracks are located less than 1 mm under the steel surface [Fore et al., 2014]. The EMAT tools should be located less than 1 mm from the pipe body, which is too close to apply a high frequency. The detection ability and efficiency of EMAT are not as good as the UT. The electromagnetic ultrasonic signal has a lower amplitude and a higher sensitivity to the noise in the surroundings.

1.3.2 Assessment

Assessment is an essential element of a well-developed pipeline integrity management program. The assessment element is a highly technical task, covering a wide range of subject topics. It includes analysis of various historical data, pipeline operating and maintenance records, and ILI data; determination of pipeline FFS; prediction of failure pressure; and evaluation of the risk associated with pipeline operation and potential failures to all relevant parties and stakeholders. This element deals with the integrity goals to be achieved and the targets that are set to ensure that such goals are accomplished. Thus, the assessment helps outline risks to the pipelines and prioritize steps such as mitigation, prevention, and control to remediate and eliminate these risks [Cheng and Norsworthy, 2017]. Furthermore, the assessment element also includes legal and regulatory requirements, and risks to community, environment, and ecology.

Generally, the frequency of conducting integrity assessments on pipelines depends on either the need for a standard periodic requirement or the need in response to instant changes in environments and operating conditions that may increase the likelihood of the pipelines failing or facing an increased severity of integrity threat. A standard assessment can be scheduled after a required period from the previous assessment, while a nonroutine assessment is conducted when required for newly observed threats such as geohazards or unusual corrosion or crack defects.

A well-established process is recommended for implementing the pipeline integrity assessment [Eiber, 2003]. This process includes steps as listed below.

- Assessment planning and scheduling, which sets assessment objectives, plans data gathering, categorizes potential failure or damage mechanisms, and identifies hazards and threats.
- Threat assessment, identifying and determining actual hazards and threats, and analyzing the implications of the hazards and threats on pipeline integrity based on predeveloped models or evaluation criteria.
- Threat assessment validation, where the identified hazards and threats and their effects on pipeline integrity are validated by historical and empirical data, and field results.
- Management review, where the management team prioritizes the integrity threats, designs threat management plans, and implements management measures.

In addition to assessment of various threats, the consequence associated with pipeline failures will be assessed and included in the assessment element. The consequence assessment process considers the severity, extent, and impact of any consequence resulting from pipeline failures which may occur in various modes. The consequence assessment will usually identify direct areas (i.e., locations that run along the pipeline's ROW) that will be impacted by a pipeline failure, and indirect areas (i.e., locations that could be impacted due to water crossings although they may be further away from the ROW areas) that can be impacted by a pipeline failure [Kishawy and Gabbar, 2010]. The assessment requires expertise in multiple areas such as engineering, management, environment and energy policy. Some basic steps in consequence assessment include [Cheng and Norsworthy, 2017]:

- Consequence assessment planning, setting goals, targets, and objectives, as well as data gathering and scheduling.
- Consequence analysis, assessing the extent of potential consequences.
- Consequence classification along the pipeline both quantitatively or qualitatively, which is often performed according to predefined consequence criteria.
- Result validation, where field verification is usually conducted for verification.
- Management review, which covers both assessment and action implementation.

Pipeline risk assessment refers to evaluation and assignment of risk levels to pipelines and determination if the risk level is acceptable. The risk assessment can enable optimization of the scheduling of pipeline inspections, maintenance, and replacement, and enhancement of pipeline reliability. Risk is defined as the multiplication of the probability of failure with the consequence induced by the failure event. A risk assessment process includes collection and analysis of individual risk factors, establishment of absolute or relative risk magnitudes, and assessment of potential consequences, as mentioned above. For risk management, the risk magnitude helps categorize pipelines into intolerable/unacceptable, tolerable, and acceptable. Three methods can be used for risk assessment, including quantitative, semiquantitative, and qualitative methods [Singh, 2017]. Selection of a risk assessment method depends on the time available to conduct the risk assessment, expected level of effort to be made for data gathering and processing, and the intended use of the resulting risk assessment results.

1.3.2.1 Determination of FFS of Pipelines

The FFS concept refers to the principle that an engineering structure performs adequately for its purpose under a condition that does not cause failure to the structure [British Standard, 2019]. Determination of the structural FFS is a multidisciplinary approach, requiring a quantitative or semiquantitative evaluation of integrity of the structure that contains a flaw (such as corrosion) or damage (such as dent). The evaluation of FFS is also called "engineering critical assessment" (ECA).

As stated, pipelines in service experience various threats to degrade the structural integrity and cause pipeline failures. The FFS determination is a crucial task that evaluates the integrity and performance condition of a pipeline under given conditions and predicts the remaining service life of the pipeline. The goal of the FFS assessment is to ensure the pipeline operates in a safe and reliable condition. The identified or perceived threats to the pipeline and their effects (usually adversely) on pipeline integrity are evaluated, either quantitatively or semiquantitatively, by various models and codes which are developed mainly based on fracture mechanics principle, as indicated by relevant parameters such as stress, strain, and/or failure pressure. The FFS assessment will also provide information about the likelihood of the threat occurrence, and associated consequences and risks [American Petroleum Institute, 2000].

FFS assessment offers a comprehensive analysis of a pipeline's current and future integrity regulations. Based on a systematic flow path, as shown in Figure 1.9, the FFS assessment provides a complete overview of the severity and potential consequences of various types of integrity threats that are already in place, the pipeline safety state under the present condition, and possible measures and actions to be taken for effective mitigation and elimination of the threats in order to resume the pipeline's sound and healthy integrity state.

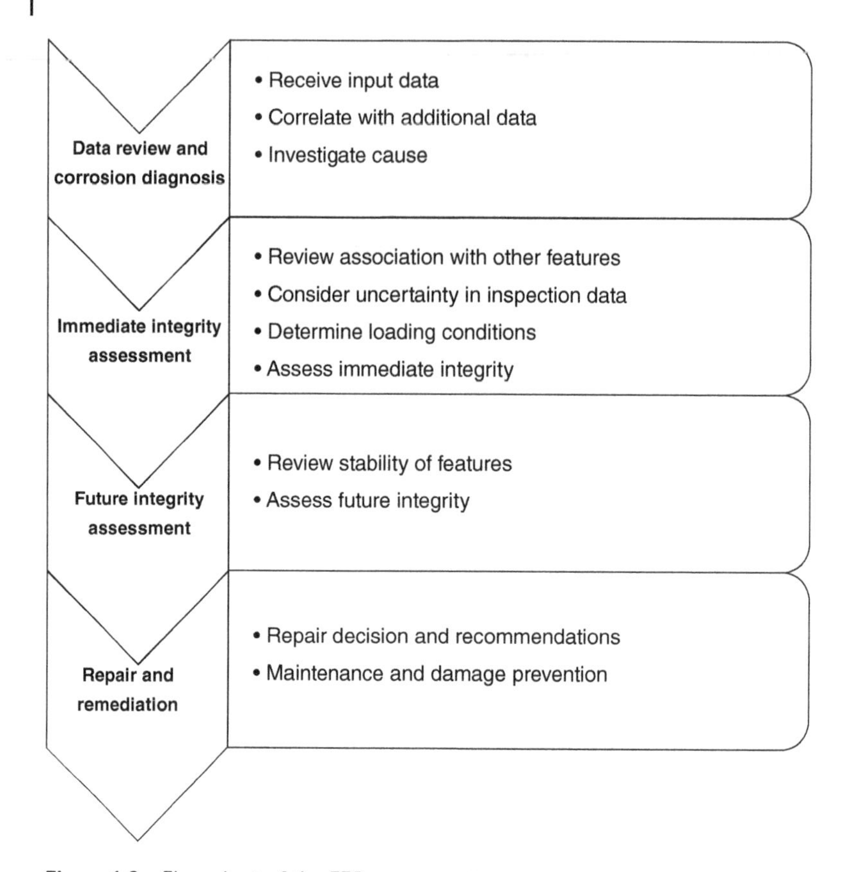

Figure 1.9 Flow chart of the FFS assessment process.

Upon completeness of ILI and collection of recorded data, the FFS assessment is often conducted by experienced experts in a wide range of subjects including materials, mechanics, corrosion, welding, computational modeling, management, and pipeline engineering. The majority of FFS assessments are for metal loss defects such as corrosion and mechanical damage on pipelines. The assessments will determine the impact of metal loss features on the immediate and future integrity of the pipelines, specifically, the failure pressure of the pipelines under the present condition, by various methods, standards and codes which are always under further improvement for enhanced assessment accuracy. The underlying causes of any metal loss features will be analyzed, and mitigation actions will be taken to stop further growth of the features. A repair plan may also be provided. In addition to metal loss, other defects such as dents, cracks, wrinkles, and buckles will be assessed in terms of their effect on pipeline integrity and, particularly, the failure pressure of the defective pipelines. Details of the assessment of various types of defect features will be given in the book.

1.3.3 Mitigation

To minimize potential risks associated with various types of threats and defects as identified and to ensure a sound operating condition through FFS assessment, mitigative measures, primarily repair work, should be planned based on the goal of maintenance of pipeline integrity and long-time service safety. Selection and conduction of an effective repair method is important, subject to considerations of multiple factors, where regulations and standards are usually used as guidelines. For example, damaged gas transmission pipelines should be either cut out and replaced or repaired by methods which are not only reliable but also engineering tested and analyzed [Farrag, 2013].

There are many methods that are appropriate for external repair of corrosion, microcracks and mechanical damages to restore the serviceability of pipelines. Some main methods are listed for further discussion.

1.3.3.1 Composite Sleeve Repair

Composite materials have been extensively used in pipeline repair with additional reinforcing capability. The materials are primarily thermoset polymers reinforced with glass or carbon fibers, such as polyester and polyurethane. The composite material sleeve is wrapped on the pipe surface to cover the damaged area. The hoop stress resulting from internal pressure is uniformly distributed on the pipe wall where the sleeve reinforcement, along with the remaining steel wall, bears the operating pressure. The repair sleeve restores the strength of the damaged pipeline. Depending on severity of the damage or defect, the thickness of the composite sleeve can be varied.

1.3.3.2 Pipe Wall Grinding and Recoating

Grinding is a common method to remove local stress raisers such as corrosion pits, microcracks, and mechanical damages to produce a smooth surface [Kiefner and Alexander, 1999]. During grinding, the operating pressure of the pipeline should be reduced by 20%. For removal of microcracks, the NDT should be used for verification after grinding. The removed thickness cannot exceed 40% of the pipe wall thickness.

1.3.3.3 Metallic Sleeve Repair

There are mainly two types of metallic sleeves, i.e., steel-reinforcing sleeves (i.e., type A) and pressure-containing sleeves (i.e., type B). The steel-reinforcing sleeves include two half pieces of steel cylinder, which are installed on the pipe exterior and then fully encircle the pipe, improving the strength of the pipe. The sleeve halves are normally welded together while the sleeve ends are not. Type B sleeve repairs a pipe with a low rate of defects and can function well without requiring high structural integrity to provide the best quality experience. The differentiating factor between type A and type B sleeves is that type B is welded directly onto the pipe. Type B sleeve can repair pipe leaks and reinforce circumferentially oriented defects.

1.3.4 Monitoring

No matter if a pipeline in operation is in its original condition or upon repair, it is always under various monitoring programs. A monitoring program can determine if a threat or consequence has occurred (e.g., new corrosion or cracking events) or is likely to occur (e.g., a new pressure cycling). It can also help identify whether the threat or consequence has changed (e.g., crack or corrosion growth). Monitored conditions affecting pipeline integrity include generation of new hazards (e.g., unauthorized ROW activities) and threats (e.g., theft), and changes in hazards (e.g., pressure cycling) and threats (e.g., crack growth) [Mora et al., 2016]. Some common and advanced monitoring techniques include aerial patrol, walking and riving the pipeline ROW, satellite, or other forms of remote monitoring (e.g., video surveillance), acoustic tools, fiber optic cables, etc. Integrity monitoring may be followed by additional measures such as mitigation and prevention, as well as reinspections. It is realized that some threats may not be able to be monitored all the time. For example, aerial or satellite monitoring of the pipeline ROW may not be available when an excavation activity or theft is taking place over the pipeline.

1.3.4.1 Internal and External Corrosion Monitoring

Performing periodic ILI and comparing data obtained from one survey to another serve as a monitoring method to estimate corrosion or defect growth rate. If a high growth rate is determined, the location will be regarded as a priority for reinspection or remediation. However, ILI tools may not be run frequently. A time interval of several years is quite common. Thus, the data comparison and result derivation may take a long time. For pipelines with disbonded coatings that shield CP current, the ILI is proper to monitor both external and internal corrosion. External corrosion of pipelines is also usually monitored with CIS of the CP potentials and other techniques such as DCVG and ACVG. These techniques also provide information about the coating performance condition. Internal corrosion is monitored by installing corrosion coupons and analysis of corrosion products taken during maintenance pigging [Cheng and Norsworthy, 2017].

1.3.4.2 Crack Monitoring

Periodic ILI surveys through proper tools are the best method for crack monitoring, including SCC, fatigue, and other manufacturing-related cracks. The locations of newly detected cracks and the cracks with an increase in size, as compared with previous ILI surveys, should be carefully documented, serving as an essential base for assessment of the crack initiation and growth. Thus, development of a plot to correlate reported crack size with previously reported sizes is useful in monitoring and management of cracks.

1.3.4.3 Welding Defect Monitoring

Welding defects are usually difficult to monitor. Most ILI tools cannot detect defects in girth welds. Instead, identification of locations with a high strain at the welds by strain measurement gauges could help refer to the welding locations that may not experience qualified quality control and should be monitored for potential crack initiation. Most cracks are generated in metallurgically imperfect regions with a high stress/strain concentration. These regions should be monitored with NDT and repaired as necessary.

1.3.4.4 Mechanical Damage Monitoring

Mechanical damage such as a dent is usually associated with a large deformation. Regions with detected deformations on the pipe body should be prioritized for integrity assessment. The damage can cause immediate failure of the pipeline or result in delayed failure owing to crack propagation over time. Thus, the damaged region should be remediated immediately when it is detected by the deformation survey.

1.3.4.5 Incorrect Operation Monitoring

A typical misoperation that is not uncommonly encountered on pipelines is pressure excursions or even unusual elevation of operating pressure. Supervisory control and data acquisition (SCADA) systems can monitor misoperational events. Frequently, an incorrect operation such as pressure excursions can be a leading reason for compromising the integrity of pipelines.

1.3.5 Prevention

Prevention of generation of various threats, hazards, and defects is key to maintaining integrity of pipelines. Prevention focuses on inhibition and elimination of threats and hazards by measures and actions that will prevent the pipelines from being exposed to conditions or environments where the integrity is compromised. The threat prevention can be applied in entire service life cycle of the pipelines. For example, quality control and quality assurance during pipeline manufacturing and construction can reduce the likelihood of various defects, such as dents, scratches, and coating damages before the pipeline is in operation. Preventive actions are usually implemented during pipeline operation to maximize the integrity of maintenance performance. During pipeline operation, preventive measures to maintain pipeline integrity include [Farrag et al., 2019]:

- Various ILI tools that inspect and identify metal loss, weld defects, and cracks.
- Real-time monitoring of the whole system with devices (e.g., flow and pressure transmitters) and operating surveillance systems that recognize and locate where leaks and failures occur to predict and prevent catastrophic failures.
- Real-time monitoring of excavation damage using sensing technologies installed along pipeline ROW.
- Threat preventive measures, damage mitigation approaches, repairs, and procedures that prevent catastrophic failures.
- Effective site selection and failure prediction models.
- Accurate inspection tools and techniques.
- Emergency response systems for automatic and timely shut-off, especially in high-consequence areas.

1.4 Plan-Do-Check-Act Integrity Management Cycle

Pipeline operators usually develop a so-called "Plan-Do-Check-Act" (PDCA) iterative cycle to implement and improve their integrity management program, ensuring sustainable pipeline operation. Figure 1.10 shows an example of the contents in a PDCA cycle. Briefly, "Plan" works ahead for changes, and analyzes and predicts results, "Do" executes the plan within a controlled environment, "Check" compares the results against the prediction, and "Act" standardizes or improves the entire process [Rees et al., 2007]. It

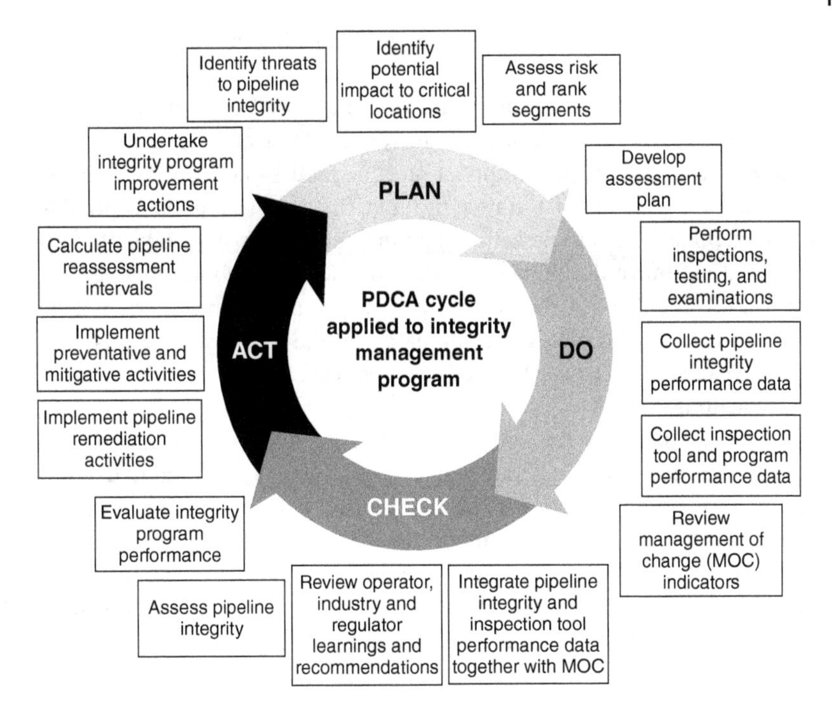

Figure 1.10 Main contents of a PDCA cycle.

has demonstrated that the PDCA cycle facilitates an efficient, structured approach toward all pipeline integrity tasks and drives an effective system.

1.4.1 Plan

As a general and obligatory guide, the pipeline integrity management PDCA cycle must be aligned with the operator's overall goals and objectives. The implementation of the developed PDCA cycle will meet the expectations of pipeline stakeholders and serve as the base for a sustainable pipeline operation in an efficient, safe, and reliable way. Strategically, the PDCA cycle must be developed to ensure seamless compatibility with other business units. Specific targets should be viewed as a series of integrated, clearly defined, and executable tasks. The managed pipelines are usually divided into segments based on service history, transported products, properties and characteristics, potential threats, consequence classes, and other criteria. The segmentation can be performed dynamically and is in response to instant changes encountered by the pipelines. Extensive data are collected and recorded based on the segmentation, but they are used for threat and consequence analysis in a consistent manner, making the results comparable from segment to segment.

Analysis of various data, primarily ILI data, based on well-established models should assess, either quantitatively to qualitatively, the susceptibility of pipelines to threats, probability of pipeline failure and associated consequences. The purpose of the integrity and risk assessments is to provide recommendations of proper actions and measures that will be taken to mitigate and minimize the threats and their impact on pipelines, meeting the goals and objectives of the company. The pipeline segments should be ranked in terms of their risk profile based on the assessment results. The ranking can serve as a basc to prioritize pipe segments for integrity maintenance actions.

If available data are not sufficient to perform integrity and risk assessments, a subject matter expert (SME) approach may be required to develop an initial risk profile [Rees et al., 2007]. Further data are obtained later, or historical data can be added to support this approach to extrapolate a relative risk model. Full probabilistic models can be operated using system-specific risk-algorithm inputs into specialist risk assessment systems. Upon validation, the assessment results are applicable to pipeline integrity management.

1.4.2 Do

Inspection is the key component of the Do stage in a PDCA cycle. In this stage, proper techniques are selected, and the inspection is deployed either along (for indirect surveys) or within (for ILI tools) each pipeline segment. Based on the inspection results and data analysis, location-specific recommendations are given, including (1) corrective responses (i.e., address inspected defects or remove detected threats so that each pipe segment is restored to a sound condition), (2) scheduled responses (i.e., address propagating or growing threats and corrosion, and sustain the pipeline's FFS over its intended life), (3) reinspection schedules contributing to a reduced probability of failure in the future service life, and (4) pipeline extension measures if a continuation is desired beyond the time when the pipe is normally considered for decommissioning.

1.4.3 Check

The Check stage closes the first iteration of a PDCA cycle. Verifications or examinations will be conducted to check if the implemented actions contribute to the defined goals and objectives. For example, to address pipeline corrosion, the Check stage includes measurements of corrosion management performance, a review of the corrosion measures, investigations of the causes of corrosion occurrence, and a summary of lessons from the corrosion-related incidents. The American Petroleum Institute (API) standard provides guiding questions to evaluate the plan conductance and effectiveness [American Petroleum Institute, 2001]. The conclusions drawn from the Check stage can be crosschecked by an SME-led performance level assessment [Colquhoun and Macpherson, 2006].

1.4.4 Act

A pipeline integrity management program, once developed and implemented, has always been in improvement, meeting the goals and objectives of sustainable pipeline operation. In the Act stage, the conclusions from the Check stage are reviewed to either validate the continuation of a consistent approach or to determine how subsequent iterations of the PDCA cycle should be continued to improve each stage and the entire cycle. When it is decided to make a new adjustment, this will be conducted in the subsequent Act stage of the cycle. To ensure the Act stage is functional as designed, three fundamental conditions are required. First, it is ensured that additional data collected during the Do stage and executed in the previous iteration are added as input for the subsequent risk assessment. Second, both outstanding and scheduled actions must be carried over. Finally, subsequent iterations of the PDCA cycle are managed by the operational expert system so that the cycle is linked with the whole integrity management system.

References

Alexander, C.R. (1999) Review of experimental and analytical investigations of dented pipelines, *Press. Vessel. Pip. Conf.*, ASME, Boston, MA, USA

Allouti, M., Schmitt, C., Pluvinage, G. (2014) Assessment of a gouge and dent defect in a pipeline by a combined criterion, *Eng. Fail. Anal.*, 36, 1–13.

American Petroleum Institute (2000) *Fitness-for-Service*. Recommended Practice 579, Ed. 1, Washington, DC, USA.

American Petroleum Institute (2001) *Managing System Integrity for Hazardous Liquid Pipelines*, National Standard 1160, Washington, DC, USA.

American Society of Mechanical Engineering (2014) *Managing System Integrity of Gas Pipelines*, ASME B31.8S, New York, NY, USA.

Banff Pipeline Workshop (2015) *Internal Corrosion in Transmission Pipelines*, Banff, Alberta, Canada.

Bauman, G. (2009) Construction issues, In: *PHMSA Workshop on New Pipeline Construction Practices*, Fort Worth, TX, USA.

Birnbaum, H.K. (1994) Hydrogen effects on deformation – relation between dislocation behavior and the macroscopic stress-strain behavior, *Scr. Metall. Mater.*, 31, 149–153.

Biro, A.L., Chenelle, B.F., Lados, D.A. (2012) Processing, microstructure, and residual stress effects on strength and fatigue crack growth properties in friction stir welding: a review, *Metall. Mater. Trans. B*, 43, 1622–1637.

British Standard (2019) *Guide to Methods for Assessing the Acceptability of Flaws in Metallic Structures*, British Standard BS 7910, London, UK.

Caleyo, F., Alfonso, L., Espina-Hernández, J.H., Hallen, J.M. (2007) Criteria for performance assessment and calibration of in-line inspections of oil and gas pipelines, *Meas. Sci. Technol.*18, 1787–1799.

Canadian Energy Pipeline Association (1996a) *Submission to the National Energy Board, Proceeding MH-2-95*. Vol. 2, Appendix D, Tab 9, Calgary, AB, Canada.

Canadian Energy Pipeline Association (1996b) *Submission to the National Energy Board, Proceeding MH-2-95*, Vol. 2, Appendix D, Tab 5, Calgary, AB, Canada.

Canadian Energy Pipeline Association (1996c) *Submission to the National Energy Board, Proceeding MH-2-95*, Vol. 2, Appendix D, Tab 4, Calgary, AB, Canada.

Canadian Energy Pipeline Association (2013) *Facilities Integrity Management Program Recommended Practice*, Ed. 1, Calgary, AB, Canada.

Canadian Energy Pipeline Association (2014) *A/C Interference Guideline Final Report*, Calgary, AB, Canada.

Canadian Energy Pipeline Association (2015) *2015 Pipeline Industry Performance Report*, Calgary, AB, Canada.

Canadian Energy Pipeline Association (2021) *Canadian Energy Evolving for Tomorrow*, Transmission Pipeline Industry Performance Report, Calgary, AB, Canada.

Canni, G., McAughey, K., Paige, D., Giese, J. (2015) *Feasibility of EMAT-Based Circumferential Crack Detection*, Report PR-437876, Pipeline Research Council International, Houston, TX, USA

Cerniauskas, S., Jose Chavez Junco, A., Grube, T., Robinius, M., Stolten, D. (2020) Options of natural gas pipeline reassignment for hydrogen: cost assessment for a Germany case study, *Int. J. Hydrogen Energy*, 45, 12095–12107.

Chen, H.F., Shu, D. (2001) Simplified limit analysis of pipelines with multi-defects, *Eng. Struct.*, 23, 207–213.

Cheng, Y.F. (2013) *Stress Corrosion Cracking of Pipelines*, John Wiley, Hoboken, NJ, USA

Cheng, Y.F. (2016) Environmental hazard: monitor safety of aged fuel pipelines, *Nature*, 529, 156.

Cheng, Y.F. (2021) *AC Corrosion of Pipelines*, AMPP, Houston, TX, USA

Cheng, Y.F., Norsworthy, R. (2017) *Pipeline Coatings*, NACE, Houston, TX, USA

Cherniavsky, A. (2018) Ratcheting analysis of “pipe–freezing soil” interaction, *Cold Regions Sci. Technol.*, 153, 97–100.

Chiodo, M.S.G., Ruggieri, C. (2009) Failure assessments of corroded pipelines with axial defects using stress-based criteria: numerical studies and verification analysis, *Int. J. Press. Vessel Pip.*, 86, 164–176.

Cole, I., Marney, D. (2012) The science of pipe corrosion: a review of the literature on the corrosion of ferrous metals in soils, *Corros. Sci.*, 56, 5–16.

Colquhoun, I., Macpherson, H. (2006) Management system approach to pipeline integrity, In: *Proc. 6th Int. Pipeline Conf.*, ASME, Calgary, AB, Canada.

Cosham, A., Hopkins, P. (2004) The effect of dents in pipelines - guidance in the pipeline defect assessment manual, *Int. J. Press. Vessels Pip.*, 81, 127–139.

Coulson, K.E.W., Worthingham, R.G. (1990) Pipe corrosion conclusion: new guidelines promise more accurate damage assessment, *Oil Gas J.*, 88, 41–44.

Department of Energy (2006) *International Energy Outlook*. Energy Information Administration Office of Integrated Analysis and Forecasting, Washington, DC, USA

Deshmukh, A.R., Venkatachalamb, G., Divekarc, H., Saraf, M.R. (2014) Effect of weld penetration on fatigue life, *Procedia Eng.*, 97, 783–789.

Det Norske Veritas (2004) *Corroded Pipelines*, RP-F10, Recommended Practice, Norway.

Djukic, M.B., Bakic, G.B., Zeravcic, V.S., Sedmak, A., Rajicic, B. (2019) The synergistic action and interplay of hydrogen embrittlement mechanism in steels and iron: localized plasticity and decohesion, *Eng. Fract. Mech.*, 216, 106528.

Dong, C.F., Liu, Z.Y., Li, X.G., Cheng Y.F. (2009) Effects of hydrogen-charging on the susceptibility of X100 pipeline steel to hydrogen-induced cracking, *Int. J. Hydrogen Energy*, 34, 9879–9884.

Eiber, B. (2003) *An Overview of Integrity Assessment Methods for Pipelines, Washington Cities and Counties Pipeline Safety Consortium*, Robert J. Eiber Consultant Inc., Columbus, OH, USA

Farrag, K. (2013) *Selection of Pipe Repair Methods*, Final Report of GTI Project Number 21087, Pipeline and Hazardous Materials Safety Administration (PHMSA), Department of Transportation, Washington, DC, USA

Farrag, K., Marean, J., Stubee, E., Gauthier, S., Olcksa, P. (2019) *Pipeline Safety and Integrity Monitoring Technologies Assessment*, Final project report prepared for California Energy Commission, Gas Technology Institute, Des Plaines, IL, USA.

Federal Energy Regulatory Commission (2011) *TransCanada Keystone Pipeline, LP, Containing Rules and Regulations Applying to the Transportation of Petroleum*, FERC No. 5.1.0, Washington, DC, USA.

Feng, Q.S., Li, R., Nie, B.H., Liu, S.C., Zhao, L.Y., Zhang, H. (2017) Literature review: theory and application of in-line inspection technologies for oil and gas pipeline girth weld defection, *Sensors*, 17, 1–24.

Focke, J. (2015) The future of pipeline integrity management, *Pipelines Int.*, Issue 3, 28–29.

Fore, T., Klein, S., Yoxall, C., Cone, S. (2014) Validation of EMAT ILI for management of stress corrosion cracking in natural gas pipelines, In: *Proc. 10th Int. Pipeline Conf.*, Calgary, AB, Canada.

Fowler, J.R., Alexander, C.R., Kovach, P.J., Connelly, L.M. (1994) *Cyclic Pressure Fatigue Life of Pipelines with Plain Dents, Dents with Gouges, and Dents with Welds.* AGA-94015627, American Gas Association, Arlington, VA, USA

Fu, A.Q., Cheng, Y.F. (2011) Characterization of the permeability of a high performance composite coating to cathodic protection and its implications on pipeline integrity, *Prog. Organ. Coat.*, 72, 423–428.

Galloway, D.L., Bawden, G.W., Leake, S.A., Honegger, D.G. (2008) Land subsidence hazards, *Landslide and Land Subsidence Hazards to Pipelines*, U.S. Geological Survey Open-File Report 2008-1164, Reston, VA, USA.

Gan, L.J., Huang, F., Zhao, Y., Liu, J., Cheng, Y.F. (2018) Hydrogen trapping and hydrogen-induced cracking of welded X100 pipeline steel in H_2S environment, *Int. J. Hydrogen Energy*, 43, 2293–2306.

Garrison Jr, W.M., Wojcieszynski, A.L. (2007) A discussion of the effect of inclusion volume fraction on the toughness of steel, *Mater. Sci. Eng. A*, 464, 321–329.

Godin, M. (2014) *Innovation Roadmap for Transmission Pipeline Transportation of Petroleum Products.* Canadian Energy Pipeline Association, Calgary, AB, Canada.

Green, K.P., Jackson, T. (2015) Canadian analysis shows pipelines safer than rail, *Pipeline Coatings*, 11, 18–20.

Hansen, M.E., Dursteler, E. (2017) *Pipelines, Rail & Trucks: Economic, Environmental, and Safety Impacts of Transporting Oil and Gas in the U.S*, Strata, Logan, UT, USA.

Harp, E.L. (2008) Landslide hazards to pipelines – regional hazard mapping, *Landslide and Land Subsidence Hazards to Pipelines*, U.S. Geological Survey Open-File Report 2008-1164, Reston, VA, USA.

Hennig, T. (2017) Removing stresses from pipeline inspection, *Pipeline Int.*, Issue 9, 40–41.

Hilvert, M., Beuker, T. (2015) High-resolution EMAT as a diagnostic tool for analysis of SCC and crack-like pipelines defects, In: *2015 India Int. Oil and Gas Pipeline Conf.*, paper no: IOGPC2015-7940, ASME, New Delhi, India.

Hirao, M., Ogi, H. (1999) An SH-wave EMAT technique for gas pipeline inspection, *NDT & E Int.*, 32, 127–132.

Hirth, J.P. (1980) Effects of hydrogen on the properties of iron and steel, *Metall. Mater. Trans. A*, 11, 861–890.

Hirth, J.P. (1984) Theories of hydrogen induced cracking of steels. *Hydrogen Embrittlement and Stress Corrosion Cracking*, R. Gibala and R.F. Hehemann, Editors, ASM, Metals Park, OH, USA, 29–41.

Hopkins, P. (2007) Pipelines: past, present, and future, In: *The 5th Asian Pacific Int. Institute Welding (IIW) Cong.* Sydney, Australia.

Hopkins, P., Jones, D.G. (1992) A study of the behavior of long and complex-shaped corrosion in transmission pipelines. Part A – pipeline technology, *Proc. 11th Int. Conf. Offshore Mechanics Arctic Eng. (OMAE 92)*, 211–217.

Jacobs Consultancy Inc. (2013) *Technical, Operational, Practical, and Safety Considerations of Hydrostatic Pressure Testing Existing Pipelines*, INGAA (i.e., Interstate Natural Gas Association of America) Foundation Final Report No. 2013.03, USA.

Jakubczak, H., Glinka, G. (1986) Fatigue analysis of manufacturing defects in weldments, *Int. J. Fatigue*, 8, 51–57.

Jin, T.Y., Cheng, Y.F. (2010) Effects of non-metallic inclusions on hydrogen-induced cracking of API5L X100 steel, *Int. J. Hydrogen Energy*, 35, 8014–8021.

Jin, T.Y., Cheng, Y.F. (2011) In-situ characterization by localized electrochemical impedance spectroscopy of the electrochemical activity of microscopic inclusions in an X100 steel, *Corros. Sci.*, 53, 850–853.

Jones, R.H. (1992) *Stress Corrosion Cracking: Materials Performance and Evaluations*, ASM, Metals Park, OH, USA

Kalisz, P. (2019) Impact of mining subsidence on natural gas pipeline failures, *IOP Conf. Ser.: Mater. Sci. Eng.* 471, 042024

Kania, R., Klein, S., Marr, J., Rosca, G., Riverol, E.S., Ruda, R., Jansing, N., Beuker, T., Ronsky, N.D., Weber, R. (2012) Validation of EMAT technology for gas pipeline crack inspection, *Proc. the 9th Int. Pipeline Conf.* ASME, Calgary, AB, Canada.

Kania, R., Rosca, G., Tandon, S., Gao, M., Krishnamurthy, R. (2014) Evaluation of EMAT tool performance and reliability, *Proc. 10th Int. Pipeline Conf.*, Calgary, AB, Canada.

Katzarov, I.H., Paxton, A.T. (2017) Hydrogen embrittlement II. Analysis of hydrogen-enhanced decohesion across (111) planes in α-Fe, *Phys. Rev. Mater.*, 1, 033603.

Kenny, S., Jukes, P. (2015) Pipeline/soil interaction modelling in support of pipeline engineering design and integrity, In: *Oil and Gas Pipelines: Integrity and Safety Handbook*, R.W. Review, Ed., Wiley, Hoboken, NY, USA, 99–142.

Kiefner, J.F., Alexander, C.R. (1999) *Repair of Line Pipe with Dents and Scratches*, Pipeline Research Council International (PRCI), Catalog No. L51788.

Kishawy, H.A., Gabbar, H.A. (2010) Review of pipeline integrity management practices, *Int. J. Press. Vessels Pip.*, 87, 373–380.

Kuang, D., Cheng, Y.F. (2015a) Study of cathodic protection shielding under coating disbondment on pipelines, *Corros. Sci.*, 99, 249–257.

Kuang, D., Cheng, Y.F. (2015b) Probing potential and solution pH under disbanded coating on pipelines, *Mater. Perf.*, 54, 40–45.

Lamontagne, M. (2002) Interaction rules – an integral factor, *Corrosion'2002*, NACE, Houston, TX, USA

Lenhart, T.R., Duncan, K.E., Beech, I.B., Sunner, J.A., Smith, W., Bonifay, V., Biri, B., Suflita, J.M. (2014) Identification and characterization of microbial biofilm communities associated with corroded oil pipeline surfaces, *Biofouling*, 30, 823–835.

Leyer, J., Sutter, P., Marchebois, H., Bosch, C., Kulgemeyer, A., Orlans-Joliet, B.J. (2005) SSC resistance of a 125 ksi steel grade in slightly sour environments, *Corros.'2005*, paper no. 05088, NACE, Houston, TX, USA.

Li, M.C., Cheng, Y.F. (2008) Corrosion of the stressed pipe steel in carbonate-bicarbonate solution studied by scanning localized electrochemical impedance spectroscopy, *Electrochim. Acta*, 53, 2831–2836.

Li, Y.M., Zhu, F.X., Cui, F.P., Fang, K. (2007) Analysis of forming mechanism of lamination defect of steel plate, *J. Northeastern Univ. Nat. Sci.*, 28, 1002–1005.

Li, Z., He, C., Liu, Z., Wu, B. (2019a) Quantitative detection of lamination defect in thin-walled metallic pipe by using circumferential lamb waves based on wavenumber analysis method, *NDT & E Int.*, 102, 56–67.

Li, H.W., Lai, Y., Wang, L., Yang, X. (2019b) Review of the state of the art: interactions between a buried pipeline and frozen soil, *Cold Regions Sci. Technol.*, 157, 171–186

Liu, H.W., Cheng, Y.F. (2017a) Mechanism of microbiologically influenced corrosion of X52 pipeline steel in a wet soil containing sulfate-reduced bacteria, *Corros. Sci.* 253, 368–378.

Liu, T., Cheng, Y.F. (2017b) The influence of cathodic protection potential on the biofilm formation and corrosion behaviour of an X70 steel pipeline in sulfate reducing bacteria media, *J. Alloy. Compd.*, 729, 180–188.

Liu, H.W., Cheng, Y.F. (2018) Microbial corrosion of X52 pipeline steel under soil with varied thicknesses soaked with a simulated soil solution containing sulfate-reducing bacteria and the associated galvanic coupling effect, *Electrochim. Acta*, 266, 312–325.

Liu, Z.Y., Li, X.G., Du, C.W., Lu, L., Zhang, Y.R., Cheng, Y.F. (2009) Effect of inclusions on initiation of stress corrosion cracks in X70 pipeline steel in an acidic soil environment, *Corros. Sci.*, 51, 895–900.

Lynch, S. (2012) Hydrogen embrittlement phenomena and mechanisms, *Corros. Rev.*, 30, 105–123.

MacDonald, K.A., Cosham, A. (2005) Best practice for the assessment of defects in pipelines – gouges and dents, *Eng. Fail. Anal.*, 12, 720–745.

MacDonald, K.A., Cosham, A., Alexander, C.R., Hopkins, P. (2007) Assessing mechanical damage in offshore pipelines – two case studies, *Eng. Fail. Anal.*, 14, 1667–1679.

Mashiri, F.R., Zhao, X.L., Grundy, P. (2001) Effects of weld profile and undercut on fatigue crack propagation life of thin-walled cruciform joint, *Thin-Wall. Struct.*, 39, 261–285.

Maxey, W.A., Mesloh, R.E., Kiefner, J.K. (1998) Use of the elastic wave tool to locate cracks along the DSAW seam welds in a 32-inch OD products pipeline, *Proc. Int. Pipeline Conf.*, vol. 1. ASME, Calgary, AB, Canada.

Mora, R., Hopkins, P., Cote, E., Shie, T. (2016) *Pipeline Integrity Management Systems: A Practical Approach*, ASME Press, New York, NY, USA

Naghipour, M., Ezzati, M., Elyasi, M. (2018) Analysis of high-strength pressurized pipes (API-5L-X80) with local gouge and dent defect, *Appl. Ocean Res.*, 78, 33–49.

National Academy of Sciences (2013) *Effects of Diluted Bitumen on Crude Oil Transmission Pipelines*, Transportation Research Board Special Report 311, Washington DC, USA

National Association of Corrosion Engineers (NACE) (2003) *Petroleum and Natural Gas Industries – Materials for Use in H_2S Containing Environments in Oil and Gas Production, Part 2: Cracking-Resistant Carbon and Low Alloy Steel, and the Use of Cast Iron*, MR0175/ISO 15156-2, NACE, Houston, TX, USA.

National Energy Board (1996) *Stress Corrosion Cracking on Canadian Oil and Gas Pipelines*, Report of the Inquiry, MH-2-95, Calgary, AB, Canada.

National Energy Board (2011) *Focus on Safety and Environment: A Comparative Analysis of Pipeline Performance 2000–2009*, Calgary, AB, Canada.

National Transportation Safety Board (2012) *Pipeline Accident Report: Enbridge Incorporated Hazardous Liquid Pipeline Rupture and Release, Marshall, Michigan, July 25, 2010*, National Transportation Safety Board/PAR-12/01, PB2012–916501, Washington, DC, USA.

Naumov, O., Moskvitin, G., Grigorieva, Y. (2019) Influence processes freezing and thawing of permafrost soils on underground pipeline's stress-strain state, *Procedia Struct. Integ.*, 20, 53–56.

Nelson, B.R. (2002) Pipeline integrity: program development, risk assessment and data management, *The 11th Annual GIS for Oil Gas Conf.*, Geospatial Information and Technology Association, Aurora, CO, USA.

Ogden, J., Jaffe, A.M., Scheitrum, D., McDonald, Z., Miller, M. (2018) Natural gas as a bridge to hydrogen transportation fuel: insights from the literature, *Energy Policy*, 115, 317–329.

Parkins, R.N. (2000) A review of stress corrosion cracking of high pressure gas pipelines, In: *Corros.'2000*, paper no. 363, NACE, Houston, USA.

Peng, X.Y., Liang, G.C., Jin, T.Y., Cheng, Y.F. (2013) Correlation of initiation of corrosion pits and metallurgical features of X100 pipeline steel, *Can. Metall. Q.*, 52, 484–487.

Peng, Z.X., Liu, J., Huang, F., Hu, Q., Cheng, C.Y., Liu, S., Cheng, Y.F. (2018) Effect of submicron-scale MnS inclusions on hydrogen trapping and HIC susceptibility of X70 pipeline steels, *Steel Res. Int.*, 89, 1700566.

Pipeline and Hazardous Materials Safety Administration (2020) *Crude Oil/ Refined Petroleum/Biofuel Accidents Impacting People or the Environment: Apparent Causes for 2010–2019*, Department of Transportation, Washington, DC, USA

Pipeline and Hazardous Materials Safety Administration (2021) *Gas Transmission Onshore Significant Incident Cause 2005–2020*. Department of Transportation, Washington, DC, USA

Psyrras, N.K., Sextos, A.G. (2018) Safety of buried steel natural gas pipelines under earthquake-induced ground shaking: a review, *Soil Dynam. Earthquake Eng.*, 106, 254–277.

Qin, G.J., Cheng, Y.F. (2021) A review on defect assessment of pipelines: principles, numerical solutions, and applications, *Int. J. Hydrogen Energy*, 191, 104329.

Rees, B., Grimes, K., Ackhurst, T. (2007) *The Value of an Integrity Management Cycle from a Business Perspective*, Pemex Producto, Villahermosa, Mexico.

Reuß, M., Grube, T., Robinius, M., Stolten, D. (2019) A hydrogen supply chain with spatial resolution: comparative analysis of infrastructure technologies in Germany, *Appl. Energy*, 247, 438–453.

Rosenfeld, M.J. (2002) Here are factors that govern evaluation of mechanical damage to pipelines, *Oil Gas J.*, 100, 64–73.

Rosenfeld, M.J., Pepper, J.W., Leewis, K. (2002) Basis of the new criteria in ASME B31.8 for prioritization and repair of mechanical damage, In: *Proc. 2002 Int. Pipeline Conf.*, IPC2002-27122, Calgary, AB, Canada.

Shuai, Y., Wang, X.H., Cheng, Y.F. (2020) Modelling of local buckling of corroded X80 gas pipeline under axial compression loading, *J. Nat. Gas Sci. Eng.*, 81, 103472.

Shuai, Y., Wang, X.H., Feng, C., Zhu, Y., Wang, C.L., Sun, T., Han, J., Cheng, Y.F. (2021) A novel strain-based assessment method of compressive buckling of X80 corroded pipelines subjected to bending moment load. *Thin-Wall. Struct.*, 167, 108172.

Singh, R. (2017) *Pipeline Integrity – Management and Risk Evaluation*, Ed. 2, Gulf Professional Publishing, Houston, TX, USA.

Sun, J.L., Cheng, Y.F. (2018) Assessment by finite element modeling of the interaction of multiple corrosion defects and the effect on failure pressure of corroded pipelines, *Eng. Struct.*, 165, 278–286

Sun, J.L., Cheng, Y.F. (2019a) Modelling of mechano-electrochemical interaction of multiple longitudinally aligned corrosion defects on oil/gas pipelines, *Eng. Struct.*, 190, 9–19.

Sun, J.L., Cheng, Y.F. (2019b) Investigation by numerical modeling of the mechano-electrochemical interaction of circumferentially aligned corrosion defects on pipelines, *Thin-Wall. Struct.*, 144, 106314.

Sun, J.L., Cheng, Y.F. (2021a) Hydrogen permeation and distribution at a high-strength X80 steel weld under stressing conditions and the implication on pipeline failure, *Int. J. Hydrogen Energy*, 46, 23100–23112.

Sun, J.L., Cheng, Y.F. (2021b) Modeling of mechano-electrochemical interaction between circumferentially aligned corrosion defects on pipeline under axial tensile stresses, *J. Petro. Sci. Eng.*, 198, 108160.

Sun, J.L., Cheng, Y.F., Woo, J., Kainat, M., Hassanien, S. (2021) Assessment of interaction between a dent and an adjacent corrosion feature on pipelines and the effect on pipeline failure pressure by finite element modeling, *J. Pipeline Sys. Eng. Pract.*, 12, 4021029.

Tang, X., Cheng, Y.F. (2009) Micro-electrochemical characterization of the effect of applied stress on local anodic dissolution behavior of pipeline steel under near-neutral pH condition, *Electrochim. Acta*, 54, 1499–1505.

TransCanada Pipelines (1996) *Response to National Energy Board Information Request # 2 of Proceeding MH-2-5*, Calgary, AB, Canada.

Trifunac, M.D., Todorovska, M.I. (2004) 1971 San Fernando and 1994 Northridge, California, earthquakes: did the zones with severely damaged buildings reoccur? *Soil Dynam. Earthquake Eng.*, 24, 225–239.

Vanaei, H.R., Eslami, A., Egbewande, A. (2017) A review on pipeline corrosion, in-line inspection (ILI), and corrosion growth rate models, *Int. J. Press. Vessel. Pip.*, 149, 43–54.

Vasseghi, A., Haghshenas, E., Soroushian, A., Rakhshandeh, M. (2021) Failure analysis of a natural gas pipeline subjected to landslide, *Eng. Fail. Anal.*, 119, 105009.

Wang, F., Chen, J. (2015) *Identification and Evaluation Technology of Magnetic Flux Leakage Detection Signals for Girth Weld Defects*, Technical Report QTSXJC-01, PetroChina Pipeline Company, Beijing, China.

Xiang, X., Jia, G. (2015) *Study on Ultrasonic Inspection Technology of Cracks in Girth Welding of Liquid Pipeline*, Technical Report MDJC-01. PetroChina Pipeline Company, Beijing, China.

Xie, M.J., Tian, Z.G. (2018) A review on pipeline integrity management utilizing in-line inspection data, *Eng. Fail. Anal.*, 92, 222–239.

Xu, L.Y., Cheng, Y.F. (2014) Experimental and numerical studies of effectiveness of cathodic protection at corrosion defects on pipelines, *Corros. Sci.*, 78, 162–171.

Xue, H.B., Cheng, Y.F. (2011) Characterization of microstructure of X80 pipeline steel and its correlation with hydrogen-induced cracking, *Corros. Sci.*, 53, 1201–1208.

Xue, H.B., Cheng, Y.F. (2013) Hydrogen permeation and electrochemical corrosion behavior of the X80 pipeline steel weld, *J. Mater. Eng. Perf.*, 22, 170–175.

Yoosef-Ghodst, N. (2015) Strain-based design of pipelines, In: *Oil and Gas Pipelines: Integrity and Safety Handbook*, R.W. Review, Ed., Wiley, Hoboken, NY, USA, 37–48.

Zarea, M., Batisse, R., Leis, B., Cardin, P., Vignal, G. (2012) Full scale experimental database of dent and gouge defects to improve burst and fatigue strength models of pipelines, In: *Proc. 2012 Int. Pipeline Conf.*, IPC2012-90620, Calgary, AB, Canada.

Zhang, G.A., Cheng, Y.F. (2009) Micro-electrochemical characterization of corrosion of welded X70 pipeline steel in near-neutral pH solution, *Corros. Sci.* 51, 1714–1724.

Zhang, C., Cheng, Y.F. (2010) Corrosion of welded X100 pipeline steel in a near-neutral pH solution, *J. Mater. Eng. Perf.*, 19, 834–840.

Zhao, W., Wang, Y., Zhang, T., Wang, Y. (2012) Study on the mechanism of high-cycle corrosion fatigue crack initiation in X80 steel, *Corros. Sci.* 57, 99–103.

Zheng, M., Luo, J.H., Zhao, X.W., Bai, Z.Q., Wang, R. (2005) Effect of pre-deformation on the fatigue crack initiation life of X60 pipeline steel, *Int. J. Press. Vessels Pip.*, 82, 546–552.

Zhou, W., Fan, J.C., Liu, S.J., Ni, J.L., Liu, X.Y. (2019) Experimental study on fatigue crack detection of marine riser based on magnetic dipole, *China Petro. Machinery*, 47, 38–44.

2

Levels I and II Assessment of Corrosion Anomalies on Pipelines

2.1 Defect Assessment for Pipeline FFS Determination

Multiple threats exist to degrade the structural integrity of pipelines in service. Particularly, the pipelines are susceptible to various damages and flaws, which are generally called anomalies, such as corrosion defects, dents, scratches, cracks, and wrinkles. The presence of these anomalies on pipelines can increase the local stress concentration and affect (usually decrease) the load-bearing (such as pressure-bearing) capability, potentially causing pipeline failures. Thus, proper measures must be taken to repair or remove the anomalies to enable pipeline resilience.

Repairing pipeline anomalies or replacement of the pipe segments containing the anomalies is costly. Moreover, the repair and replacement process can introduce new flaws and/or residual stress, which would further elevate the failure risk. The anomalies, once generated on pipelines, do not necessarily fail the pipelines immediately. A pipeline containing the anomalies is still operatable if the anomalies can pass a proper safety assessment criterion using carefully developed technical programs. Therefore, the pipeline integrity management program often conducts assessment on anomalies such as corrosion defects to understand their impact on structural integrity of the pipelines and to determine their FFS.

The pipeline FFS determination is based on quantitative evaluation of the structural integrity of a pipeline that contains anomalies under operating conditions. Along with the FFS determination, the defect assessment can also provide information about the failure pressure of the pipeline, and the local stress and strain concentrations at the anomalies. While the FFS determination by defect assessment is a multidisciplinary task, involving materials, mechanics, corrosion, numerical computation, etc., the basic principle of the defect assessment techniques is based on fracture

Defect Assessment for Integrity Management of Pipelines, First Edition. Y. Frank Cheng.

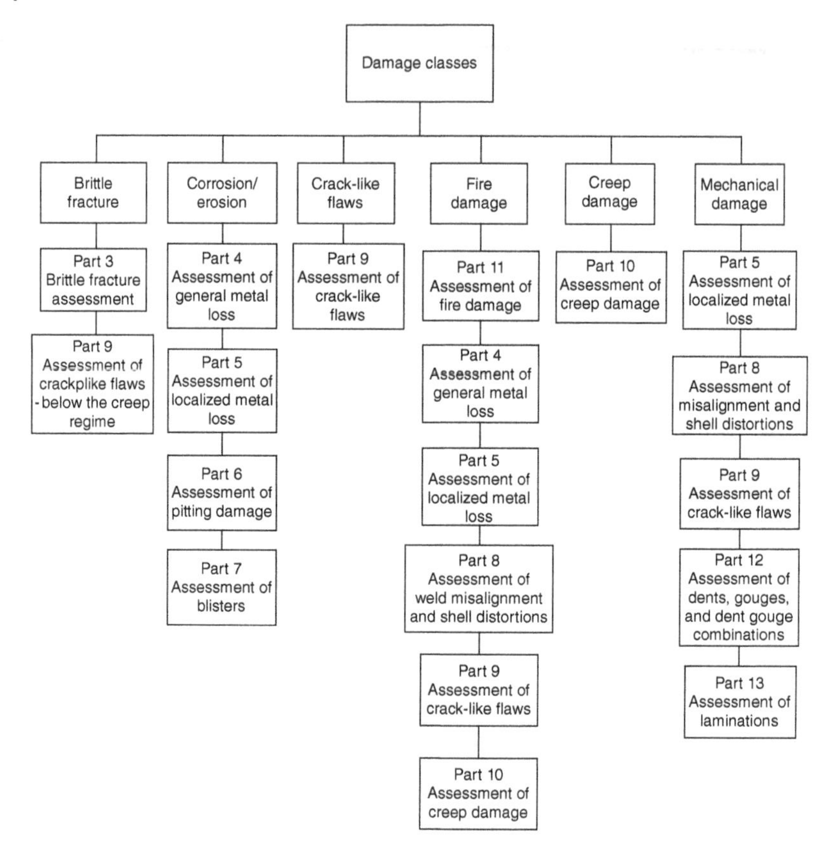

Figure 2.1 Example of the FFS determination process for various types of damage mechanisms on pipelines and the corresponding orders and contents for the damage assessment. *Source:* From American Petroleum Institute [2000].

mechanics. Due to the diversity of anomalies on pipelines and each type of anomalies is associated with a specific mechanism causing pipeline failure, the FFS assessment and determination is complicated. Figure 2.1 shows an example of the FFS determination process for various types of damage mechanisms on pipelines and the corresponding orders and contents for the damage assessment contained in American Petroleum Institute [2000].

For all types of defects and flaws, the general procedure to conduct the FFS assessment and determination is similar. First, the defect type will be characterized by relevant tools such as ILI tools and the formation mechanism will be analyzed. Proper assessment techniques will then be selected for specific defects, along with collection of all relevant condition data and operating parameters. The assessment will be performed on the pipelines containing defects to determine the FFS condition. The assessment results will be used to provide recommendations on whether the pipelines are acceptable for continuous operation. In addition, the failure pressure of the assessed pipelines will be predicted under the given condition. Next, the remaining service life of the pipelines will be estimated based on time-dependent algorithms for defect growth on the pipelines under normal operating conditions. If the defect growth does not depend on time, the remaining life assessment may not be conducted. Moreover, the assessment results will recommend whether the defective area on pipelines should be repaired or replaced. If the assessment is performed under Level I or Level II, whether a further assessment at a higher level is conducted will be considered.

Determination of FFS by defect assessment on pipelines brings tremendous benefits. It can help save the budget for pipeline integrity management through reduced inspections. The inclusion of the FFS determination as a part of the maintenance program can prevent the extension of shutdowns and potentially avoid unplanned shutdowns. For example, an FFS assessment can assist pipeline operators in quickly making decisions based on ILI data on whether a pipe segment containing defects needs to be repaired or can continue in service.

2.2 Evolution of Defect Assessment Techniques

Anomalies on pipelines threatening the pipeline integrity can be categorized into three groups, i.e., three-dimensional (3D) volumetric metal loss such as corrosion and gouging, geometric deformation such as dents and wrinkles, and crack-like flaws [Holiday et al., 2022]. Corrosion is one of the primary mechanisms degrading structural integrity to cause pipeline failures. A statistical analysis was performed on the database managed by the Pipeline and Hazardous Materials Safety Administration (PHMSA) about pipeline failures occurring from 2010 to 2015 that contains 432 crude oil pipeline failures and 238 natural gas pipeline failures [Dai et al., 2017]. The three major failure causes were corrosion, pipe/weld material failure, and equipment failure for oil pipelines, while three major failure causes for natural gas pipelines include pipe/weld material failure, excavation damage, and corrosion. It is thus seen that corrosion is one of the major causes of failures of both oil and gas pipelines. Moreover, corrosion can initiate cracks such as stress corrosion cracks. Therefore, corrosion defect assessment is critical to manage the pipeline integrity.

Of various inspection methods for pipeline integrity management, the ILI tools are the most used, as compared with direct assessment or hydrostatic testing, due to their inspection accuracy, practicality, and cost-effectiveness. Upon completion of ILI tool runs, the collected data will be analyzed to detect, locate, and size the defects including corrosion defects. Defect assessment is then performed to determine the FFS and predict failure pressure of the pipelines. Generally, a pipeline containing defects can continue to operate if a maximum allowable operating pressure (MAOP) passes the reliability assessment by determining failure pressure of the pipelines [American Petroleum Institute, 2000].

2.2.1 Historical Background of Defect Assessment on Pipelines

The defect assessment on pipelines started from evaluation of welding defects contained in pressure vessels in the manufacturing industry. To control the weld quality, standards and codes were proposed to control the weld defects on permissible levels. It was realized that these standards could, sometimes in specific cases, be conservative, especially for materials with great ductility under low stress levels. To determine the number of weld repairs at a more accurate level, it was proposed to develop a procedure for the industry to assess the FFS of equipment containing welding defects. As a result, testing was conducted to characterize the fracture behavior of welds containing defects using the crack tip opening displacement (CTOD). Built upon the testing program, a standard, i.e., British Standard (BS) PD6493 (now BS 7910), was issued for the assessment of defects in fusion-welded structures in 1991 [British Standard, 1991].

Later, similar requirements by the oil and gas industry were made to assess offshore jacket platforms in the North Sea reserves, where the platforms were of large tubular construction and contained a great number of welded joints between plates and nodal connections. The assessment was expected to perform a minimum number of weld repairs while maintaining safety of the structures to resist fatigue cracking in the service environment. Thus, new rules for the assessment of fatigue cracks to determine the FFS of the structures were introduced in the initial version of BS PD6493. In addition, the FFS assessment was also required by the nuclear industry to ensure a high integrity and tolerance to welding defects of critical pressure vessels. A wider application of FFS assessment procedures on pressure equipment in the petrochemical industry has been proposed in the recent years to extend the life of aging equipment, justify reduced inspections, and lower the cost of repairs and replacement associated with shutdowns. All of them drove the further development of defect assessment for FFS determination. In 2000, the API reviewed and compiled relevant procedures, and published the widely used recommended practice (RP), i.e., API 579 [American Petroleum Institute, 2000]. In addition to welding defects, corrosion, locally thinned areas, and mechanical damage such as dents and gouges were also included for assessment.

In the United Kingdom, along with the application of BS PD6493, further work on assessment of corrosion and locally thinned areas on pipelines was incorporated into the PD6493. With extensive application experiences, consolidations, and revisions, BS updated the PD6493 as BS 7910 for FFS assessment [British Standard, 2019]. Procedures for FFS assessment and determination have also been developed in other countries.

In the last half century, defect assessment on pipelines has evolved in significant developments of the modeling methods, geometrical definition, computational solutions and improved prediction accuracy, and reliability. The defect assessment methods can generally be grouped into two categories, i.e., generic methods and pipeline-specific methods [Cosham and Hopkins, 2004]. The generic methods include various technical procedures for assessing the impact of defects on engineering structures based on fracture mechanics principles [American Petroleum Institute, 2000, 2007; British Standard, 2013]. Generic standards could be overly conservative when applied to specific structures such as pipelines. As a comparison, the pipeline-specific methods are usually based on experimental testing such as burst tests, with limited theoretical validation. They are thus called semiempirical methods, which may become invalid or unreliable when applied outside the empirical limits.

2.2.2 Level-by-level Defect Assessment Approach

As stated, the FFS determination of pipelines containing defects by defect assessment is based on fracture mechanics, providing both scientific understanding and quantitative evaluation of the impact of defects on the pipelines. At present, there are three levels of methods for corrosion defect assessment on pipelines [Qin and Cheng, 2021]. Level I methods mainly assess a single corrosion defect and determine an allowable pressure capacity of corroded pipelines by taking a longitudinal area of metal loss as the primary geometrical parameter of the defect [Cosham and Hopkins, 2002]. Sometimes, the Level I defect assessments simply use tables and screening curves as bases while conducting limited computations. The assessment results can be quite conservative, like design code calculations.

When the Level I assessment does not meet acceptance criteria, a Level II assessment can be performed. The Level II assessment requires much more simulations and computations to reduce the conservatism associated with the Level I methods, generating more accurate results. The computations are so complex that certain software is often required. There are two types of Level II methods available for defect assessment [Benjamin et al., 2016a, b]. The Level IIa method uses an effective area to assess a corrosion defect with a complex shape, and the Level IIb method is developed to evaluate multiple defects in interaction.

Generally, if a pipeline passes the Level II assessment criteria, it can be returned to service. However, in severe cases, a Level III assessment may be required. The Level III defect assessment requires more detailed data, a well-developed numerical model with a sound physical meaning, a considerable amount of academic knowledge, and a certain computational software platform to solve numerical functions. Nowadays, Level III analysis has mainly been performed in laboratories and is rarely conducted in the industry. Various Level III defect assessment methods solve multiple nonlinear problems such as defect geometric nonlinearity, material nonlinearity, and boundary nonlinearity through finite element (FE) modeling and analysis [Caleyo et al., 2002]. By considering the dynamic nature of corrosion defect and its growth on pipelines in service environments, Cheng and coworkers integrate electrochemical corrosion reactions with the mechanical stress field, proposing the so-called "mechano-electrochemical interaction" theoretical concept [Xu and Cheng, 2013; Wang et al., 2021]. Based on the concept, an FE-based multiphysics field coupling model was developed, assessing corrosion defects and the resulting effect on failure pressure of pipelines, and quantifying the defect growth rate on the pipelines for prediction of the remaining service life. The Level III method has so far been the most accurate technique for defect assessment on pipelines.

Selection and use of the three levels of defect assessment techniques depend on specific task requirements, accuracy of the prediction results, and knowledge preparation for the assessment. Generally, the simplified assessment methods such as the Level I methods are more conservative than sophisticated and detailed engineering analyses such as Level III methods which can produce more realistic results that are more accurate than those from Levels I and II assessments. It is thus expected that a pipeline that fails the Level I assessment could pass the Level II or Level III assessment due to different conservatisms and simplifying assumptions in each of the assessment methods. Moreover, Level I assessments usually require specific procedures to be followed, with a minimum flexibility to analyze the modeled target. However, Level II methods permit a reasonable evaluation to determine the engineering reliability of the pipeline structure. For Level III assessments, a big flexibility exists, where the industry standards provide some guidelines only, and the assessment processes including the model development and validation are fully open to the integrity specialists. Thus, a step-by-step procedure guiding the advanced Level III analyses does not exist considering the varied and complex engineering conditions and scenarios.

2.3 Level I Defect Assessment on Pipelines

2.3.1 Principle and Codes

Corrosion defects are 3D metal loss features with longitudinal (i.e., length, L), circumferential (i.e., width, W), and radial (i.e., depth, d) dimensions present on pipe body. Accurate definition of the shape and geometric dimensions of a corrosion defect is always challenging due to irregularity of the defect. Moreover, the defect keeps growing by corrosion reaction with time. It is not uncommon that multiple corrosion defects have not separated from each other. Instead, they are linked or overlapped, making it more difficult to have an accurate dimensional determination. According to the Level I defect assessment method, the maximum metal loss is used to estimate residual strength of the corroded pipelines, where only the longitudinal and radial dimensions (i.e., L and d) of the defect are considered. In other words, the corrosion defect is simplified as a two-dimensional (2D) feature with a depth and a length only, while the defect width (W) is ignored [Kiefner and Vieth, 1990a]. Nowadays, numerous Level I-based standards and codes have been developed to assess corrosion defects and predict failure pressure of the corroded pipelines in industry. These include NG-18 equations [Kiefner et al., 1973], ASME B31G [ASME, 1991; 2012], modified B31G (i.e., RSTRENG-0.85) [Kiefner and Vieth, 1990b], SHELL-92 [Klever et al., 1995], Canadian Standardization Association (CSA) Z662 [Canadian Standardization Association, 2007], Det Norske Veritas (DNV)-RP-F101 [Det Norske Veritas, 2010], API RP 579 [American Petroleum Institute, 2000; Shdid and Hajali, 2015], FITNET FSS [Seib et al., 2008; Cicero et al., 2009], BS 7910 [British Standard, 2013; Cosham and Andrews, 2019], etc. In addition, existing Level I assessment models have been modified, and new models have been proposed in academic institutions and governmental research agencies for further improvement of the defect assessment accuracy [Cronin and Pick, 2000; Benjamin and Andrade, 2003; Choi et al., 2003; Mustaffa and Van Gelder, 2010; Keshtegar and Miri, 2014; Wang and Zarghamee, 2014; Ghani et al., 2016; Phan et al., 2017; Shuai et al., 2017].

As stated, the Level I defect assessment mainly considers the longitudinal and radial dimensions of the defect, as well as the internal operating pressure of pipelines [Fu, 1999; Bjørnøy and Marley, 2001]. Generally, the contour of a corrosion defect is described as rectangular, parabolic, or mixed shape. For example, according to NG-18 equations [Kiefner et al., 1973],

$$P_F = \sigma_{flow} \frac{2t}{D} \left[\frac{1-(A/A_0)}{1-A/A_0 \cdot M^{-1}} \right] \tag{2.1}$$

$$M = \sqrt{1 + \frac{2.51(L/2)^2}{Dt} - \frac{0.54(L/2)^4}{(Dt)^4}} \tag{2.2}$$

where P_F is failure pressure, σ_{flow} is flow stress, which is defined as a decrease in strength of pipeline steels due to corrosion, t is pipe wall thickness, D is pipe outer diameter, A_0 is cross-sectional area of the pipe before corrosion occurs (i.e., $A_0 = tL$), A is area of a corrosion defect which is a function of the depth (d) and length (L) of the defect, and M is Folias factor, i.e., a geometric correction factor to describe bulging effect of a shell surface with a thinned wall thickness than surrounding shells [Folias, 1956, 2000]. It is noted that the flow stress concept was initially proposed by Battelle to model complex plastic flow and the work-hardening phenomenon associated with structural collapse. Flow strength is a notional material property with a value between yield strength (σ_y) and ultimate tensile strength (σ_u) [Kiefner et al., 1973].

Table 2.1 lists some typical Level I defect assessment models [Qin and Cheng, 2021]. The difference in prediction of pipeline failure pressure by the various assessment models is mainly attributed to flow stress, defect profile, and Folias factor [Escoe, 2006]. An accurate definition of the geometric dimension of a corrosion defect is critical to the prediction results, affecting the assessment criteria. For ASME B31G standard, the corrosion defect on pipelines is generally regarded as parabolic, with a shape factor of 2/3. The value of $\sqrt{20Dt}$ is used to define the length of the corrosion defect. The parabolic shape is applied for short corrosion defects, which is defined as $L \leq \sqrt{20Dt}$. The long corrosion defects, i.e., $L > \sqrt{20Dt}$, are treated as rectangular. As a comparison, the modified ASME B31G code regards corrosion defect as a mixed type of geometric shape, where the critical geometric value is defined as $\sqrt{50Dt}$, with a shape factor of 0.85. The flow stress is set as (σ_y + 69) MPa, which is close to the conventional fracture mechanics definition of flow stress, i.e., the average of yield strength and ultimate tensile strength, $(\sigma_y + \sigma_u)/2$, as used in BS 7910 [British Standard, 2013]. Most Level

Table 2.1 Level I defect assessment models for prediction of failure pressure of pipelines containing single corrosion defect.

Method	σ_{flow}	M (Q for DNV)	Defect shape	P_F	Publishing time	Source
ASME B31G	$1.1\sigma_y$	$M=\sqrt{1+\frac{0.8L^2}{Dt}}$	Parabolic	$\begin{cases} P_F=\sigma_{flow}\frac{2t}{D}\left[\frac{1-\frac{2}{3}\cdot\frac{d_{max}}{t}}{1-\frac{2}{3}\cdot\frac{d_{max}}{tM}}\right], & L\le\sqrt{20Dt} \\ P_F=\sigma_{flow}\frac{2t}{D}\left[1-\frac{d_{max}}{t}\right], & L>\sqrt{20Dt} \end{cases}$	1984	ASME [2012]
Modified ASME B31G	σ_y+69	$M=\begin{cases}\sqrt{1+\frac{2.51\left(\frac{L}{2}\right)^2}{Dt}-\frac{0.54\left(\frac{L}{2}\right)^4}{(Dt)^4}}, & L\le\sqrt{50Dt} \\ 0.0032\frac{L^2}{Dt}+3.3, & L>\sqrt{50Dt}\end{cases}$	Mixed	$P_F=\sigma_{flow}\frac{2t}{D}\left[\frac{1-\frac{2}{3}\cdot\frac{d_{max}}{t}}{1-\frac{2}{3}\cdot\frac{d_{max}}{tM}}\right]$	1989	Kiefner and Vieth [1990b]
SHELL-92	σ_u	$M=\sqrt{1+\frac{0.8L^2}{Dt}}$	Rectangular	$P_F=\sigma_{flow}\frac{2t}{D}\left[\frac{1-\frac{d_{max}}{t}}{1-\frac{d_{max}}{tM}}\right]$	1995	Klever et al. [1995]

(Continued)

Table 2.1 (Continued)

Method	σ_{flow}	M (Q for DNV)	Defect shape	P_F	Publishing time	Source
CSA Z662	$\begin{cases} 1.15\sigma_y, \ \sigma_y \leq 241 \ \text{MPa} \\ 0.9\sigma_u, \ \sigma_y > 241 \ \text{MPa} \end{cases}$	$M = \begin{cases} \sqrt{1 + \dfrac{2.51\left(\dfrac{L}{2}\right)^2}{Dt} - \dfrac{0.54\left(\dfrac{L}{2}\right)^4}{(Dt)^4}}, & L \leq \sqrt{50Dt} \\ 0.0032\dfrac{L^2}{Dt} + 3.3, & L > \sqrt{50Dt} \end{cases}$	Rectangular	$P_F = \sigma_{flow}\dfrac{2t}{D}\left[\dfrac{1 - \dfrac{d_{ave}}{t}}{1 - \dfrac{d_{ave}}{tM}}\right]$	2002	CSA [2007]
DNV-RP-F101	σ_u	$Q = \sqrt{1 + \dfrac{0.31L^2}{Dt}}$	Rectangular	$P_F = \sigma_{flow}\dfrac{2t}{D\text{-}t}\left[\dfrac{1 - \dfrac{d_{max}}{t}}{1 - \dfrac{d_{max}}{tQ}}\right]$	1999	Det Norske Veritas [2010]
API RP 579	$\sigma_y/0.9$	$M = \sqrt{1 + \dfrac{0.31L^2}{Dt}}$	Rectangular	$P_F = \sigma_{flow}\dfrac{2t}{D}\left[\dfrac{1 - \dfrac{d_{max}}{t}}{1 - \dfrac{d_{max}}{tM}}\right]$	2000	American Petroleum Institute [2000] and Shdid and Hajali [2015]
FITNESS FSS	$\sigma_u\left(\dfrac{1}{2}\right)^{65/\sigma_y}$	$M = \sqrt{1 + \dfrac{0.8L^2}{Dt}}$	Rectangular	$P_F = \sigma_{flow}\dfrac{2t}{D\text{-}t}\left[\dfrac{1 - \dfrac{d_{max}}{t}}{1 - \dfrac{d_{max}}{tM}}\right]$	2008	Seib et al. [2008] and Cicero et al. [2009]

RPA	$\sigma_y + 69$	$M = \begin{cases} \sqrt{1 + \frac{2.51\left(\frac{L}{2}\right)^2}{Dt} - \frac{0.54\left(\frac{L}{2}\right)^4}{(Dt)^4}}, & L \leq \sqrt{20Dt} \\ 2.1 + 0.7\left(\frac{L}{Dt}\right), & L > \sqrt{20Dt} \end{cases}$	Rectangular + Parabolic	$P_F = \sigma_{flow}\frac{2t}{D}\left[\frac{1 - a\cdot\frac{d_{\max}}{t}}{1 - a\cdot\frac{d_{\max}}{tM}}\right]$	2003	Benjamin and Andrade [2003]
BS 7910: 2013 Annex G[a]	$\frac{\sigma_u + \sigma_y}{2}$	$M = \sqrt{1 + \frac{0.31L^2}{Dt}}$	Rectangular	$P_F = \sigma_{flow}\frac{2t}{D\text{-}t}\left[\frac{1 - \frac{d_{\max}}{t}}{1 - \frac{d_{\max}}{tM}}\right]$	2013	British Standard [2013] and Cosham and Andrews [2019]

[a] *Note:* BS 7910 applies for defects in all types of structures and components, while Annex G of BS 7910 is applicable for pressured structures and components. Annex G is intended to be simple for pipeline defect assessment [Cosham and Andrews, 2019].

Source: From Qin and Cheng [2021].

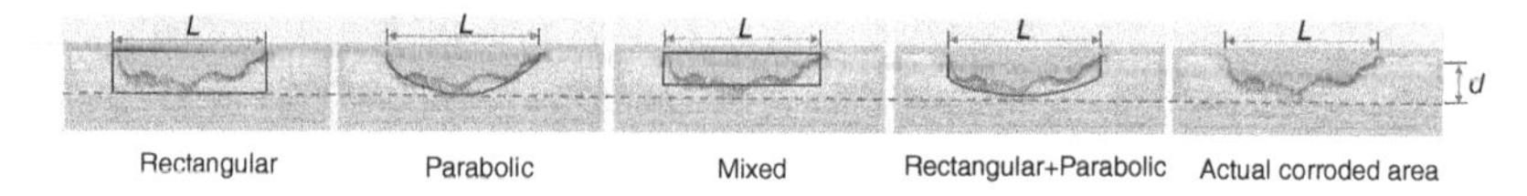

Figure 2.2 Illustration of the methods used in various Level I defect assessment codes and models to define the geometric dimension of a corrosion defect on pipelines.
Source: From Qin and Cheng [2021].

I assessment models use the maximum depth of an irregularly shaped corrosion defect, d_{max}, to define the defect depth for simplification. However, in CSA model, an average defect depth, d_{ave}, is used to describe the defect contour. Furthermore, a rectangular parabola area (RPA) method was proposed to modify the defect profile [Benjamin and Andrade, 2003; Amaya-G'omez et al., 2019]. Figure 2.2 illustratively shows the methods of various Level I assessment codes and models to define and calculate the geometrical dimension of a corrosion defect on pipelines [Qin and Cheng, 2021].

It is also seen that flow stress is used to define failure pressure of pipelines with specific defect dimensional conditions. Generally, the flow stress defined in most of the Level I defect assessment methods is between yield strength and ultimate tensile strength of the pipeline steels. The exceptions include SHELL-92 code and DNV-RP-F101 code, where the ultimate tensile strength was used to represent the flow stress [Klever et al., 1995; Batte et al., 1997]. Moreover, the Folias factor, M, was replaced with a length correction factor, Q, as the geometric correction factor.

2.3.2 Applications of Level I Defect Assessment for Pipeline FFS Determination and Failure Pressure Prediction

The Level I defect assessment methods determine failure pressure of corroded pipelines by a combination of Folias factor, flow stress and corrosion defect profiles. However, the applicability of the assessment codes and models is different. For example, the ASME B31G method, which is the benchmark standard for assessment of corrosion defect on pipelines, works properly for the pipelines under an operating temperature lower than 120 °C when the flow stress is $1.1\sigma_y$ [American Society of Mechanical Engineering, 2012]. When the flow stress is about $\sigma_y + 69$ MPa, the modified ASME B31G and RPA codes are suggested for defect assessment on carbon and alloy steel pipelines with a yield strength below 483 MPa and the operating temperature lower than 120 °C [Kiefner and Vieth, 1990b].

Table 2.2 summarizes the applicability of various Level I defect assessment methods in terms of the defect dimension and steel grade. The ASME B31G

Table 2.2 Applicability of the Level I defect assessment methods in terms of the defect dimension and steel grade.

		ASME B31G	Modified ASME B31G	SHELL-92	CSA	RPA	DNV (part B)	API RP 579	FITNET FSS	BS 7910 Annex G	Source
Defect geometry	Length	Both short and long defects	Short defects	Not reported	Not reported	Both short and long defects	Long defects	Not reported	Not reported	Not reported	Filho et al. [2014]
	Depth	10–80% pipe wall thickness	20–80% pipe wall thickness	10–85% pipe wall thickness	10–85% pipe wall thickness	10–85% pipe wall thickness	10–85% pipe wall thickness	10–85% pipe wall thickness	10–85% pipe wall thickness	20–80% pipe wall thickness	British Standard [2013], Motta et al. [2017], and Keshtegar and Seghier [2018]
Pipe steel property		Below X56	Below X65	Not reported	Below X65	Below X65	Below X65	Below X65	Not reported	Below X80[a]	British Standard [2013], Motta et al. [2017], and Keshtegar and Seghier [2018]

[a] *Note:* BS 7910 Annex G does not apply for materials with a specified minimum yield strength exceeding 555 MPa (equivalent to X80 grade).
Source: From Qin and Cheng [2021].

and RPA codes are recommended for assessment of both short and long corrosion defects, while the Modified B31G is appropriate for short defects and the DNV RP-F101 for long defects only [Filho et al., 2014]. The ASME B31G considers corrosion defects on pipelines that are under internal pressure only, while external loads are not included. In addition to the longitudinal length, the Level I methods also consider other geometrical parameters, primarily the defect depth. Principally, the models listed in Table 2.1 are applicable for shallow to deep corrosion defects (i.e., the depth of the corrosion defect ranges from 10% to 85% of pipe wall thickness), except for ASME B31G, modified B31G and BS 7910 Annex G. In terms of the constraint of pipe steels, the Level I methods are limited to materials with properties and performance behavior that have been validated in modeling development. It is noted that some models such as FITNET FSS and SHELL are not available in public literature. As a result, the limitations of these models in the materials aspect are not defined.

Furthermore, Level I defect assessment methods can be used as probabilistic approaches to corroded pipelines. According to the Level I assessment-based deterministic approaches, such as ASME B31G, modified B31G and DNV-RP-F101, the time dependence of geometrical parameters of the corrosion defect should be included in the models. Therefore, these models can be further improved to develop approaches for evaluation and prediction of the time-dependent failure probability [Zhang et al., 2019a] and remaining service life [Ahammed, 1998] of pipelines.

In a recent work, Zhu [2021] reviewed some main Level I defect assessment methods, including ASME B31G, Modified ASME B31G, Shell-92, CSA Z662, and a Level II method RSTRENG, comparing the burst pressures of X52 steel pipelines containing a corrosion defect with a fixed uniform depth of $d/t = 0.5$ obtained from the methods, as shown in Figure 2.3. In the figure, the x-axis is the normalized defect length (i.e., $\frac{L}{\sqrt{Dt}}$), and the y-axis refers to the normalized burst pressure by P_b, i.e., the modeled burst pressure, to P_0, i.e., Tresca strength solution as given by [Tresca, 1864]:

$$P_0 = \frac{2t}{D}\sigma u \tag{2.3}$$

It is seen that the modified ASME B31G model predicts the highest burst pressures for all the given defect lengths. The CSA Z662 and Shell-92 give identical results which are slightly smaller than the RSTRENG results. The ASME B31G predicts comparable results to Modified B31G for short corrosion defects, but the smallest results for long corrosion defects.

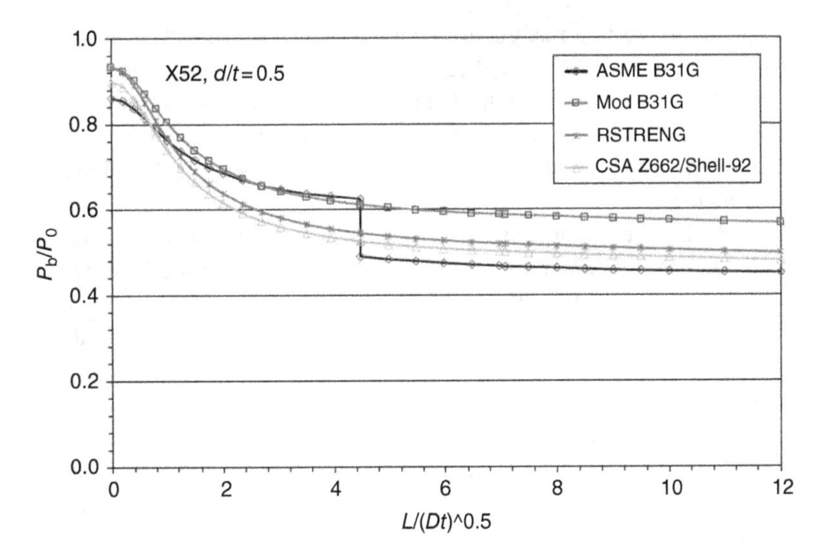

Figure 2.3 Comparison of the burst pressures of X52 steel pipe containing a corrosion defect obtained from some main Level I defect assessment methods. *Source:* From Zhu [2021].

Considering the improved level of the RSTRENG method, both ASME B31G and Modified B31G predict results deviating from the RSTRENG results more than the results from CSA Z662 and Shell-92.

2.3.3 Commentary Remarks for Level I Defect Assessment Methods

Generally, the Level I defect assessment models are semiempirical in nature. The Level I assessment requires the least amount of information and data and involves the fewest computations of all levels of defect assessment techniques. The Level I defect assessment is easy to use, and the modeling results can be conservative, just like design code calculations. Complex stress conditions at the corrosion defect and irregular defect geometries limit the modeling accuracy in prediction of failure pressure of the pipelines.

First, various Level I models and codes were developed and derived from full-scale burst tests and numerical simulations on pipelines under internal pressure, i.e., the only source of stress applied on the pipelines, while other sources of stress were ignored. Underground pipelines are always affected by various natural and environmental forces, which can produce mechanical responses at corrosion defects and the defect-free pipe segments. For example, when a pipeline is buried in geotechnically unstable regions, such as permafrost and semipermafrost areas and slopes, a significant longitudinal stress or strain would be exerted on the pipeline due to ground movement [Cheng, 2020]. Obviously, the buried pipelines are under stress conditions generated from multiple sources, where the internal pressure is one of them. However, stress sources other than internal pressure are not included in the Level I methods.

Second, all Level I defect assessment models and codes were developed from testing data obtained on low and moderate grades of pipeline steels. For example, the ASME B31G code was validated through tests conducted on steels below API X56 grade. The DNV-RP-F101 standard was developed and verified by 138 burst tests on X42 to X65 steels, while BS 7910 Annex G was based on a large parametric database of failure pressures and validated against over 100 full-scale burst tests. The codes and methods did not consider pipelines made of high grades of steel, such as X80 steel, which were not available at the time when the Level I methods were developed. High-strength pipeline steels possess unique metallurgical microstructure, alloying treatment, and mechanical properties, making them distinguishable in strength, toughness, and weldability from the low grades of steel [Barsanti et al., 2002; Zhang et al., 2019b]. To date, large-diameter, thin-wall, and long-distance transmission pipelines made of high-strength steels have been widely used in natural gas transportation due to significant economic benefits [Kalwa et al., 2002; Zhang et al., 2019b]. It is expected that more high-strength steel pipeline projects will be launched in the next 5–10 years in the world with rapidly growing energy demand globally [Gray and Siciliano, 2009]. The Level I defect assessment methods should include the effect of mechanical strength and fracture toughness of steels on pipeline failure, making the methods applicable to the pipelines made of high-strength grades of steels. Nowadays, burst tests have been performed on high-strength steel pipes to provide testing data [Netto et al., 2005; Yeom et al., 2015; Zhu, 2015, 2021], serving as a firm foundation to improve the Level I methods, making the defect assessment applicable for high-strength steel pipelines.

Third, the Level I defect assessment methods use the longitudinal dimension of the metal loss area to define the defect geometry and determine residual strength of the corroded pipelines, while ignoring the circumferential dimension of the defect. Moreover, the Level I models and codes simplify the corrosion defects with a smooth contour, which apparently deviates from the real complex geometry of the defects. Without an accurate definition of the defect geometry, it is impossible to obtain true stress and strain values, and thus, the failure pressure of the corroded pipelines.

Finally, a single and isolated corrosion defect is the object the Level I methods assess for pipeline failure prediction. However, colonies of corrosion defects, rather than isolated, single defect, are often present on pipelines in the field. Generally, multiple, adjacent corrosion defects can result in a mutual interaction between them, making it more complicated to evaluate the FFS of the corroded pipelines, as compared with the pipelines containing the same number of defects, but they are isolated and thus treated separately [Benjamin and Cunha, 2007; Benjamin et al., 2016a, b]. Moreover, the interaction between the corrosion defects can decrease failure pressure of the corroded pipelines [Sun and Cheng, 2018].

2.4 Level II Defect Assessment on Pipelines

A Level II assessment can be conducted if the Level I assessment on a pipeline containing corrosion defect does not pass acceptance criteria. The Level II defect assessment requires many more computations to replace the simple, conservative computations in the Level I assessment so that fewer conservative calculations may produce more accurate results. However, the computations are often complex. The Level II defect assessments include two types of methods. The Level IIa method is developed upon the Level I method for further enhancement of the assessment accuracy for single corrosion defect by an improved definition and characterization of the defect geometrical dimension. Moreover, the developed level IIb method can assess multiple corrosion defects and their interaction on pipeline failure, making the targeted defects more representative of the reality encountered on pipelines in the field.

2.4.1 Principle and Codes

2.4.1.1 The Level IIa Method

The Level IIa method was originally proposed to improve the assessment accuracy of failure for pipelines containing a single corrosion defect with complex geometry. The method includes RSTRENG [American Society of Mechanical Engineering, 2009], KAPA program [Kiefner, 2000], API 579 Level II [Janelle, 2005], and DNV RP-F101 [Det Norske Veritas, 2010]. Generally, a wide range of defect and damage types which are detected during in-service inspection of pipelines can be defined and assessed, while corrosion defect and local metal loss are the main assessment targets.

The Level IIa defect assessment method determines the failure stress at the corrosion defect based on estimation of a so-called effective area of metal

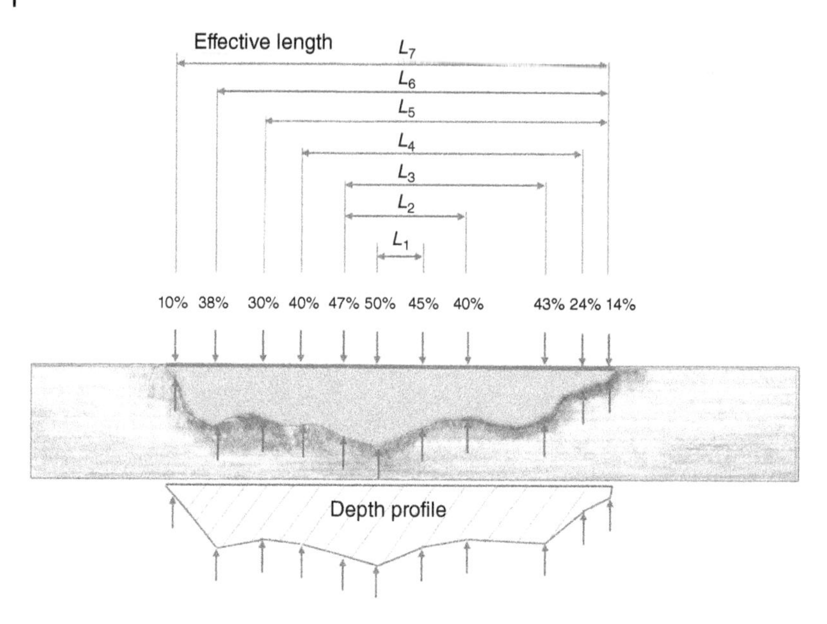

Figure 2.4 Illustration of the length and depth profiles of a corrosion defect to derive the effective area (% means the percentage of pipe wall thickness). *Source:* From McNealy et al. [2008].

loss through an iterative process in grid mode, representing the longitudinal region of the metal loss area according to a detailed depth profile at the defect [McNealy et al., 2008]. The process remarkably decreases the conservativeness of the defect assessment through an improved definition of the defect dimension. Figure 2.4 shows the illustrative diagram defining the length and depth profiles of a corrosion defect to derive the effective area [McNealy et al., 2008]. A corrosion defect is labeled with different length grids in the longitudinal direction (i.e., L_1, L_2, ..., L_n), which correspond to specific depths along the pipe wall thickness direction. The total length of the defect is called an effective length (i.e., L_{eff}), and the area associated with L_{eff} is the effective area (i.e., A_{eff}). Table 2.3 lists some main Level IIa models and codes for assessment of a complex-shaped corrosion defect on pipelines, including flow stress, geometric factor, failure pressure, and the effective length criteria.

For most locally thinning areas, the circumferential direction is limited because hoop stresses are typically twice of longitudinal stresses. As a result, assessment of the area should be concentrated on the circumferential stress direction if internal pressure is the primary source of stress. When secondary stress sources such as ground movement are included in the assessment, the

Table 2.3 Some main Level IIa models and codes for assessment of a complex-shaped corrosion defect on pipelines.

Method	σ_{flow}	M (Q for DNV)	Shape	P_F	Source
RSTRENG	$\sigma_y + 69$	$M = \begin{cases} \sqrt{1 + \frac{0.6275 L_{eff}^2}{Dt} - \frac{0.03375 L_{eff}^4}{(Dt)^4}}, & L_{eff} \le \sqrt{50Dt} \\ 0.0032 \frac{L_{eff}^2}{Dt} + 3.3, & L_{eff} > \sqrt{50Dt} \end{cases}$	Effective area	$P_F = \sigma_{flow} \frac{2t}{D} \left[\frac{1 - (A_{eff}/A_0)}{1 - A_{eff}/A_0 \cdot M^{-1}} \right]$	McNealy et al. [2008]
DNV for complex-shaped defect (Part B)	σ_u	$Q = \sqrt{1 + \frac{0.31 L_{total}^2}{Dt}}$	Effective area	$P_F = \sigma_{flow} \frac{2t}{D\text{-}t} \left[\frac{1 - \frac{d_{ave}}{t}}{1 - \frac{d_{ave}}{tQ}} \right]$	Det Norske Veritas [2010]
API 579	$\sigma_y + 69$	$M = \sqrt{\frac{1.02 + 0.4411(\lambda)^2 + 0.0006124(\lambda)^4}{1 + 0.02642(\lambda)^2 + 1.533(10)^{-6}(\lambda)^4}}$ where $\lambda = 1.285 L_{eff}/\sqrt{Dt}$	Effective area	$P_F = \sigma_{flow} \frac{2t}{D} \left[\frac{1 - (A_{eff}/A_0)}{1 - A_{eff}/A_0 \cdot M^{-1}} \right]$	American Petroleum Institute [2000]

Note: The original area of corrosion defect is $A_0 = L_{eff} t$.
Source: From Qin and Cheng [2021].

longitudinal stress direction should also be considered. In Level II assessment, an iterative process that slices the critical thickness profile of the metal loss area into subsections, as mentioned above, is conducted to accurately define the dimension of the defect. Each subsection is evaluated, and acceptance is based on the limiting subsection. The remaining strength of the defect is then determined, and the failure pressure of pipelines is predicted.

2.4.1.2 The Level IIb Method

Generally, multiple corrosion defects coexist on pipelines. When they are in adjacency, a mutual interaction is induced between the defects, affecting (usually adversely) the failure pressure of the pipelines [Sun and Cheng, 2018]. Principally, each corrosion defect is associated with a specific mechanical stress/strain field, which may interact with the stress/strain field generated by another defect if the two defects are located within a critical spacing. It is widely accepted that the interaction between corrosion defects can further decrease failure pressure of the pipelines, as compared with the pipelines containing two corrosion defects but they are isolated [Benjamin et al., 2016a]. Therefore, to accurately assess the impact of corrosion defects on FFS of pipelines, a critical task is to determine whether there is an interaction between the adjacent defects.

Interaction rules can be used to define a limiting space between adjacent corrosion defects. The interaction effect can be ignored if the spacing between the defects exceeds a critical value. As a result, the defects can be assessed individually. Most interaction rules can be expressed as $S_L \leq S_L^{Lim}$ and $S_C \leq S_C^{Lim}$, where S_L and S_C are longitudinal and circumferential spacings, respectively, between two adjacent corrosion defects, and S_L^{Lim} and S_C^{Lim} refer to the limiting longitudinal and circumferential spacings, respectively. Table 2.4 shows various interaction rules developed for defining mutual interaction between adjacent corrosion defects in the past three decades [Qin and Cheng, 2021]. It is seen that the interaction rules are dependent on geometries of both the pipe and the corrosion defect. Most rules are associated with either the defect geometry or the pipe geometry, while the POF rule considers both.

Previously, both the Level I assessment method and interaction rule were combined to predict the failure pressure of pipelines containing multiple corrosion defects [Benjamin et al., 2016a]. The interaction rules are used to determine whether an interaction occurs between adjacent corrosion defects. If an interaction does not exist, the two defects will be assessed separately. One of the available Level I defect assessment models or codes is chosen to calculate failure pressure of the pipelines at individual corrosion

Table 2.4 Various interaction rules developed for defining mutual interaction between adjacent corrosion defects in the past three decades.

Interaction rules	Limit longitudinal spacing, S_L^{Lim}	Limit circumferential spacing, S_C^{Lim}	Source
CW	$S_L^{Lim} = \min(L_1, L_2)$	$S_C^{Lim} = \min(W_1, W_2)$	Coulsen and Worthingham [1990]
Kiefner and Vieth	$S_L^{Lim} = 25.4mm$	$S_C^{Lim} = 6t$	Kiefner and Vieth [1990a]
3WT	$S_L^{Lim} = 3t$	$S_C^{Lim} = 3t$	Hopkins and Jones [1992]
POF	$S_L^{Lim} = \min(6t, L_1, L_2)$	$S_C^{Lim} = \min(6t, W_1, W_2)$	Pitchford [1999]
DNV-RP-F101	$S_L^{Lim} = 2\sqrt{Dt}$	$S_C^{Lim} = \pi\sqrt{Dt}$	Det Norske Veritas [2010]
6WT	$S_L^{Lim} = 6t$	$S_C^{Lim} = 6t$	Lamontagne [2002]
API 579	$S_L^{Lim} = (L_1 + L_2)/2$	$S_w^{Lim} = (W_1 + W_2)/2$	American Petroleum Institute [2007]
ASME B31G	$S_L^{Lim} = 3t$	$S_C^{Lim} = 3t$	American Society of Mechanical Engineering [2012]
BS 7910	$S_L^{Lim} = 2\sqrt{Dt}$	$S_C^{Lim} = 3\sqrt{Dt}$	British Standard [2013]
Al-Owaisi	$S_L^{Lim} = 3t$	$S_C^{Lim} = t$	Al-Owaisi et al. [2016]
Li et al.	$S_L^{Lim} = 2\sqrt{Dt}\left(L \le \sqrt{20Dt}\right)$	$S_C^{Lim} = 0.1\pi D\left(L \le \sqrt{20Dt}\right)$	Li et al. [2016]
	$S_L^{Lim} = \sqrt{Dt}\left(\sqrt{20Dt} < L \le \sqrt{50Dt}\right)$	$S_C^{Lim} = 0.05\pi D\left(\sqrt{20Dt} < L \le \sqrt{50Dt}\right)$	
	$S_L^{Lim} = t\left(L > \sqrt{50Dt}\right)$	$S_C^{Lim} = t\left(L > \sqrt{50Dt}\right)$	
Sun and Cheng	$S_L^{Lim} = 2.5\sqrt{Dt}$	$S_C^{Lim} = 5.3t$	Sun and Cheng [2018]
BS 7910 Annex G	$S_L^{Lim} = \min(L_1, L_2)$	$S_C^{Lim} = \max(L_1, L_2)$	British Standard [2013]

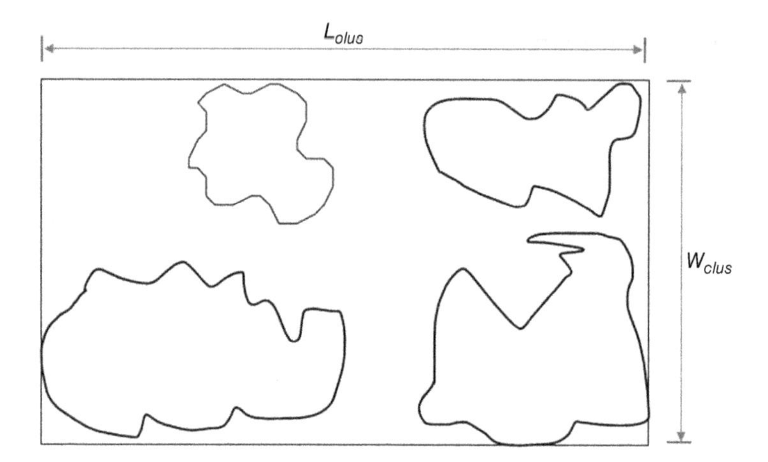

Figure 2.5 "Envelope rectangle" concept used for defect assessment of multiple corrosion defects on a pipeline. *Source:* From Benjamin et al. [2016a].

defects. The lower failure pressure determined from the specific defect is taken as the final value of the failure pressure. If the interaction rules show existence of a mutual interaction between the defects, the final failure pressure will be determined based on the Level I assessment method while considering the interaction, which usually further decreases the failure pressure of the pipelines. This methodology uses a so-called "envelope rectangle" concept to increase the size of actual area of the corrosion defects, as shown in Figure 2.5 [Benjamin et al., 2016a], where the total size under defect assessment includes both the summation of the areas of individual defects and the nearby area. The depth of the defect cluster, d_{clus}, is defined as the maximum depth of the deepest defect. The overall length of the defect cluster, L_{clus}, and the overall width of the defect cluster, W_{clus}, refer to the length and width, respectively, of the envelope rectangle. Obviously, the sizing method used in the "envelope rectangle" concept for assessment of multiple corrosion defects will give conservative results [Al-Owaisi et al., 2016].

To overcome the conservatism of the previous assessment methods for multiple corrosion defects, the Level IIa method is used to calculate the pipeline's failure pressure at individual corrosion defects, ignoring the mutual interaction between the defects. This apparently simplifies the defect assessment, making the results deviate from the realistic service condition. The Level IIb method is developed to calculate the failure pressure of pipelines containing multiple defects, while including the potential interaction between the defects into consideration. A commonly used Level IIb method is DNV-RP-F101 [Det Norske Veritas, 2010]. Figure 2.6 shows how the defect

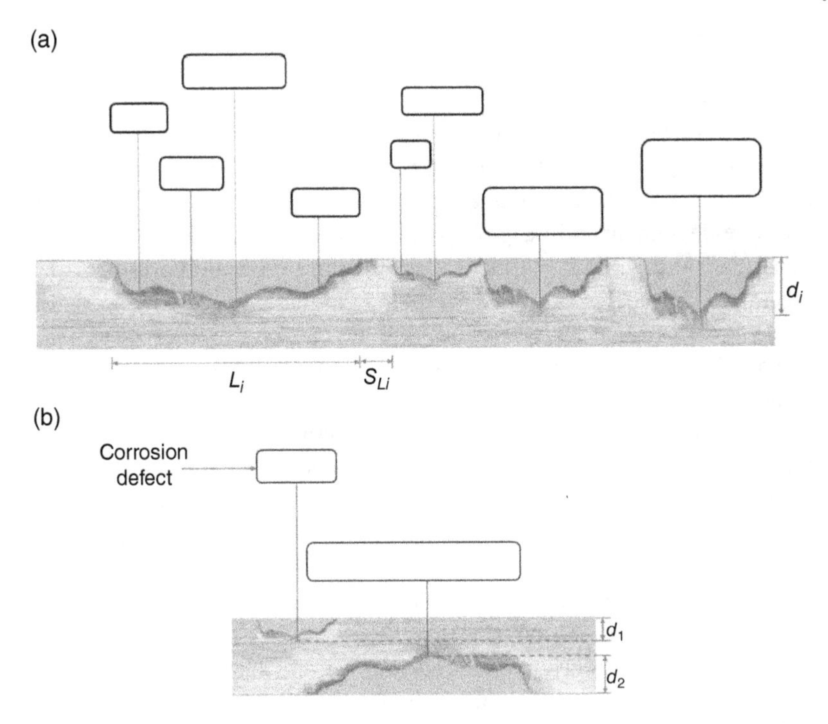

Figure 2.6 "Envelope rectangle" concept used for defect assessment of multiple corrosion defects on a pipeline (a) multiple external defects, (b) an internal defect and an external defect. *Source:* From Det Norske Veritas [2010].

dimension is defined, where the corrosion defects present on a pipe surface are projected to a longitudinal plane to generate a 2D shape featuring of a length and a depth. The defects are regarded as a composite defect if the projections of the defects are overlapped (Figure 2.6a). The failure pressure of the pipelines considers the total length (L_i) and the maximum depth (d_i) of the composite defect, as well as the longitudinal spacing between adjacent defect projections (S_{Li}). When the composite defect consists of overlapped projections of both the internal and external defects, the total depth is the sum of the maximum depths of the internal and external projections, i.e., $d_1 + d_2$, as shown in Figure 2.6b [Dann and Maes, 2018].

2.4.2 Commentary Remarks for Level II Defect Assessment Methods

The Level II defect assessment methods are applicable for failure prediction of pipelines containing a single corrosion defect, multiple isolated corrosion defects, and multiple interacting corrosion defects with improved definition of the defect geometry. The Level II methods use the so-called effective area to define the defect dimension by projecting real contours of the defects onto a longitudinal plane. The sizing method is appropriate for any type of defect geometry. The Level II defect assessment is not limited by material types [Amandi et al., 2019]. Like Level I methods, the Level II assessment methods can include corrosion defect growth models to assess the time-dependent failure probability of the pipelines. Specifically, the RSTRENG and API RP 579 models are the most accurate Level II methods to predict pipeline failure as verified by burst tests [Chauhan, 2004], while the DNV RP-F101 method is too conservative for assessment of complex-shaped defects, as compared with API-579 and RSTRENG standards.

Although the Level II defect assessment methods overcome the assessment conservatism compared with the Level I methods, problems still exist to affect the modeling and prediction accuracy. For example, in sizing the corrosion defects, the contour of each defect is projected on a longitudinal plane, ignoring the circumferential growth of the defect. It was reported that such a simplification caused a conservative result in assessing pipelines containing interacting corrosion defects [ANSYS, 2013].

In addition, overlapped corrosion defects on pipelines are usually more common than longitudinally and circumferentially aligned defects [Benjamin et al., 2005]. The Level IIb methods project the overlapped corrosion defects on the longitudinal plane to form a composite defect, where the maximum depth and the total length of the composite defect are used to calculate the failure pressure. This simplification ignores the real geometry of the defect located on the top layer. Thus, the FFS determination of pipelines containing overlapped corrosion defects is conservative when assessed by the Level II methods.

Finally, the interaction rules used for adjacent corrosion defects are empirical and do not consider the influence of steel grade on the interaction between corrosion defects. This practice is supported by Sun and Cheng's work [Sun and Cheng, 2018], where FE modeling was conducted on low-, middle-, and high-grade pipe steels. It was found that the interaction effect between corrosion defects was not apparently affected by the steel grade. However, the defect geometry greatly affected the interaction between the defects, although there have been few interaction rules considering the geometrical features of the defects. Recently, an interaction rule considering the critical length of corrosion defect has been proposed [Sun and Cheng, 2018]. Apparently, further work is required to improve accuracy when defining the interaction rules for adjacent corrosion defects on pipelines.

References

Ahammed, M. (1998) Probabilistic estimation of remaining life of a pipeline in the presence of active corrosion defects, *Int. J. Press. Vessel Pip.*, 75, 321–329.

Al-Owaisi, S.S., Becker, A.A., Sun, W. (2016) Analysis of shape and location effects of closely spaced metal loss defects in pressurized pipes, *Eng. Fail. Anal.*, 68, 172–186.

Amandi, K.U., Diemuodeke, E.O., Briggs, T.A. (2019) Model for remaining strength estimation of a corroded pipeline with interacting defects for oil and gas operations, *Cogent Eng.*, 6, 1663682.

Amaya-G'omez, R., S'anchez-Silva, M., Bastidas-Arteaga, E., Schoefs, F., Munoz, F. (2019) Reliability assessments of corroded pipelines based on internal pressure – a review, *Eng. Fail. Anal.*, 98, 190–214.

American Petroleum Institute (2000) *Fitness-for-Service*, API Recommended Practice 579, API Publishing Services, Ed. 1, Washington, DC, USA.

American Petroleum Institute (2007) *Fitness-for-Service*, API Recommended Practice 579-1, API Publishing Services, Ed. 2, Washington DC, USA.

American Society of Mechanical Engineering (1991) *Manual for Determining the Remaining Strength of Corroded Pipelines, A Supplement to ASME B31 Code for Pressure Piping*, ASME B31G-1991 (Revision of ANSI/ASME B31G-1984), ASME, New York, NY, USA

American Society of Mechanical Engineering (2009) *Manual for Determining the Remaining Strength of Corroded Pipelines – A Supplement to ASME B31 Code for Pressure Piping*, B31G-2009, ASME, New York, NY, USA

American Society of Mechanical Engineering (2012) *Manual for Determining the Remaining Strength of Corroded Pipelines: A Supplement to ASME B31 Code for Pressure Piping*, ASME, New York, NY, USA

ANSYS (2013) *ANSYS Fluent Theory Guide*, ANSYS Inc., Canonsburg, PA, USA

Barsanti, L., Bruschi, R., Donati, E. (2002) From X80 to X100: know-how reached by ENI group of high strength steel, In: *Proc. Appl. Evaluation High Grade Linepipe in Hostile Environ. Conf.*, Yokohama, Japan.

Batte, D., Fu, B., Kirkwood, M.G., Vu, D. (1997) Advanced methods for integrity assessment of corroded pipelines, *Pipes Pipelines Int.*, 42, 5–11.

Benjamin, A.C., Andrade, E.D.Q. (2003) Modified method for the assessment of the remaining strength of corroded pipelines, In: *Proc. 2003 Rio Pipeline Conf.*, Rio de Janeiro, Brazil.

Benjamin, A.C., Freire, J.L.F., Vieira, R.D., Diniz, J.L., De Andrade, E.Q. (2005) Burst tests on pipeline containing interacting corrosion defects, in: *Proc. Int. Conf. Offshore Mechanics Arctic Eng.*, Paper no: OMAE2005-67059, Halkidiki, Greece.

Benjamin, A.C., Cunha, D.J.S. (2007) New method for the prediction of the failure pressure of interacting corrosion defects. In: *Proc. the 17th Int. Offshore Polar Eng. Conf.*, ISOPE, p. 3456–3465.

Benjamin, A.C., Freire, J.L.F., Vieira, R.D., Cunha, D.J.S. (2016a) Interaction of corrosion defects in pipelines – part 1: fundamentals, *Int. J. Press. Vessel Pip.*, 144, 56–62.

Benjamin, A.C., Freire, J.L.F., Vieira, R.D., Cunha, D.J.S. (2016b) Interaction of corrosion defects in pipelines – part 2: MTI JIP database of corroded pipe tests, *Int. J. Press. Vessel Pip.*, 145, 41–59.

Bjørnøy, O.H., Marley, M. (2001) Assessment of corroded pipelines/past, present and future, In: *Proc. 11th Int. Conf. Offshore Polar Eng. (ISOPE 2001)*, Stavanger, USA.

British Standard (1991) *Guidance on Methods for Assessing the Acceptability of Flaws in Fusion Welded Structures*, BS PD 6493, British Standard Institute, London, UK

British Standard (2013) *Guide to Methods for Assessing the Acceptability of Flaws in Metallic Structures*, British Standard Institute, London, UK

Caleyo, F., Gonzalez, J.L., Hallen, J.M. (2002) A study on the reliability assessment methodology for pipelines with active corrosion defects, *Int. J. Press. Vessel Pip.*, 79, 77–86.

Canadian Standard Association (2007) *Limit State Equation for Burst of Large Leaks and Rupture for Corrosion Defect: Oil and Gas Pipeline Systems*, CSA-Z662-07, Ottawa, ON, Canada.

Chauhan, V. (2004) *Experimental Validation of Methods for Assessing Closely Spaced Corrosion Metal Loss Defects in Pipelines*, PRCI Report no. L52007, Pipeline Research Council International, USA.

Cheng, Y.F. (2020) Technical insights into the long-term integrity and sustainability of China-Russia eastern gas pipeline, *Oi/Gas Storage Transport.*, 39, 1–8.

Choi, J.B., Goo, B.K., Kim, J.C., Kim, Y.J., Kim, W.S. (2003) Development of limit load solutions for corroded gas pipelines, *Int. J. Press. Vessel Pip.*, 80, 121–128.

Cicero, S., Lacalle, R., Cicero, R., Ferreno, D. (2009) Assessment of local thin areas in a marine pipeline by using the FITNET FFS corrosion module, *Int. J. Press. Vessel Pip.*, 86, 329–334.

Cosham, A., Andrews, R. (2019) The assessment of locally thinned areas: background to the guidance given in annex G of BS 7910:2013, *Int. J. Press. Vessel Pip.*, 169, 177–187.

Cosham, A., Hopkins, P. (2002) The pipeline defect assessment manual. In: *Proc. 2002 Int. Pipeline Conf.*, Calgary, AB, Canada.

Cosham, A., Hopkins, P. (2004) An overview of the pipeline defect assessment manual (PDAM), In: *Proc. 4th Int. Pipeline Technol. Conf.*, Oostende, Belgium.

Coulsen, K.E.W., Worthingham, R.W. (1990) New guidelines promise more accurate damage assessment, *Oil Gas J.*, 88, 16.

Cronin, D.S., Pick, R.J. (2000) A new multi-level assessment procedure for corroded line pipe, In: *Proc. 2000 Int. Pipeline Conf.*, ASME, Calgary, AB, Canada.

CSA (2007) *Limit State Equation for Burst of Large Leaks and Rupture for Corrosion Defect: Oil and Gas Pipeline Systems*, CSA-Z662–07. Ottawa, ON, Canada.

Dai, L., Wang, D., Wang, T., Feng, Q.S., Yang, X.Q. (2017) Analysis and comparison of long-distance pipeline failures, *J. Petro. Eng.*, 2017, 3174636.

Dann, M.R., Maes, M.A. (2018) Stochastic corrosion growth modeling for pipelines using mass inspection data, *Reliability Eng. Sys. Safety*, 180, 245–254.

Det Norske Veritas (2010) *Corroded Pipelines – Recommended Practice,* RPF101, Norway.

Escoe, K. (2006) *Piping and Pipelines Assessment Guide*, Gulf Professional Publishing, London, UK

Filho, J.E., Machado, R.D., Bertin, R.J., Valentini, M.D. (2014) On the failure pressure of pipelines containing wall reduction and isolated pit corrosion defects, *Comput. Struct.*, 132, 22–33.

Folias, E.S. (1956) *The Stresses in a Cylindrical Shell Containing an Axial Crack*, Aerospace Research Laboratories, Office of Aerospace Research, United States Air Force, Washington DC, USA

Folias, E.S. (2000) Predicting failures in cylindrical pressurized vessels from tests carried out on flat plates, In: *Proc. 9th Int. Conf. Press. Vessel Technol.*, Sydney, Australia.

Fu, B. (1999) Advanced engineering methods for assessing the remaining strength of corroded pipelines, *IMechE Conf. Trans.*, 8, C571.

Ghani, M.A., Tewfik, G., Djahida, D. (2016) Determination of limit load solution for the remaining load-carrying capacity of corroded pipelines, *J. Press. Vessel Technol.*, 138, 051701.

Gray, J.M., Siciliano, F. (2009) High strength microalloyed linepipe: half a century of evolution, in: *Proc. Int. Symp. High-Strength Steel Development*, Beijing, China.

Holiday, C., Wilde, A., Clyne, A. (2022) Understanding complex anomalies, *World Pipelines* 2, 35–39.

Hopkins, P., Jones, D.G. (1992) A study of the behavior of long and complex-shaped corrosion in transmission pipelines, part A – pipeline technology, In: *Proc. 11th Int. Conf. Offshore Mechanics Arctic Eng.*, Vol. 5, ASME, pp. 211–217.

Janelle, J. (2005) *An Overview and Validation of the Fitness-for-Service Assessment Procedures for Local Thin Areas*, PhD thesis, Ohio University, USA.

Kalwa, C., Hillenbrand, H.G., Graf, M. (2002) High-strength steel pipes: new development applications, in: *Proc. Onshore Pipeline Conf.*, Houston, TX, USA.

Keshtegar, B., Miri, M. (2014) Reliability analysis of corroded pipes using conjugate HL–RF algorithm based on average shear stress yield criterion, *Eng. Fail. Anal.*, 46, 104–117.

Keshtegar, B., Seghier, M.E.A.B. (2018) Modified response surface method basis harmony search to predict the burst pressure of corroded pipelines, *Eng. Fail. Anal.*, 89, 177–199.

Kiefner, J.F., Maxey, W.A., Eiber, R.J., Duffy, A.R. (1973) *The Failure Stress Levels of Flaws in Pressurised Cylinders*, ASTM STP 536, Philadelphia, PA, USA.

Kiefner, J.F., Vieth, P.H. (1990a) Evaluating pipe conclusion: PC program speeds new criterion for evaluating corroded pipe, *Oil Gas J.*, 88, 91–93.

Kiefner, J.F., Vieth, P.H. (1990b) Evaluating pipe – 1. New method corrects criterion for evaluating corroded pipe, *Oil Gas J.*, 88, 56–59.

Kiefner, J.F. (2000) *Kiefner & Associates Pipe Assessment (KAPA)*, Kiefner and Associates Inc., Columbus, OH, USA

Klever, F.J., Stewart, G., Van der Valk, C.A. (1995) *New Developments in Burst Strength Predictions for Locally Corroded Pipelines*, ASME, New York, NY, USA

Lamontagne, M. (2002) Interaction rules – an integral factor, In: *Corrosion' 2002*, NACE, Houston, TX, USA.

Li, X., Bai, Y., Su, C., Li, M. (2016) Effect of interaction between corrosion defects on failure pressure of thin wall steel pipeline, *Int. J. Pres. Vessel Pip.*, 138, 8–18.

McNealy, R., Gao, M., Limo, T., Deaton, B. (2008) Defect assessment using effective area method from in-line inspection data, In: *Proc. 2008 Int. Pipeline Conf.*, Calgary, AB, Canada.

Motta, R.S., Cabral, H.L., Afonso, S.M., Willmersdorf, R.B., Bouchonneau, N., Lyra, P.R., De Andrade, E.Q., (2017) Comparative studies for failure pressure prediction of corroded pipelines, *Eng. Fail. Anal.*, 81, 178–192.

Mustaffa, Z., Van Gelder, P. (2010) A review and probabilistic analysis of limit state functions of corroded pipelines, In: *Proc. 20th Int. Offshore Polar Eng. Conf., June 20–25*, International Society of Offshore and Polar Engineers (ISOPE), Beijing, China.

Netto, T.A., Ferraz, U.S., Estefen, S.F. (2005) The effect of corrosion defects on the burst pressure of pipelines, *J. Constr. Steel Res.*, 61, 1185–1204.

Phan, H.C., A.S. Dhar, Mondal, B.C. (2017) Revisiting burst pressure models for corroded pipelines, *Can. J. Civ. Eng.*, 44, 485–494.

Pitchford, J. (1999) Specification and requirements for the intelligent pig inspection of pipelines, *Pipelines Int.*, 44, 17–27.

Qin, G.J., Cheng, Y.F. (2021) A review on defect assessment of pipelines: principles, numerical solutions, and applications, *Int. J. Press. Vessel Pip.*, 191, 104329.

Seib, E., Uz, M.V., Koçak, M. (2008) European fitness-for-service procedure fitness and its application to welded aerospace structures, In: *Proc. 1st Int. Cong. Welding Joining Technol. & 17th Tech. Sessions*. Spanish Welding Society, Madrid, Spain.

Shdid, A.C., Hajali, M. (2015) The effect of the size and orientation of corrosion damage on the residual strength of steel pipes: analytical validation of a numerical model, In: *Proc. 2015 Press. Vessels Pip. Conf.*, ASME, New York, NY, USA

Shuai, Y., Shuai, J., Xu, K. (2017) Probabilistic analysis of corroded pipelines based on a new failure pressure model, *Eng. Fail. Anal.*, 81, 216–233.

Sun, J.L., Cheng, Y.F. (2018) Assessment by finite element modeling of the interaction of multiple corrosion defects and the effect on failure pressure of corroded pipelines, *Eng. Struct.*, 165, 278–286.

Tresca, H. (1864) Memoire sur iecoulement des corps solides soumis a de fortes pressions, *C. R. Acad. Sci. Paris*, 59, 754–758.

Wang, N., Zarghamee, M.S. (2014) Evaluating fitness-for-service of corroded metal pipelines: structural reliability bases, *J. Pipeline Syst. Eng.*, 5, 04013012.

Wang, Y.C., Xu, L.Y., Sun, J.L., Cheng, Y.F. (2021) Mechano-electrochemical interaction for pipeline corrosion: a review, *J. Pipeline Sci. Eng.*, 1, 1–16.

Xu, L.Y., Cheng, Y.F. (2013) Development of a finite element model for simulation and prediction of mechano-electrochemical effect of pipeline corrosion, *Corros. Sci.*, 73, 150–160.

Yeom, K.J., Lee, Y.K., Oh, K.H., Kim, W.S. (2015) Integrity assessment of a corroded API X70 pipe with a single defect by burst pressure analysis, *Eng. Fail. Anal.*, 57, 553–561.

Zhang, P., Su, L., Qin, G.J., Kong, X., Peng, Y. (2019a) Failure probability of corroded pipeline considering the correlation of random variables, *Eng. Fail. Anal.*, 99, 34–45.

Zhang, S.Z., Cheng, Y.F., Feng, X.D., Wei, L.J., Wang, R.J., Duo, Y.Q., Han, Y.X. (2019b) Performance characteristics and technical challenges of X80 pipeline steel – a review, *Oil Gas Storage Transport.*, 38, 481–495.

Zhu, X.K. (2015) A new material failure criterion for numerical simulation of burst pressure of corrosion defects in pipelines, In: *Proc. Press. Vessel Piping Conf.*, ASME, New York, NY, USA.

Zhu, X.K. (2021) A comparative study of burst failure models for assessing remaining strength of corroded pipelines, *J. Pipeline Sci. Eng.*, 1, 36–50.

3

Level III Assessment of Corrosion Anomalies on Pipelines

3.1 Introduction

Nonlinearities exist in pipelines containing corrosion defects in terms of the defect geometry, pipe material, and boundary conditions of the pipe segment. The Level III assessment methods determine pipeline FFS and predict the failure pressure by considering the nonlinearities with available commercial software programs such as ANSYS [2013], ABAQUS [Systemes, 2015], and COMSOL Multiphysics [Comsol, 2018], where FE modeling is conducted to solve the nonlinear problems associated with the corroded pipelines.

During modeling, the corrosion defects, depending on their geometric features, are usually simplified as either a 3D equal-depth shape (i.e., approximately cuboid), a semi-ellipsoid, or a hemisphere [Al-Owaisi et al., 2018; Cosham and Hopkins, 2004; Liu et al., 2009]. The boundary conditions of the pipe segment to be modeled should not affect modeling of the corrosion defect by setting a sufficient length of the pipe segment [Benjamin et al., 2006]. To save computational time, a quarter or a half of the pipe segment containing a single or multiple corrosion defects is usually used for modeling due to geometrical symmetry.

3.2 Principle and Methods

3.2.1 Stress Conditions of Pipelines

Buried pipelines are under complex stress conditions. Figure 3.1 shows schematically various types of stress that apply to a pipeline in place [Qin and Cheng, 2021a]. The primary stress, i.e., hoop stress, is resulted from

Defect Assessment for Integrity Management of Pipelines, First Edition. Y. Frank Cheng.

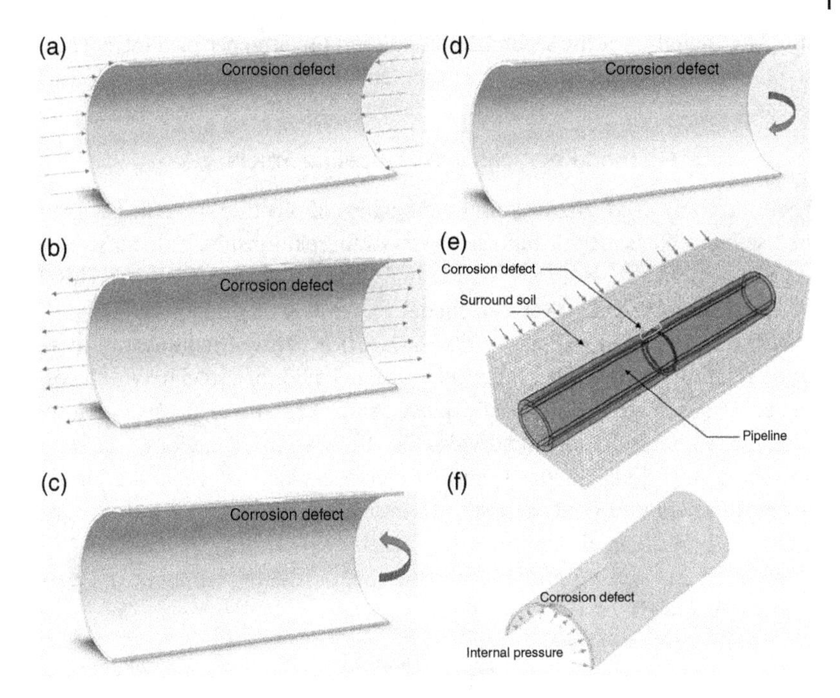

Figure 3.1 Schematic diagram of a buried pipeline under (a) an axial compressive stress, (b) an axial tensile stress, (c) a closing bending moment, (d) an opening bending moment, (e) surface loading in pipe–soil interaction, and (f) internal pressure. *Source:* (a and b) Adapted from Xu and Cheng [2012]; (c and d) adapted from Mondal and Dhar [2019]; (e) adapted from Wu and Li [2019]; (f) adapted from Sun and Cheng [2019].

internal pressure (Figure 3.1f). In addition, the pipeline is subject to additional stresses, which are usually called secondary stresses. For example, ground movement can exert either longitudinal tensile or compressive stress on the pipe in geotechnically unstable regions, as shown in Figure 3.1a,b [Cheng, 2020]. A bending moment can apply to the pipe if it is in suspension due to soil erosion (Figure 3.1c,d) [Cunha et al., 2014; Qin and Cheng, 2021b]. Temperature differences also cause an axial force and a bending moment on the pipe wall [Mondal and Dhar, 2019]. The underground pipeline is sometimes under surface load, as seen in Figure 3.1e, caused by the gravity of soil, construction trucks, or vehicles, especially when the pipe is across a highway [Warman and Hart, 2005]. All the stresses affect failure pressure of the pipeline. Frequently, nonlinear boundary conditions and external stresses are applied on a pipe along the longitudinal direction to simulate the effect of the secondary stresses on pipeline performance [Thesi et al., 2010; Wu and Li, 2019; Wu et al., 2019].

3.2.2 Stress–Strain Relationships of Pipeline Steels

The commonly used stress–strain relationships of pipeline steels in Level III assessment models include bilinear stress–strain relationship, trilinear stress–strain relationship, isotropic hardening model, Ramberg-Osgood (R-O) stress–strain rule, and power–law model.

Bilinear and trilinear stress–strain relationships. The bilinear stress–strain relationship describes the mechanical behavior of a steel in two stages, i.e., an elastic stage and an elastic–plastic stage [Yatabe et al., 2004]. However, the trilinear relationship divides the stress–strain behavior of the steel into three stages, i.e., an elastic stage, an elastic–plastic stage, and a plastic stage [Ramberg and Osgood, 1943]. In each stage of the two relationships, the stress and the strain are linearly related.

Isotropic hardening model. The isotropic hardening behavior of steels follows [Al-Owaisi et al., 2018]:

$$\sigma_{\text{yhard}} = \sigma_{\text{exp}}(\varepsilon_{\text{eff}}) - \sigma_{\text{y}} = \sigma\left(\varepsilon_{\text{p}} + \frac{\sigma_{\text{e}}}{E}\right) - \sigma_{\text{y}} \tag{3.1}$$

where σ_{yhard} is stress enhancement during plastic deformation, σ_{exp} is an experimental stress function, ε_{eff} is effective strain, σ is stress, ε_{p} is the sum of plastic strain, σ_{e} is effective stress, and E is Young's modulus.

The R-O stress–strain rule. According to R-O rule [Ramberg and Osgood, 1943],

$$\varepsilon = \frac{\sigma}{E} + K\left(\frac{\sigma}{\sigma_{\text{y}}}\right)^{n}\left(\frac{\sigma}{E}\right) \tag{3.2}$$

$$\varepsilon = \frac{\sigma}{E} + 0.079\left(\frac{\sigma}{\sigma_{\text{u}}}\right)^{12.64} \tag{3.3}$$

where ε is strain, and K and n are R-O material parameters, which are constants that depend on the modeled steel. Eq. (3.2) is used for low-grade pipe steels, such as X46 and X60 steels, and Eq. (3.3) is applicable for high-grade steels, such as X70 and X80 steels [Benjamin et al., 2006].

Power–law model. According to the power–law model [Liu et al., 2017],

$$\begin{cases} \sigma = E\varepsilon \; \sigma < \sigma_y \\ \sigma = K\varepsilon^n \; \sigma \geq \sigma_y \end{cases} \tag{3.4}$$

3.2.3 Pipeline Failure Criteria

3.2.3.1 Stress-based Criteria

According to traditional strength design theories, the maximum stress on a material should be below an allowable stress level to maintain safety of the material [Fu and Kirkwood, 1995]. Generally, a pipeline is assumed to fail when the equivalent stress on the pipe body (for defect-free pipeline) or at a corrosion defect (for defect-containing pipeline) exceeds a reference stress. The equivalent stress is either the Tresca yield stress (σ_{Tresca}) or von Mises stress (σ_{Mises}), which are defined as follows:

$$\sigma_{\text{Tresca}} = \sigma_1 - \sigma_3 \tag{3.5}$$

$$\sigma_{\text{Mises}} = \frac{1}{\sqrt{2}}\sqrt{(\sigma_1 - \sigma_2)^2 + (\sigma_2 - \sigma_3)^2 + (\sigma_3 - \sigma_1)^2} \tag{3.6}$$

where σ_1, σ_2, σ_3 are three principal stresses of a pipeline.

In terms of the reference stress level, the ultimate tensile strength of pipe steels is usually proposed as the failure criterion for a pipeline. The pipeline fails when the von Mises stress along the ligament thickness direction at a defect reaches σ_u [Det Norske Veritas, 2004]. In addition, since a plastic strain occurring at the corrosion defect can cause pipeline failure, sometimes, the yield strength of pipe steels is also used as the reference level, especially in geohazard regions where a significant axial strain is generated due to pipe–soil interactions [Hale et al., 1991].

3.2.3.2 Strain-based Criteria

The strain-based failure criteria require that the maximum deformation of a pipeline should not exceed a limit value [Det Norske Veritas, 2007]. The criteria include limit tensile strain, limit compressive strain, and limit elliptical deformation.

Limit tensile strain. A significant amount of tensile strain applied on a pipeline can cause its failure, especially at a corrosion defect where a high tensile strain level is usually developed locally. It is thus critical to analyze if the tensile strain is excessive when using the strain-based criteria. According to CSA Z662 [Canadian Standardization Association, 2007], the limit tensile strain is set as 2.5%. For both DNV-OS-F101 [Det Norske Veritas, 2007] and American Society of Civil Engineering (2005), the limit tensile strain is 2%.

Limit compressive strain. When a pipeline is subject to a compressive strain, the pipeline is prone to bend, causing buckling and wrinkling. In general, the compressive strain limit can be estimated by an empirical formula [Canadian Standardization Association, 2007]:

$$\varepsilon_{\mathrm{lim}} = 0.5\frac{t}{D} - 0.0025 + 3000\left(\frac{(P_i - P_e)D}{2tE}\right)^2 \tag{3.7}$$

where $\varepsilon_{\mathrm{lim}}$ is compressive strain limit of a pipe, P_i is maximum design internal pressure, and P_e is minimum external hydrostatic pressure.

Limit ovalization. Upon extrusion, the cross-section of a pipe tends to be elliptical. The ovalization deformation will affect safety of the pipeline. The elliptization deformation rate is defined as:

$$\Delta = 2\left(\frac{D_{\mathrm{max}} - D_{\mathrm{min}}}{D_{\mathrm{max}} + D_{\mathrm{min}}}\right) \tag{3.8}$$

where Δ is pipe ovality, D_{max} is maximum pipe outer diameter, and D_{min} is minimum pipe outer diameter. The limit ovalization criteria are set as 3% by CSA Z662 [Canadian Standardization Association, 2007] and DNV-OS-F101 [Det Norske Veritas, 2007], and 5.5–6.2% by API [American Petroleum Institute, 2000].

3.3 Applications for FFS Determination and Failure Pressure Prediction of Pipelines

Some typical examples of using Level III assessment for pipelines containing a single corrosion defect are listed in Table 3.1. It is seen that the FE models and codes are applicable for a wide variety of defect geometries and steel grades. The corrosion defects to be modeled are mainly external defects. The affecting factors include defect geometry, internal pressure, soil strain, mechanical vibration, bending moment, debris flow impact, etc. Generally, the defect geometry affects local stress and strain distributions, and thus, failure pressure of the pipelines. Moreover, the failure pressure usually decreases when the pipelines are under combined stresses resulting from multiple sources.

In addition to assessment of a single corrosion defect on pipelines, the Level III methods are also applicable for assessment of multiple corrosion defects which are adjacent to enable mutual interactions. As widely recognized, the interaction between defects can further reduce the failure pressure of pipelines, as compared with the pipelines containing identical

Table 3.1 Some typical examples of the application of Level III assessment for pipelines containing a single corrosion defect.

Method	Load	Material	Stress–strain rule	Failure criterion	Defect geometry	Source
Xu and Cheng	Soil-induced strain (tensile and compressive) and internal pressure	X65, X80 and X100	Isotropic-hardening plasticity model	Stress-based	Semi-ellipsoid	Xu and Cheng [2012]
Shuai et al.	Axial compressive and internal pressure	X80	R-O	Stress-based	Cuboid	Shuai et al. [2020]
Oh et al.	Pressure, bending moments and internal pressure	X65	Bilinear	Stress-based	Cuboid and semi-sphere	Oh et al. [2009]
Arumugam et al.	Axial compressive stress and internal pressure	X52	Power–law model	Stress-based	Cuboid	Arumugam et al. [2020]
Dewanbabee	Axial compression and internal pressure	X46	Power–law model	Stress-based	Cuboid	Dewanbabee [2009]
Zhou et al.	Longitudinal tensile and compressive strain	X70	Isotropic hardening plasticity model	Stain-based	Cuboid	Zhou et al. [2016]
Kim et al.	Axial displacement and internal pressure	X70	Power–law model	Strain-based	Cuboid	Kim et al. [2020]
Mondal and Dhar	Axial forces, bending moments and internal pressure	X65	Isotropic hardening plasticity model	Stress-based	Cuboid	Mondal and Dhar [2019]
Chegeni et al.	Bending moment and internal pressure	X52 and X46	Power–law model	Stress-based	Cube and cuboid	Chegeni et al. [2019]
Shim et al.	Bending moment and internal pressure	ASTM A333 Grade 6	Power–law model	Stress-based	Semi-ellipsoid	Shim et al. [2003]
Wu et al.	Debris flows and internal pressure	X80	Bilinear	Stress-based	Cuboid	Wu et al. [2019]
Qin and Cheng	Cyclic loading and internal pressure	X60 and X80	R-O	Stress-based	Cuboid	Qin and Cheng [2020]
Zhang and Zhou	Internal pressure	X52 and X80	Power–law model	Stress-based	Semi-ellipsoid and Cuboid	Zhang and Zhou [2020]
Jin et al.	Internal pressure	X70	Bilinear	Stress-based	Semi-sphere	Jin et al. [2020]
Shuai et al.	Internal pressure	X46, X52, X60	Power–law model	Stress-based	Cuboid	Shuai et al. [2017]

Source: From Qin and Cheng [2021a].

number of defects, but they are isolated and do not interact with each other [Sun and Cheng, 2018, 2019, 2020, 2021]. Interaction rules were proposed to define whether an interaction exists between adjacent corrosion defects [Lamontagne, 2002]. Table 3.2 shows typical applications of the FE-based Level III assessment for multiple corrosion defects with mutual interactions and the effect on failure pressure of the pipelines. The modeled defects are oriented in various ways, including longitudinal, circumferential, and radial alignments, as well as mixed forms. The radial alignment may occur on the inner surface of pipelines. Pipe materials range from low to high grades of steel. The stresses under consideration include hoop stress resulting from internal pressure, axial stress induced by ground movement, etc. The geometry of corrosion defects and their distributions remarkably affect the mutual interaction between adjacent defects and thus the failure pressure of the pipelines [Lee et al., 2005; Shuai et al., 2020]. In general, the failure pressure of a pipeline containing multiple corrosion defects is lower compared with the pipeline containing a single defect. Moreover, the mutual interaction of adjacent defects further decreases the failure pressure of the pipeline. The interaction is enhanced as the distance between the defects decreases.

3.3.1 A Single Corrosion Defect on Pipelines

Compared with the available industry standards and codes for Levels I and II defect assessment, as reviewed in Chapter 2, the FE-based Level III assessment methods show an improved accuracy in defect sizing, a strong capability in solving nonlinear problems for pipelines containing various shapes of corrosion defects, and thus a reliable prediction of failure pressure and remaining strength of the pipelines. This section includes an FE-based Level III model for assessment of a single corrosion defect on pipelines made of various grades of steels, i.e., X65, X80, and X100 steels, under combined stresses resulting from internal pressure and ground movement [Xu and Cheng, 2012]. As a comparison, results obtained from three commonly used Levels I and II industry standards, i.e., ASME B31G [American Society of Mechanical Engineering, 2009, 2012], RSTRENG [Kiefner and Vieth, 1990a, 1990b], and DNV RP F101 [Det Norske Veritas, 2010], are included to evaluate the applicability and accuracy of the developed FE model.

3.3.1.1 Failure Pressure Prediction and Evaluation of the Accuracy of Existing Industry Models

Three grades of pipeline steel (i.e., X65, X80, and X100 steels) and four corrosion depths (i.e., 20%, 40%, 60%, and 80% of pipe wall thickness) were chosen for FE modeling while assuming the defect length and width remained

Table 3.2 Typical examples of the applications of FE-based Level III assessment for multiple corrosion defects with mutual interactions and the effect on failure pressure of the pipelines.

Method	Defect orientation	Pipe steel	Stress–strain rule	Stress	Source
Kuppusamy et al.	Longitudinal alignment, circumferential alignment	X46	Power–law model	Internal pressure, ground movement (i.e., axial compressive stress)	Kuppusamy et al. [2016]
Han et al.	Longitudinal alignment, overlapped on internal pipe surface	X65	R-O	Internal pressure	Han et al. [2016]
Sun and Cheng	Longitudinal alignment, circumferential alignment, overlapped on external pipe surface	X46, X60 & X80	R-O	Internal pressure	Sun and Cheng [2018]
Chen et al.	Longitudinal alignment, circumferential alignment	X80	Power–law model	Internal pressure	Chen et al. [2015]
De Andrade et al.	Longitudinal alignment, circumferential alignment, mixed type of interaction	X80	R-O	Internal pressure	De Andrade et al. [2006]
Silva et al.	Longitudinal alignment, circumferential alignment	X52	Power–law model	Internal pressure	Silva et al. [2007]
Capula Colindres et al.	Overlapped internal and external surface	X42, X52, X60, X70, X80, X100	Power–law model	Internal pressure	Capula Colindres et al. [2020]
Li et al.	Longitudinal alignment, circumferential alignment, mixed type of interaction	X80	R-O	Internal pressure	Li et al. [2016]

Source: From Qin and Cheng [2021a].

Table 3.3 Failure pressure of a pipe (19 mm in wall thickness) made of X65, X80, or X100 steel containing a corrosion defect with various depths and 200 mm in length determined by various models.

	Failure pressure (MPa)		
Model	**X65 steel**	**X80 steel**	**X100 steel**
Defect depth is 20% of pipe wall thickness			
ASME B31G	22.1	31.3	38.9
RSTRENG	22.9	31.2	38.0
DNV-RP-F101	25.8	34.4	40.3
Xu and Cheng	26.1	35.0	42.2
Defect depth is 40% of pipe wall thickness			
ASME B31G	20.4	28.9	35.9
RSTRENG	20.6	28.1	34.2
DNV-RP-F101	23.5	31.3	36.7
Xu and Cheng	23.0	29.3	36.0
Defect depth is 60% of pipe wall thickness			
ASME B31G	18.0	26.0	32.3
RSTRENG	17.7	24.1	29.4
DNV-RP-F101	19.9	26.5	31.0
Xu and Cheng	18.9	24.0	28.8
Defect depth is 80% of pipe wall thickness			
ASME B31G	15.8	22.4	27.8
RSTRENG	13.7	18.1	22.7
DNV-RP-F101	13.6	18.1	21.3
Xu and Cheng	13.9	18.0	20.2

Source: From Xu and Cheng [2012] / with permission from Elsevier.

unchanged for research purposes. A quarter of the pipe segment was modeled due to geometrical symmetry. The corrosion defect was treated as an elliptical shape. Table 3.3 shows the failure pressure of a pipe (19 mm in wall thickness) made of X65, X80, or X100 steel containing a corrosion defect with various depths and 200 mm in length determined by various models. As expected, at specific defect depths, the failure pressure of the pipe predicted by all models increases with increased steel grade. This is due to the enhanced strength of a high grade of steel, providing a great resistance to deformation and rupture at the defect. For the shallow corrosion defect with a thickness of 20% of the pipe wall thickness, the DNV model obtains results slightly smaller than the results from Xu and Cheng model, while the ASME B31G and RSTRENG models generate much lower failure pressures. With increased corrosion defect depth, the failure pressure of the pipe decreases. The failure pressures obtained by the three industry models are greater than the results of the Xu and Cheng model, especially for deep corrosion defects with a depths of 60% and 80% of the pipe wall thickness, except the results on X65 steel pipe. Particularly, the ASME B31G predicts the greatest failure pressure for X100 steel.

Generally, the three industry models for prediction of failure pressure of pipelines containing a corrosion defect give results greater than the FE-based Xu and Cheng model for shallow (or small) corrosion defects. In other words, the industry models tend to underestimate the failure pressure of the corroded pipelines. However, overestimation occurs with increased steel grade and corrosion defect depth. Moreover, the prediction accuracy decreases as the corrosion depth and steel grade increase. This is expected as the industry models (i.e., ASME B31G and RSTRENG) were developed based on data from burst testing conducted on low-grade pipe steels such as X52 and X65 steels. The RSTRENG model is a modification of ASME B31G by changing the flow stress, σ_{flow}, to specified minimum yield strength (SMYS) + 69 MPa, and the effective area of corrosion defect factor to 0.85. In ASME B31G, the S_{flow} is 1.1 × SMYS and the defect area factor is 2/3. Thus, the flow stress calculated by RSTRENG model is increased and close to the ultimate tensile strength of steels, compared to the ASME B31G model. The effective area of corrosion defect is also improved, and close to the real value used in FE model. Thus, the RSTRENG model gives a better prediction accuracy than ASME B31G. However, the model is still applicable for low grades of steel only such as X65 steel. Of the three industry models, the DNV model is the most accurate, and the prediction results are close to the FE-based Xu and Cheng model. The DNV model was developed and validated based on full-scale testing on high-grade steels up to X80 grade [Amirat et al., 2006]. In principle, the X100 steel pipeline steel is beyond the predictive scope of DNV model. The DNV model provides the failure pressure close to the value obtained from the FE model for X100 steel. The primary reason is that DNV model uses directly the ultimate tensile strength as flow stress.

In summary, the Levels I and II industry models and methods are not appropriate for defect assessment for the high-strength steel pipelines, e.g., X80 and X100 steels. For the low-grade steel pipelines, these models and methods are not applicable for deep corrosion defects with a depth exceeding 60% of the pipe wall thickness. Level III methods must be used for defect assessment to provide accurate results for high-strength steel pipelines and deep corrosion defects.

3.3.1.2 Local Stress and Strain Distributions at the Corrosion Defect

Figure 3.2 shows the von Mises stress of the inner and outer surfaces of an X80 steel pipe in the presence and absence of a corrosion defect with the

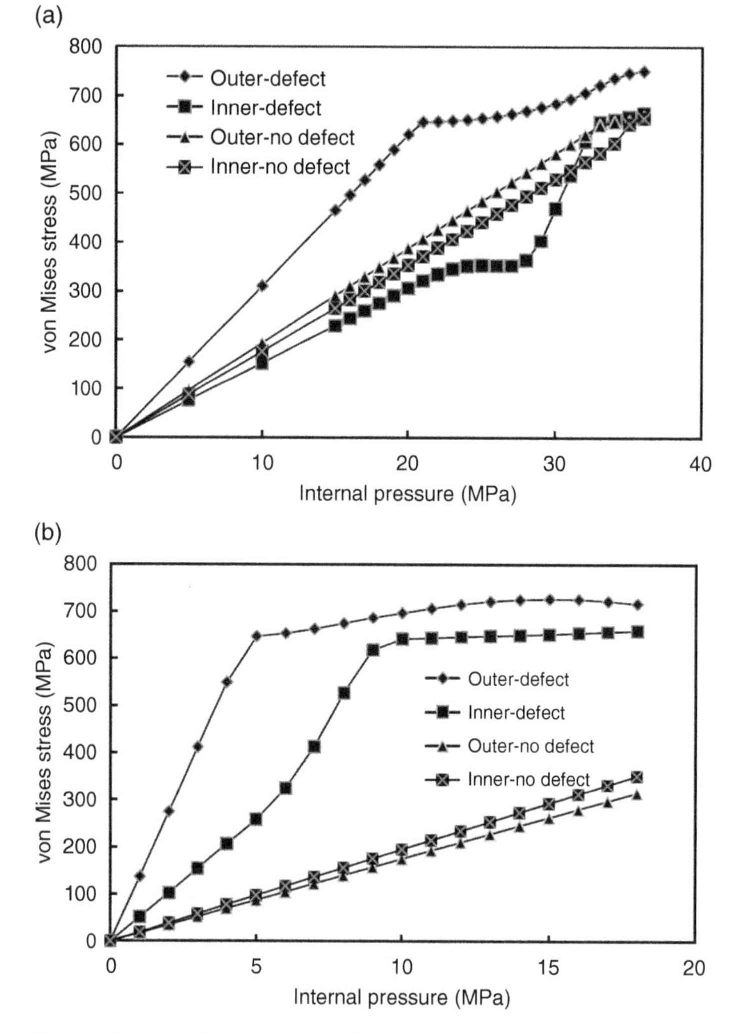

Figure 3.2 von Mises stresses of the inner and outer surfaces of X80 steel pipe as a function of internal pressure in the presence and absence of a corrosion defect with (a) 20% and (b) 80% of wall loss depth. *Source:* From Xu and Cheng [2012] / with permission from Elsevier.

depth of 20% and 80% of pipe wall thickness, respectively, where "outer-defect" and "inner-defect" refer to the location of the defect bottom for the corrosion defect present on the outer and inner pipe surfaces, respectively. It is seen that, in the absence of corrosion defect, there is little difference of the von Mises stress on the inner and outer surfaces. When a corrosion defect is present, the von Mises stress at the defect base of the outer surface is much greater than that of the inner surface. When the corrosion depth is up to 80% of wall thickness, the von Mises stresses of both the inner and outer surfaces are much greater than the stresses when the corrosion defect is 20% of wall loss depth. Apparent yielding points are observed in both curves.

Effective plastic strain is usually used to describe the intensity of plastic deformation. A comparison of the effective plastic strain on the outer and inner surfaces of the X80 steel pipe in the absence and presence of a corrosion defect with depths in 20% and 80% of pipe wall thickness, respectively, is shown in Figure 3.3. It is seen that there is no plastic deformation on the pipe over the measured pressure range when a corrosion defect is absent. When the corrosion defect is as deep as 20% of the pipe wall depth, the effective plastic strain of the outer surface becomes apparent when the internal pressure exceeds 25 MPa. After that, the plastic strain increases rapidly, indicating a high plastic deformation. When the defect is up to 80% of wall loss depth, plastic deformation is appreciable at an internal pressure of 5 MPa only. A small effective plastic strain is observed on the inner face of the pipe as well.

Apparently, the depth of a corrosion defect is a major factor affecting local stress and strain distributions. Generally, in the absence of a corrosion defect, the von Mises stress of the inner surface is slightly greater than that of the outer surface owing to the primary stress source of internal pressure. Even at a small corrosion depth (e.g., 20% of pipe wall thickness), the von Mises stress at the bottom of the defect is much greater than the inner surface of the pipe. With the corrosion depth increasing to 80% of pipe wall thickness, the stress concentration at the defect is further enhanced. It is thus concluded that the internal corrosion defect, especially the deep corrosion defect, could be the location to fail due to a high stress concentration. Another interesting phenomenon is that an increased corrosion depth affects von Mises stress of the inner surface of the pipe but does not affect the effective plastic strain. The plastic deformation mainly occurs on the outer defect, although the inner surface also suffers from plastic deformation, but at a low level.

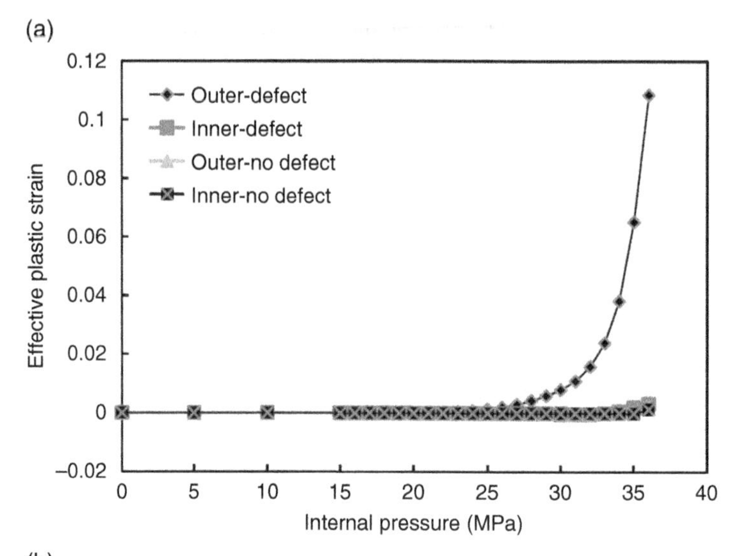

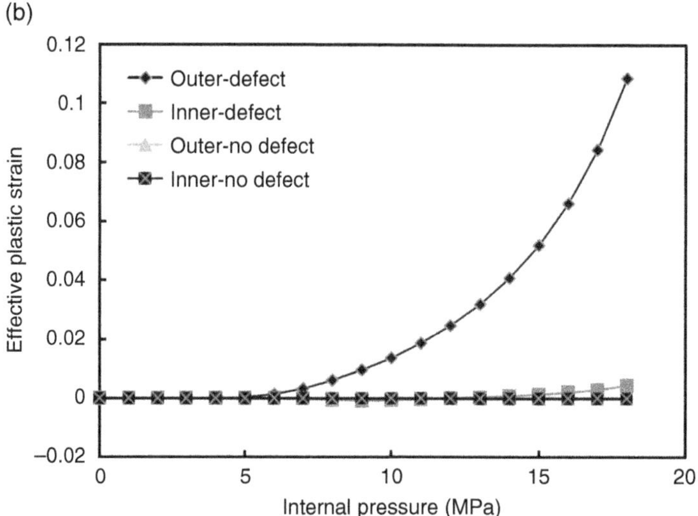

Figure 3.3 Effective plastic strain of the inner and outer surfaces of the X80 steel pipe as a function of internal pressure in the presence and absence of a corrosion defect with (a) 20% and (b) 80% of pipe wall thickness. *Source:* From Xu and Cheng [2012] / with permission from Elsevier.

3.3.1.3 Failure Pressure of Pipelines Containing a Corrosion Defect Under a Combined Internal Pressure and Axial Strain

Pipelines experiencing significant axial strain due to pipe–soil interactions are usually named pre-strained pipelines. The pre-strain may affect failure

Table 3.4 Failure pressures determined by the FE model for an X80 steel pipe containing a corrosion defect with a depth of 80% pipe wall thickness under various axial tensile or compressive pre-strains.

Tensile pre-strain	**0**	**+0.1%**	**+0.2%**	**+0.25%**
Failure pressure (MPa)	18.0	17.0	15.3	14.4
Compressive pre-strain	0	−0.1%	−0.2%	−0.25%
Failure pressure (MPa)	18.0	17.3	16.5	15.0

Source: From Xu and Cheng [2012] / with permission from Elsevier.

pressure of the pipelines, especially when corrosion defects are present on the pipe body. The failure pressures determined by the FE model for an X80 steel pipe containing a corrosion defect with the depth of 80% of pipe wall thickness under various axial tensile or compressive pre-strains are shown in Table 3.4, where the signs "+" and "−" indicate the tensile and compressive pre-strains, respectively. It is seen that the failure pressure of the defective pipe is reduced by applied axial pre-strain, no matter whether tensile or compressive. Moreover, as the pre-strain increases, the failure pressure further decreases. Under specific pre-strains, the failure pressure under a compressive pre-strain is greater than that under a tensile pre-strain. Thus, an axial tensile strain would decrease the pressure-bearing capability more significantly than an axial compressive strain. Figure 3.4 shows the distribution of plastic deformation of the X80 steel pipe containing a corrosion defect with a depth of 80% pipe wall thickness under different pre-strain conditions. Apparently, the plastic deformation area increases when either a tensile or a compressive pre-strain is applied. However, the plastic deformation area under the tensile strain expands along the hoop direction and is away from the defect, while the plastic area under the compressive pre-strain is located at the defect. Therefore, the axial tensile strain results in an expansion of the plastic area along the hoop direction, potentially causing circumferential cracking. On the contrary, the axial compressive strain generates a local buckling or wrinkling effect at the corrosion defect.

In summary, a pipeline containing a corrosion defect with pre-strain due to ground movement will have a decreased failure pressure compared with the pipeline without pre-strain experience. Depending on the nature of the pre-strain, the failure mode of the pipeline will be different. An axial tensile strain will cause circumferential cracking starting at the corrosion defect, while an axial compressive strain can cause buckling or wrinkling of the pipeline at the defect.

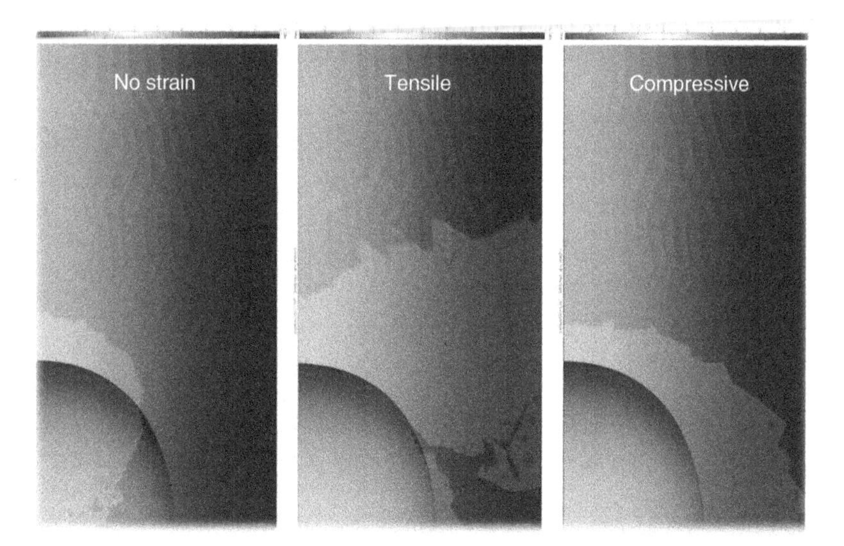

Figure 3.4 Distribution of plastic deformation of the X80 steel pipe containing a corrosion defect with a depth of 80% pipe wall thickness under different pre-strain conditions. *Source:* From Xu and Cheng [2012] / with permission from Elsevier.

3.3.2 Multiple Corrosion Defects on Pipelines

Most corrosion defects on pipelines are in proximity so that mutual interactions exist between the adjacent defects [Det Norske Veritas, 2004]. While extensive efforts have been made to develop models and numerical codes to assess the single corrosion defect (i.e., the defect is isolated from any other defects so that an interaction does not happen between the defects), technical gaps still exist in accurate assessment of multiple defects with mutual interactions for pipeline FFS determination and failure pressure prediction.

Analysis of the interaction between adjacent corrosion defects is expected to provide more accurate evaluation of the pipeline FFS as compared with the assessment where the interaction is ignored [Chen and Shu, 2001; Benjamin et al., 2016a, b]. It has been confirmed that the interaction between corrosion defects influenced remarkably the failure pressure of corroded pipelines [Chiodo and Ruggieri, 2009]. Investigations on multiple corrosion defects on pipelines have resulted in establishment of the so-called interaction rules, such as CW rule [Coulsen and Worthingham, 1990], DNV-RP-F101 code [Det Norske Veritas, 2004], 6WT rule [Lamontagne, 2002], and 3WT rule [Hopkins and Jones, 1992]. These rules and codes are used to determine whether an interaction exists between adjacent corrosion defects based on calculation of the strength of pipeline steels at the corrosion defects with known geometrical factors. When the interaction effect is identified between two adjacent corrosion defects, they will not be treated as two single defects. As a result, conventional methods for defect assessment and FFS determination, such as ASME B31G and modified B31G, cannot be used for assessment of these defects. Generally, the failure pressure of pipelines decreases when multiple corrosion defects interact with each other [Chen and Shu, 2001].

The methods available today for assessment of multiple corrosion defects on pipelines suffer from some problems and limitations. For example, the modeling results are validated with data obtained from a specific grade of pipeline steel, and thus, are not applicable for the steel strength under which an interaction rule is to be applied. Therefore, the influence of steel grade on the interaction between corrosion defects has rarely been studied. Moreover, in addition to circumferential and longitudinal orientations, the corrosion defects usually overlap with each other on pipelines, which is a common scenario found in the field [Han et al., 2016]. However, to date, there has been limited investigation on assessment of overlapped corrosion defects and their effect on failure pressure of the pipelines. To fill the gap, numerical models should be developed to define the interaction between adjacent corrosion defects, and the effect on pipeline FFS determination. Upon validation by either burst testing or comparison of the defect growth over time, as detected by periodic ILI, the developed models can serve as a novel methodology for assessment of multiple corrosion defects on pipelines.

3.3.2.1 The Model

Sun and Cheng [2018] developed an FE-based model for assessment of X46, X60, and X80 steel pipes containing multiple corrosion defects, which were either longitudinally aligned, circumferentially aligned, or overlapped with each other. While considering the effects of defect size and steel grade, the critical spacing between the corrosion defects was defined so that it was able to assess if an interaction effect existed between the defects to affect failure pressure of the pipelines. For the selected pipe steels, X46 steel pipelines were generally built in the 1950s, representing many pipeline networks in North America. A typical example of the "old" X46 steel pipelines is Enbridge Pipelines' Line 5, which was built in 1953 and of 645 miles in length, transporting up to 540,000 barrels per day of light crude oil, light synthetic crude, and natural gas liquids, constituting an important link in Enbridge's US Mainline System [Enbridge, 2014]. The X60 steel possesses an improved mechanical strength with a good ductility, representing the mediate grade of steel for pipelines. The X80 steel is categorized as high-strength steel, which has been extensively used mainly in Europe and Asia for pipeline construction, with improved transporting capacity, a high internal pressure, and reduced steel weight for significant economic benefits in operation [Zhang et al., 2019].

Figure 3.5 shows the 3D model of a steel pipe containing corrosion defects which are oriented either longitudinally, circumferentially, or overlapped with each other. Symmetrical constraints were applied on the planes to be modeled. Displacement in Z direction of the uncorroded end was also constrained. The length of the pipe was sufficient to avoid the effect of boundary conditions on the corroded area. The uncorroded area on the pipe was under coarse meshing to reduce computation. The appropriate meshing sizes for different defects were determined by mesh sensitivity analysis, and the mesh sizes of 8 and 10 mm were chosen in circumferential and longitudinal directions, respectively. The maximum and minimum element sizes of the corroded region were 3 and 1 mm, respectively. The corrosion defects have a smooth edge, with a radius equaling the defect depth, to avoid a high stress concentration.

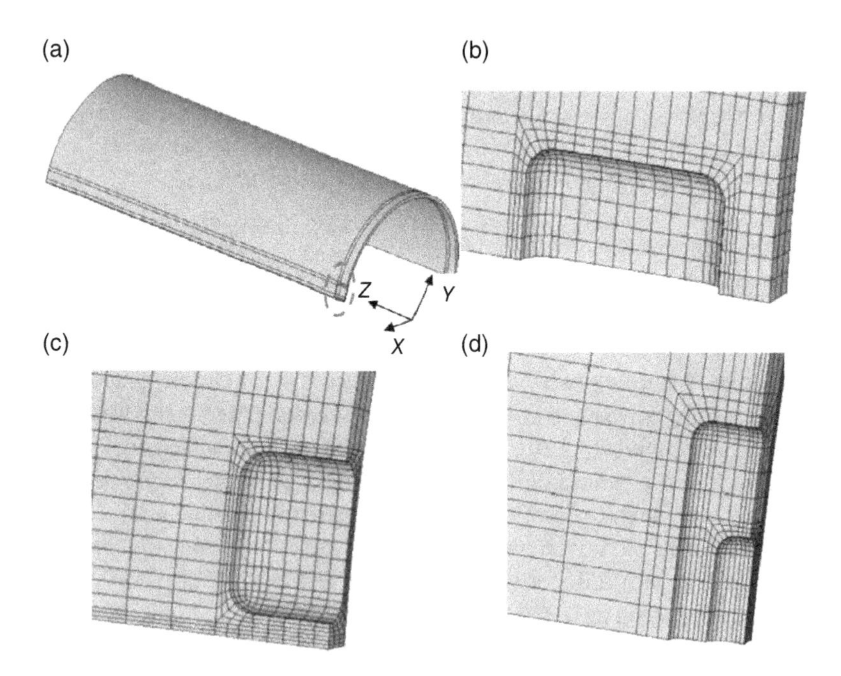

Figure 3.5 The 3D model for a steel pipe containing corrosion defects (a) a quarter model and meshes of (b) a longitudinal and (c) a circumferential corrosion defect, and (d) two overlapping corrosion defects. *Source:* From Sun and Cheng [2018] / with permission from Elsevier.

The R-O stress–strain rule [Ramberg and Osgood, 1943] was used to define the strain-hardening behavior of pipeline steels. The mechanical properties of X46, X60, and X80 pipeline steels are included in Sun and Cheng [2018]. The ultimate tensile strength of the steel is used as the failure criterion for pipelines, as recommended by DNV RP F101 standard [Det Norske Veritas, 2010]. The criterion has been verified to enable prediction of the failure pressure of corroded pipelines with a better accuracy compared to ASME B31G and the modified B31G standards [American Society of Mechanical Engineering, 1991, 2009].

3.3.2.2 Interaction of Longitudinally or Circumferentially Aligned Corrosion Defects on Pipelines

Since the interaction between corrosion defects results in a decreased failure pressure, to quantify the interaction effect, the failure pressure of a pipeline containing a single corrosion defect, i.e., P_{single}, is used as a reference. The criterion used to evaluate the interaction between corrosion defects is defined as:

$$\begin{cases} P_{\text{multiple}}/P_{\text{single}} \geq 0.99, & \text{no interaction} \\ P_{\text{multiple}}/P_{\text{single}} < 0.99, & \text{interaction exists} \end{cases} \tag{3.9}$$

where P_{multiple} is the failure pressure of a pipeline containing multiple corrosion defects. When the ratio of $P_{\text{multiple}}/P_{\text{single}}$ equals or is larger than 0.99, there is no interaction effect between the adjacent corrosion defects. If the ratio is smaller than 0.99, an interaction exists between the defects to decrease the failure pressure. The smaller the ratio, the more significant the interaction on decreased failure pressure of the pipeline.

The effect of the longitudinal spacing of corrosion defects located in the axial direction of a steel pipe on the ratio of failure pressures (i.e., $P_{\text{multiple/Psingle}}$) of the pipes made of X46, X60, and X80 steels is shown in Figure 3.6a. It is seen that the ratio increases with increased longitudinal spacing between the corrosion defects, indicating that the interaction of the corrosion defects decreases as the longitudinal spacing increases. There is little effect of the steel grade on the $P_{\text{multiple}}/P_{\text{single}}$ ratio at specific longitudinal spacings. The dashed line in the figure denotes the failure pressure ratio of 0.99, where a critical longitudinal spacing of 152.4 mm is identified for the three grades of pipeline steel.

Figure 3.6b shows the effect of the circumferential spacing of two corrosion defects which are located on the circumferential direction on the ratio of $P_{\text{multiple}}/P_{\text{single}}$ of pipelines made of X46, X60, and X80 steels. The ratio also increases with increased circumferential spacing for the three grades

(a)

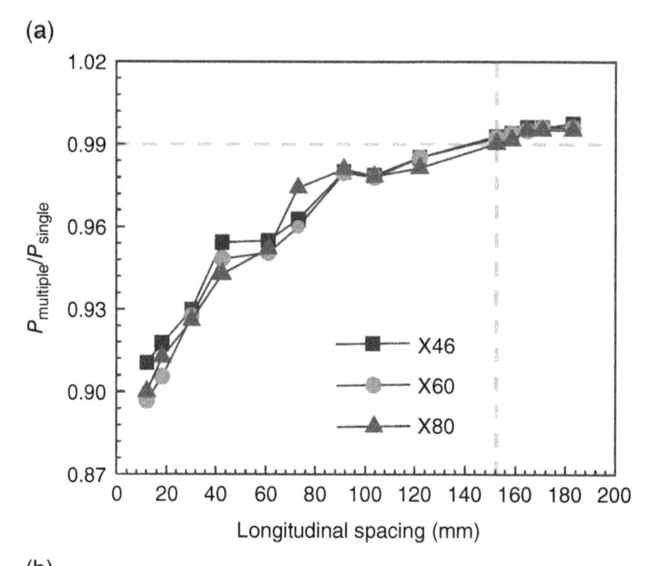

(b)

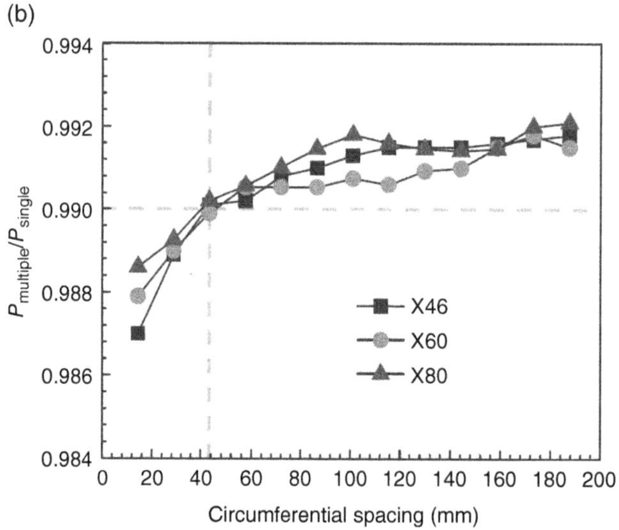

Figure 3.6 Effect of (a) the longitudinal spacing and (b) circumferential space of corrosion defects on the ratio of failure pressures (i.e., $P_{multiple}/P_{single}$) of pipelines made of X46, X60, and X80 steels. *Source:* From Sun and Cheng [2018] / with permission from Elsevier.

of steel. The steel grade slightly affects the dependence of the failure pressure ratio on circumferential spacing. The three curves reach a critical ratio at the circumferential spacing of about 43.2 mm, which is much smaller than the critical spacing for longitudinally oriented defects.

For pipelines containing longitudinally aligned corrosion defects, when the spacing between them is larger than 152.4 mm, the interaction effect does not exist, and the multiple defects can be treated separately. The steel grade does not affect this conclusion. As a result, the ratio of the failure pressures is not consistent with the behavior of a single corrosion defect with doubled length. When the corrosion defects are oriented circumferentially, the critical spacing is 43.2 mm. Thus, the longitudinal corrosion defects are associated with a larger spacing where the interaction between the defects exists and affects the failure pressure of the pipeline. The circumferential spacing of the defects has a smaller influence on failure pressure compared with the longitudinal spacing of two corrosion defects with identical dimensions. This difference is attributed to the fact that the defect length affects the failure pressure of pipelines at a more significant level than the defect width while their depths are kept constant.

3.3.2.3 Overlapped Corrosion Defects on Pipelines

Overlapped corrosion defects are aligned along the radial direction of a pipe with the lengths of the defects in axial direction. Figure 3.7 shows the distributions of von Mises stress on an X46 steel pipe containing two overlapped corrosion defects, where L_1 is a half of the axial length of the top defect, and d_1 and d_2 are depths of the top and bottom defect, respectively, under an operating pressure of 20 MPa. It is seen that the stress distribution is not uniform. The stress of the defects is greater than that of the pipe surface. With increased d_2, the stress of the bottom defect increases apparently. Thus, a stress concentration occurs mainly at the bottom defect as its depth increases, with the maximum von Mises stress observed at the defect side.

Table 3.5 lists the failure pressures of X46, X60, and X80 steel pipes containing two overlapped corrosion defects with L_1 of 19.8 mm and various d_1 and d_2/d_1 ratios. With other variables fixed, the failure pressure decreases with the increasing d_1 and d_2/d_1 ratio (i.e., the increasing d_2 at a fixed d_1). The failure pressure increases with the increased steel strength.

The effect of corrosion defect overlapping on failure pressure of a pipeline is defined as the ratio of $P_{overlapped}/P_{single}$, where $P_{overlapped}$ and P_{single} are failure pressures of the pipeline containing overlapped corrosion defects and a single defect, respectively. Generally, the smaller the ratio, the more significant the overlapped defects causing reduction in failure pressure. Figure 3.8 shows the failure pressure ratio, i.e., $P_{overlapped}/P_{single}$, of an X46 steel pipe containing two overlapped corrosion defects as a function of the defect depth ratio, i.e., d_2/d_1. Linear relationships are observed between the $P_{overlapped}/P_{single}$ ratio and d_2/d_1. Generally, the width of a corrosion defect does not affect greatly the failure pressure, as compared with

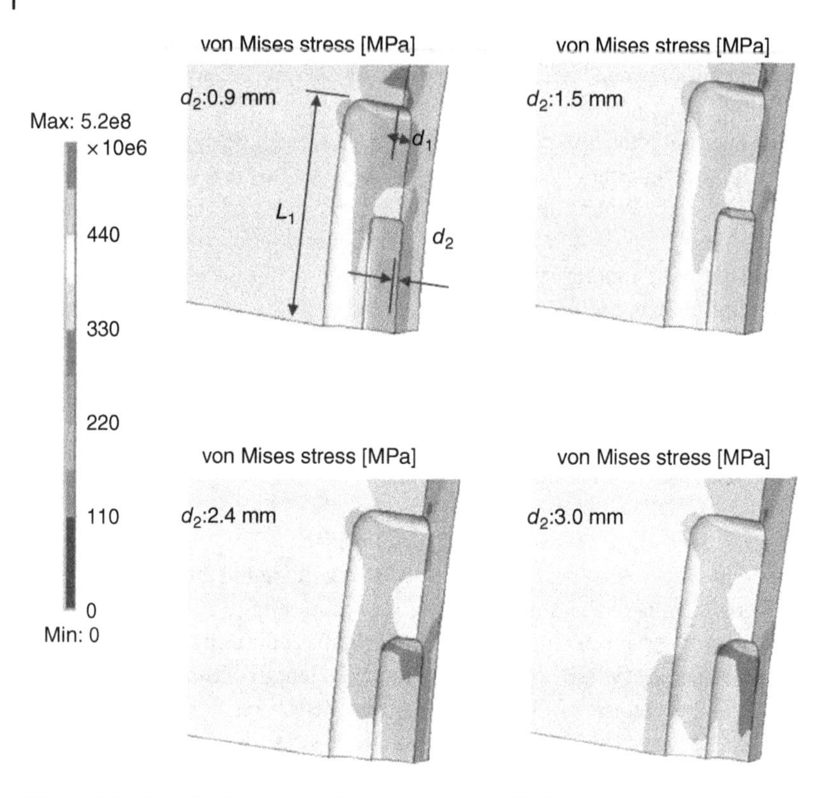

Figure 3.7 Distributions of von Mises stress on an X46 steel pipe containing two overlapped corrosion defects (where L_1 is a half of the axial length of the top defect, and d_1 and d_2 are depths of the top and bottom defect, respectively) under an operating pressure of 20 MPa. *Source:* From Sun and Cheng [2018] / with permission from Elsevier.

the defect depth (d) and length (L). This applies for both a single defect and multiple defects oriented either longitudinally or circumferentially [Chandra and Mondal, 2016]. The presence of the bottom defect lowers the failure pressure, compared to that with a single top defect only. The $P_{overlapped}/P_{single}$ ratio linearly decreases with the increasing d_2/d_1 ratio, indicating that the failure pressure decreases with increased d_2 (i.e., the depth of the bottom defect) at fixed d_1. The slope of the lines obtained by linear regression analysis is used to quantify the relationship between the $P_{overlapped/Psingle}$ ratio and the d_2/d_1 ratio. Moreover, at individual d_2/d_1 ratios, there is a smaller $P_{overlapped}/P_{single}$ ratio at a larger d_1, i.e., the depth of the top layer defect. Thus, an increase in the depth of either top or bottom corrosion defect would reduce the failure pressure of pipelines. A further

Table 3.5 Failure pressures of X46, X60, and X80 steel pipes containing two overlapped corrosion defects with L_1 of 19.8 mm and various d_1 (i.e., 2, 3 and 4 mm) and d_2/d_1 ratios.

d_2/d_1	0.3	0.4	0.5	0.6	0.7	0.8	0.9	1.0
X46 steel								
2	21.86	21.84	21.80	21.76	21.71	21.68	21.65	21.62
3	21.40	21.31	21.20	21.11	21.02	20.96	20.89	20.85
4	20.54	20.41	20.29	20.19	20.04	19.93	19.75	19.67
X60 steel								
2	26.00	25.98	25.92	25.90	25.84	25.82	25.76	25.68
3	25.27	25.17	25.06	25.01	24.88	24.77	24.67	24.59
4	24.11	23.98	23.82	23.69	23.54	23.38	23.23	23.07
X80 steel								
2	31.80	31.77	31.70	31.67	31.57	31.54	31.48	31.44
3	31.03	30.90	30.74	30.61	30.49	30.39	30.30	30.23
4	29.73	29.54	29.38	29.23	29.01	28.85	28.60	28.47

Source: From Sun and Cheng [2018] / with permission from Elsevier.

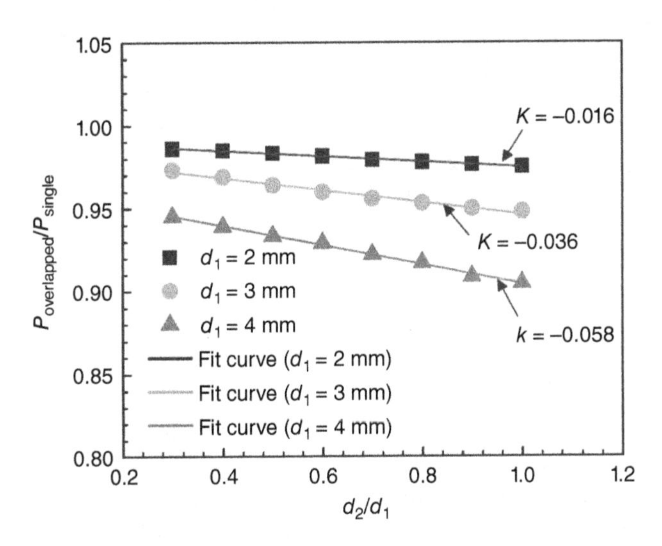

Figure 3.8 The failure pressure ratio, i.e., $P_{overlapped}/P_{single}$, of an X46 steel pipe containing two overlapped corrosion defects as a function of the defect depth ratio, i.e., d_2/d_1. *Source:* From Sun and Cheng [2018] / with permission from Elsevier.

comparison of absolute value of the linear slope shows that the absolute value of the slope increases with the increasing d_1. Thus, with the top defect becoming deeper, the $P_{\text{overlapped}}/P_{\text{single}}$ ratio decreases more rapidly with the bottom layer defect depth.

3.3.2.4 Quantification of the Interaction of Multiple Corrosion Defects

The interaction of multiple corrosion defects on pipelines is quantified by the ratio of $P_{\text{multiple}}/P_{\text{single}}$. In particular, the failure pressure of pipelines containing a single defect is determined by [Chen et al., 2015]:

$$P_{\text{single}} = \sigma_u \frac{2t}{(D-t)} f\left(\frac{d}{t}, \frac{L}{\sqrt{Dt}}, \frac{\beta}{\pi}\right) \tag{3.10}$$

where β is the width angle of a defect and $f()$ is a reduction function of the defect geometry with respect to d/t, $L/\sqrt{Dt}$, and β/π [Su et al., 2016]. The equation can also be used to calculate the failure pressure of pipelines containing longitudinal or circumferential corrosion defects if the parameters d, L, and β are replaced with equivalent depth d_e, equivalent length L_e, and equivalent width angle β_e, respectively, of the multiple defects [Chandra and Mondal, 2016]. Thus, the expression $P_{\text{multiple}}/P_{\text{single}}$ is written as:

$$\frac{P_{\text{multiple}}}{P_{\text{single}}} = \frac{\sigma_u \dfrac{2t}{(D-t)} f_2\left(\dfrac{d_e}{t}, \dfrac{L_e}{\sqrt{Dt}}, \dfrac{\beta_e}{\pi}\right)}{\sigma_u \dfrac{2t}{(D-t)} f_1\left(\dfrac{d}{t}, \dfrac{L}{\sqrt{Dt}}, \dfrac{\beta}{\pi}\right)} \tag{3.11}$$

Since the models for multiple corrosion defects and a single defect are based on the same pipeline, the steel's ultimate tensile strength σ_u, and the pipe dimensions (i.e., t and D) are identical for the numerator and denominator in Eq. (3.11). It is thus simplified as:

$$\frac{P_{\text{multiple}}}{P_{\text{single}}} = \frac{f_2\left(\dfrac{d}{t}, \dfrac{L}{\sqrt{Dt}}, \dfrac{\beta}{\pi}\right)}{f_1\left(\dfrac{d}{t}, \dfrac{L}{\sqrt{Dt}}, \dfrac{\beta}{\pi}\right)} \tag{3.12}$$

The failure pressure ratio, $P_{\text{multiple}}/P_{\text{single}}$, then depends on the defect geometry only. Neither the material properties nor the pipe geometry influences the interaction effect between multiple corrosion defects.

In summary, the maximum longitudinal interaction spacing is determined as 152.4 mm (i.e., $2.5\sqrt{Dt}$), which is well consistent with the value of $2\sqrt{Dt}$ in DNV-RP-F101 [Det Norske Veritas, 2004]. The maximum circumferential interaction spacing is 43.2 mm (i.e., 5.3 t), which is approximately identical to the value of 6 t as defined in 6WT rule [Lamontagne, 2002]. The circumferential spacing has less effect on the failure pressure compared with the longitudinal spacing. While the defect depth is kept constant, the defect width is not as important as the length to affect the failure pressure of pipelines [De Andrade et al., 2006]. Since the longitudinal and circumferential spacing of corrosion defects is normalized, with considerations of the outer diameter and pipe wall thickness, the conclusions apply for pipelines with other dimensions. As the geometry and orientation of corrosion defects affect failure pressure of the pipelines, the conclusions are subject to varied dimensions of the corrosion defects. For overlapped corrosion defects on pipelines, a stress concentration occurs mainly at the bottom defect. The stress level increases as the defect depth increases. Generally, the existence of the bottom corrosion defect reduces the failure pressure of the pipeline, as compared to the pipe containing a single defect only. The increase in the depth of either top or bottom corrosion defect would reduce the failure pressure. Finally, the steel grade does not apparently affect the interaction between adjacent corrosion defects which are oriented either longitudinally, circumferentially, or overlapped with each other.

3.3.3 Defect Assessment Under Mechanical Vibration Induced by ILI Operation

In-line inspection by MFL or UT tool can accurately detect and size various types of defects on pipelines [Vanaei et al., 2017]. During ILI operation, the tool moves inside a pipeline and may cause vibration, especially when it encounters obstacles such as rough inner wall surface, dents, girth weld between pipe segments, and corrosion pits [Zhang et al., 2015, 2020]. Vibrations could also happen on pipelines that are suspended on erosive soil supports. The vibration can generate a cyclic loading on inner pipe wall [Zhang et al., 2015], affecting local stress and strain distributions at the defects and thus the failure pressure of the pipeline. Of various defect assessment models and methods developed to date, none of them has considered the mechanical response due to the ILI-induced vibration and its effect on the local stress and strain concentrations, as well as the resulting pipeline failure. Qin and Cheng [2020], for the first time of its kind, developed a new FE-based model attempting to assess corrosion defects and predict failure pressure of pipelines under cyclic loading resulting from the ILI tool operation.

3.3.3.1 The Model

The FE model developed for a corrosion defect on the interior of a steel pipe is shown in Figure 3.9, where t, L, W, and d were defined previously. The

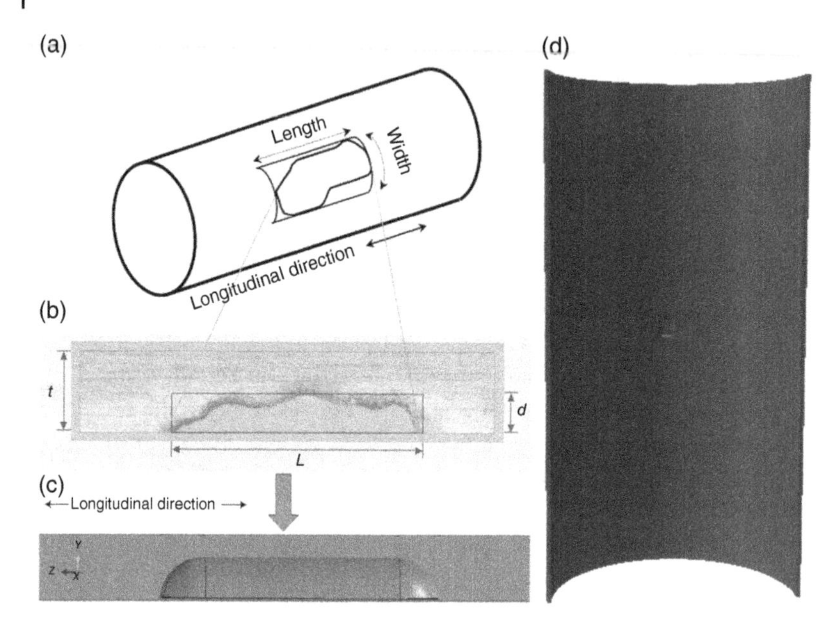

Figure 3.9 Schematic illustration of the geometry of a corrosion defect on the interior of a steel pipe and the 3D model of the pipe (a) A corrosion defect on a pipe, (b) Dimension of the corrosion defect, (c) A half of the 3D model of the defect, (d) An internal corrosion defect on the pipe. *Source:* From Qin and Cheng [2020] / with permission from Elsevier.

dimension selection of the modeled corrosion defect is based on the industry's corroded pipe database [Det Norske Veritas, 2004]. It is assumed that both ends of the pipe segment are under a full consolidation constraint, ensuring that both
the displacements and rotation of the pipe in three principal directions, i.e., x-, y-, and z-directions, are zero [Systemes, 2015].

The FE model with multiple numbers of elements is tested to obtain a reasonable mesh setting so that the modeling results are sufficiently stable and reliable. From the mesh density analysis, the mesh size for the pipe interior is 10 mm in both x- and z-directions, as shown in Figure 3.10a. To improve the modeling accuracy, the mesh refinement is performed at the corrosion defect, with details shown in Figure 3.10b. To model the effect of cyclic loading induced by vibration of ILI tools on failure pressure of the pipeline, a 3D model is developed, as shown in Figure 3.10c, where the sealing cup of the ILI tool advances along the axial direction (i.e., z-axis) of the pipe. The diagram of the FE model and the interface between the sealing cup and the pipe interior are shown in Figures 3.10d,e, respectively, where R_p is pipe outer

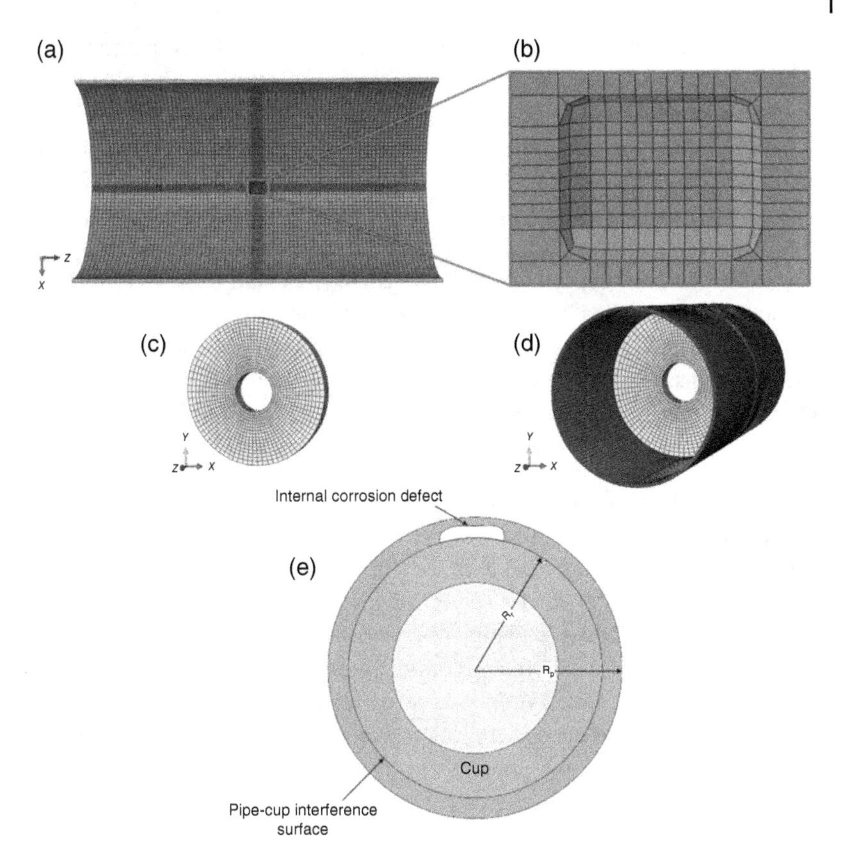

Figure 3.10 A 3D model illustrating a steel pipe containing an internal corrosion defect (a) meshes of the internal pipe surface, (b) meshes of the corrosion defect, (c) meshes of the sealing cup of the ILI tool, (d) assembly model of the pipe and the sealing cup, and (e) axial view of the interface between the sealing cup and the corroded pipe. *Source:* From Qin and Cheng [2020] / with permission from Elsevier.

radius and R_r is outer radius of the sealing cup of the ILI tool. The contact between the cup and the pipe internal surface is a nonlinear surface-to-surface contact, with a friction coefficient setting as 0.4 [Zhang et al., 2020].

For modeling of pipeline steels, the R-O relationship is used to describe the strain-hardening behavior of the steels [Ramberg and Osgood, 1943]. The failure criterion of pipelines is usually stress based. Specifically, a pipeline fails when the equivalent von Mises stress exceeds the ultimate tensile strength of the steel at the corrosion defect.

To model the cyclic loading induced by ILI tool operation in pipelines, the movement speed of the tool is selected as 1 m/s. The vibration of the tool is implemented by applying an acceleration on the pipe interior as shown in Figure 3.10, and the cyclic loading is applied on y-axis, i.e., the pipe wall thickness direction. Depending on acceleration and load change of the ILI tool [Mohammad et al., 2007], the frequency, f, of the cyclic loading is set as 10, 15, 20, and 25 Hz [Zhang et al., 2015, 2017]. The stress ratio of the cyclic loading, R_c, is determined by the ratio of the minimum stress, σ_{min}, to the maximum stress, σ_{max}. When $R_c = -1$, σ_{min} and σ_{max} have the same absolute value, but opposite directions, where the mean value is zero. The cyclic loading is called symmetrical relative to the zero-stress level. When R_c is not equal to -1, σ_{min} and σ_{max} have different absolute values and opposite directions. The mean value is not zero and the cyclic loading is not symmetrical about the zero-stress level.

3.3.3.2 Distributions of von Mises Stress at Corrosion Defect Under Cyclic Loading: Effect of *R*-ratio

The maximum von Mises stress at the corrosion defect on pipe segments made of X60 and X80 steels, respectively, with different *R*-ratios during cyclic loading are shown in Figure 3.11. Generally, as the *R*-ratio decreases, i.e., the difference between the minimum and maximum stresses increases, the maximum von Mises stress at the corrosion defect increases. At all *R*-ratios, the local stress concentration exceeds yield stress of the steels,

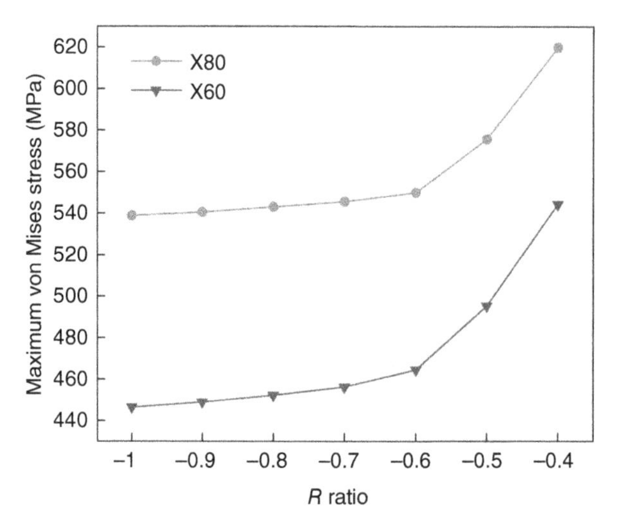

Figure 3.11 The maximum von Mises stress at the corrosion defect on X60 and X80 steel pipes as a function of *R*-ratio of cyclic loading. *Source:* From Qin and Cheng [2020] / with permission from Elsevier.

indicating that local plastic deformation occurs at the corrosion defect. It is noted that the stress increases rapidly when the *R*-ratio is below −0.6, i.e., a critical *R*-ratio governing pipeline failures under cyclic loading. This can cause the pressure-bearing capacity of the pipe to drop rapidly. The stress distribution at the corrosion defect on X80 steel pipe under an internal pressure of 8 MPa and cyclic loading with various *R*-ratios is shown in Figure 3.12. A stress concentration occurs at the defect. As the *R*-ratio decreases, the stress level at the defect increases, and the stress concentration area spreads along the defect length direction. The stress level is always the greatest at the two sides of the defect, as shown in dark red.

The effect of *R*-ratio on failure pressure of the X80 steel pipe containing a corrosion defect under cyclic loading with various *R*-ratios is shown in

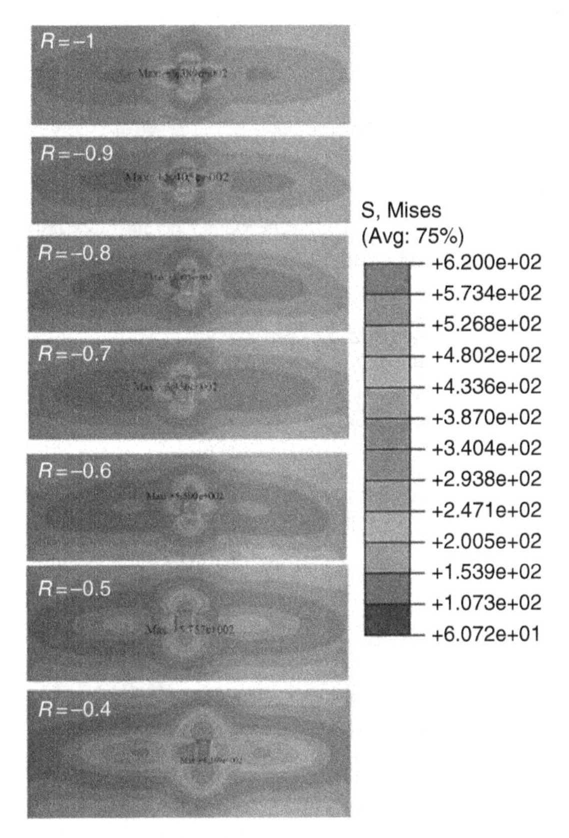

Figure 3.12 Distribution of von Mise stress (MPa) at the corrosion defect under an internal pressure of 8 MPa and cyclic loading with various *R*-ratios. *Source:* From Qin and Cheng [2020] / with permission from Elsevier.

Table 3.6 Failure pressure of X80 steel pipe containing a corrosion defect under cyclic loading with various *R*-ratios.

***R*-ratio**	**−1**	**−0.9**	**−0.8**	**−0.7**	**−0.6**	**−0.5**	**−0.4**
Failure pressure (MPa)	20.94	20.68	20.32	19.85	19.09	17.88	15.86
Frequency (Hz)	**15**	**20**	**25**	**30**			
Failure pressure (MPa)	19.09	19.83	20.57	20.93			

Source: From Qin and Cheng [2020] / with permission from Elsevier.

Table 3.6. The results show that the failure pressure decreases gradually with the decreased *R*-ratio, and then drops rapidly when the *R*-ratio is below −0.6, i.e., the critical *R*-ratio resulting in pipeline failures during cyclic loading under the given condition.

3.3.3.3 Distributions of von Mises Stress and Strain at Corrosion Defect Under Cyclic Loading: Effect of Cyclic Frequency

The maximum von Mises stress at the corrosion defect on X60 and X80 steel pipes under cyclic loading with various frequencies is shown in Figure 3.13. It is seen that the von Mises stress decreases with increased frequency. The maximum von Mises stresses on both steels exceed the yield stress under the combined internal pressure and cyclic loading, resulting in local plastic

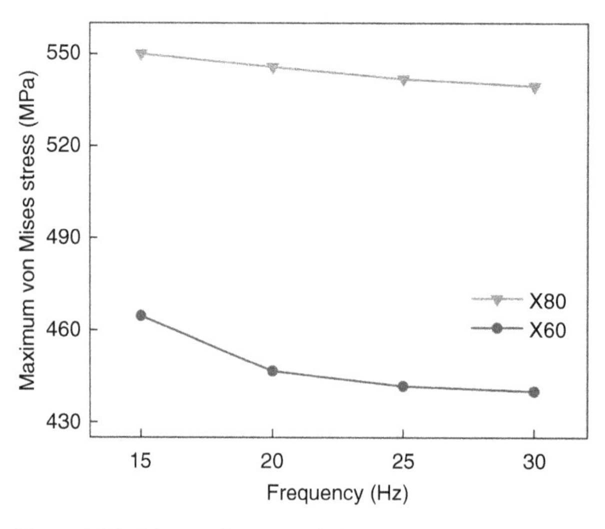

Figure 3.13 The maximum von Mises stress at the corrosion defect on X60 and X80 steel pipes as a function of cyclic frequency of the loading. *Source:* From Qin and Cheng [2020] / with permission from Elsevier.

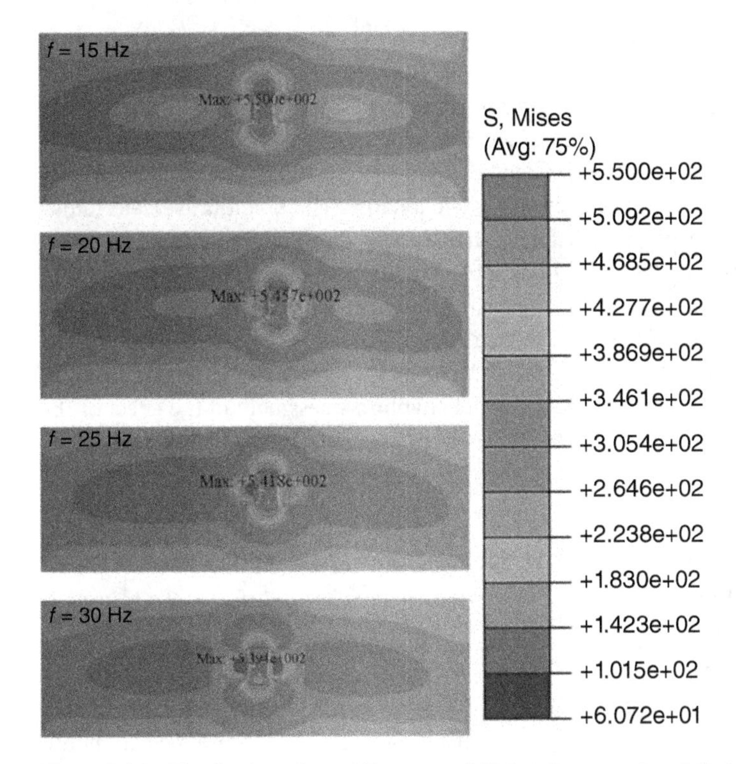

Figure 3.14 Distribution of von Mise stress (MPa) at the corrosion defect under an internal pressure of 8 MPa and cyclic loading with various frequencies. *Source:* From Qin and Cheng [2020] / with permission from Elsevier.

deformation at the defect. Figure 3.14 shows the von Mises stress distribution at the corrosion defect on X80 steel pipe under an internal pressure of 8 MPa and cyclic loading with various frequencies. The stress concentration mainly occurs at the defect bottom, and the stress level decreases with increased frequency.

The failure pressure of X80 steel pipe containing a corrosion defect as a function of the cyclic frequency is shown in Table 3.6. Generally, the failure pressure increases with increased cyclic frequency. Thus, a low-frequency cyclic stress is more likely to compromise the integrity of pipelines than high-frequency cyclic loading.

3.3.3.4 ILI Operation and Its Potential Effect on Integrity of Pipelines Containing Corrosion Defect

The ILI provides a reliable method for corrosion defect detection, location, and sizing. The ILI data analysis and assessment have always been integral to pipeline integrity management program. To date, none of the available defect assessment methods has considered the effect of cyclic loading induced by operational vibration of ILI tools on integrity of the pipelines. The developed model can determine the local stress distribution at corrosion defect under cyclic loading as a function of R-ratio and cyclic frequency. The failure pressure of the corroded pipelines is predicted under the combined action of internal pressure and cyclic loading. The quantitative determination of the relationship between the pipeline's failure pressure and the cyclic loading parameters would provide the industry recommendations about safe operating conditions (e.g., operating pressure) of the corroded pipelines during operation of ILI tools, which can generate an important effect on the stress concentration at corrosion defect and thus failure pressure of the pipelines. Therefore, an accurate model enabling assessment of the effect of ILI operation on pipeline integrity must be included in the defect assessment program.

3.3.4 Corrosion Defect at Pipeline Elbow and the Burst Pressure Determination

3.3.4.1 Burst Pressure Prediction of Pipeline Elbow Containing Corrosion Defect

Elbow is an important component in changing pipe routes and directions in a pipeline system. According to the bending angle, there are three commonly used elbows, i.e., 45°, 90°, and 180° elbows. In some special applications, elbows with other abnormal bending angles such as 60° are also available. Of various angled elbows, the 90° elbow is the most widely used type in upstream gathering pipelines. It is noted that such a high angle of elbow is rarely found in long-distance transmission pipelines. Generally, there is no limitation required for the bending radius of an elbow, where the selection is based on actual requirements and the pipe diameter. Owing to the bending curvature, the local stress at the pipe elbow is nonuniformly distributed under internal operating pressure. External coatings applied on the elbow are more likely to peel off than the coatings on a straight pipe. As a result, corrosion can occur to generate defects on the elbow [Khalajestani et al., 2015; Lam and Zhou, 2016]. Although many standards and codes have been developed for defect assessment on pipelines, such as those listed in Chapters 2 and 3, these methods are mostly applicable for straight pipes. Limitations exist when applied to elbows.

Recently, efforts have been made to investigate integrity of intact and corroded pipe elbows, but most of the work focused on bending capacity [Oh et al., 2007; Kirn et al., 2013; Li et al., 2014] or fatigue performance [Liu et al., 2018; Kim et al., 2020] of the elbows. For modeling determination of burst pressure of pipe elbows, Goodall [1978] developed a model to calculate the burst pressure of intact pipe elbows subject to internal pressure in 1978. Based on the model, Li et al. [2001] and Khalajestani et al. [2015] proposed empirical formulas to assess the failure pressure of pipe elbows containing corrosion defect. However, Zhang et al. [2014] thought that the results obtained from Goodall model were more conservative than full-scale burst tests and FE modeling. Wang and Zhou [2019] improved the Goodall model by using von Mises stress criterion, which was combined with the DNV RP-F101 code, a commonly used method for prediction of burst pressure of straight corroded pipes. As a result, a new prediction model was proposed for the burst pressure of corroded elbows. However, the applicability of the model to high-strength steel pipelines remains to be verified.

This section introduces the latest FE-based model developed to determine the burst pressure of X80 steel pipe elbows containing corrosion defect [Shuai et al., 2022]. While the effect of defect geometry on elbow burst is considered, the modeling results are verified by burst testing data.

3.3.4.2 Development of the FE Model

A 3D model of a 90° elbow on X80 steel pipe containing an external corrosion defect and the toroidal coordinate system illustrating the structure of the pipe elbow and the defect location is shown in Figure 3.15. The corrosion defect is modeled as a rectangular shape with smooth sides [Zhang and Zhou, 2020; Shuai et al., 2021] on the outer surface of the elbow by generating a local reduction in pipe wall thickness, although many actual corrosion defects on pipelines are usually irregular. It is expected that the burst capacity of a pipe containing an actual corrosion defect is greater than the burst capability when a rectangular corrosion defect is present. The latter has a larger area of reduced wall thickness at the maximum defect depth than the former. Thus, the model considers a situation more dangerous to pipeline safety than the reality. In Figure 3.15b, θ represents the clock position of corrosion defect in the circumferential direction of the pipe elbow, Φ is defined as the axial routing angle of the pipe, R_p is pipe outer radius, and R_b is bending radius of the elbow.

To determine burst pressure of the pipe elbow containing a corrosion defect, the ultimate tensile strength criterion for burst assessment of corroded pipelines is used. According to the criterion, burst failure occurs when the von Mises equivalent stress throughout the ligament thickness at the corrosion defect area reaches the ultimate tensile strength of the steel. The burst behavior of pipelines is a multi-nonlinear process. The RIKS algorithm, which is also called arc-length algorithm, is used to solve the FE model. A grid sensitivity study is conducted to obtain an exact

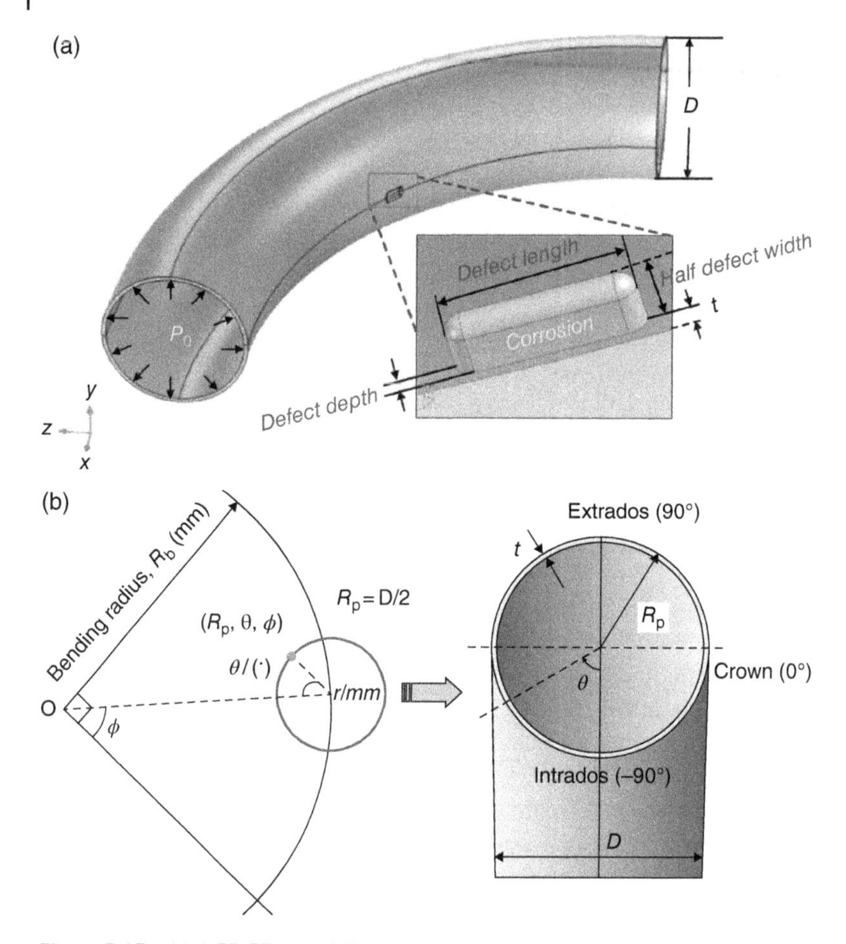

Figure 3.15 (a) A 3D FE model illustrating the dimensional parameters of a corrosion defect on an elbow of X80 steel pipe. (b) Location of the corrosion defect on pipe elbow in an annular coordinate system. *Source:* From Shuai et al. [2022] / with permission from Elsevier.

convergence solution using several trial calculations. With an increased mesh density, the burst pressure increases gradually and then remains almost unchanged, indicating that the mesh has converged. The minimum number of elements corresponding to the stable solution is selected as the optimal mesh condition. The element size of the corrosion defect is obtained as 1.5 mm, while the grid size of the pipe away from the defect area is approximately 12 mm. The total number of the model elements was 89,681. Figure 3.16 shows the cloud diagram of the von Mises

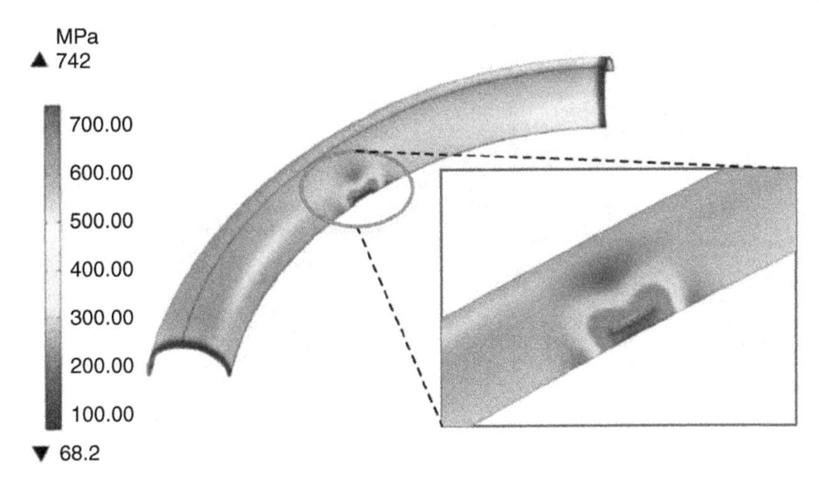

Figure 3.16 Distribution of von Mises stress at the X80 steel pipe elbow at the burst moment. *Source:* From Shuai et al. [2022] / with permission from Elsevier.

equivalent stress distribution of a pipe elbow at the burst moment. It is seen that there is a high stress concentration at the corrosion defect, while the maximum stress reaches the ultimate tensile strength, i.e., 742 MPa, for X80 steel.

3.3.4.3 Effects of Corrosion Defect Dimension on Burst Capacity of Pipe Elbow

The effect of corrosion depth on failure pressure of the X80 steel pipe elbow (18.4 mm in wall thickness) subject to internal pressure is shown in Figure 3.17, where the corrosion defect (150 mm in width) is located at intrados, extrados, and crown of the elbow, respectively. The burst pressures predicted by the FE model, P_{FE}, are normalized by burst pressure of a defect-free elbow, P_{GM}. It is seen that the failure pressure of the corroded elbow decreases linearly with increased corrosion depth. The corrosion length is also important in failure pressure determination. The longer the corrosion defect, the faster the decrease in failure pressure with the increased defect depth.

To determine the effect of corrosion length on burst pressure of the corroded pipe elbow, the FE modeling is performed with various normalized defect lengths, $L/\sqrt{Dt}$. Figure 3.18 shows the burst pressure of the pipe elbow as a function of corrosion length, $L/\sqrt{Dt}$, at various depths (i.e., d/t). Three typical locations, i.e., intrados, extrados, and crown, where the corrosion

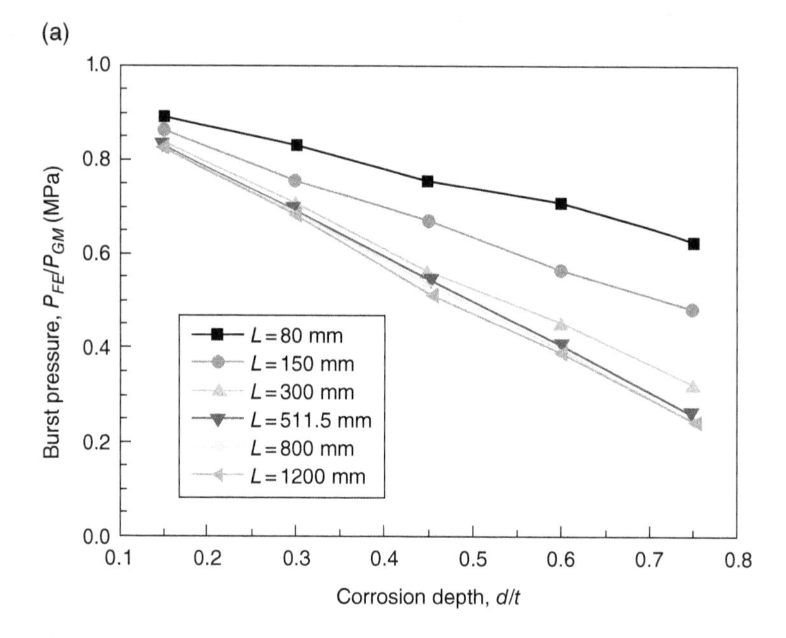

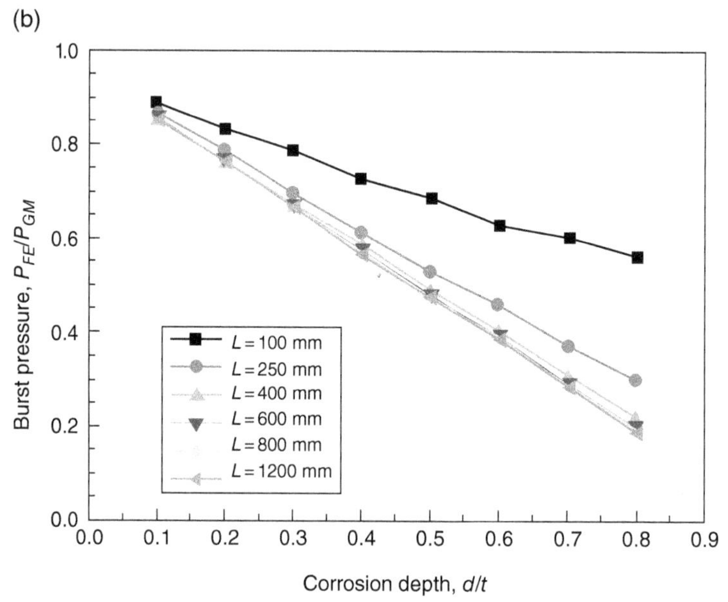

Figure 3.17 Effect of corrosion depth on burst pressure of X80 steel pipe elbow while considering the corrosion defect length, where the defects are located at (a) intrados, (b) extrados, and (c) crown of the elbow, respectively. *Source:* From Shuai et al. [2022] / with permission from Elsevier.

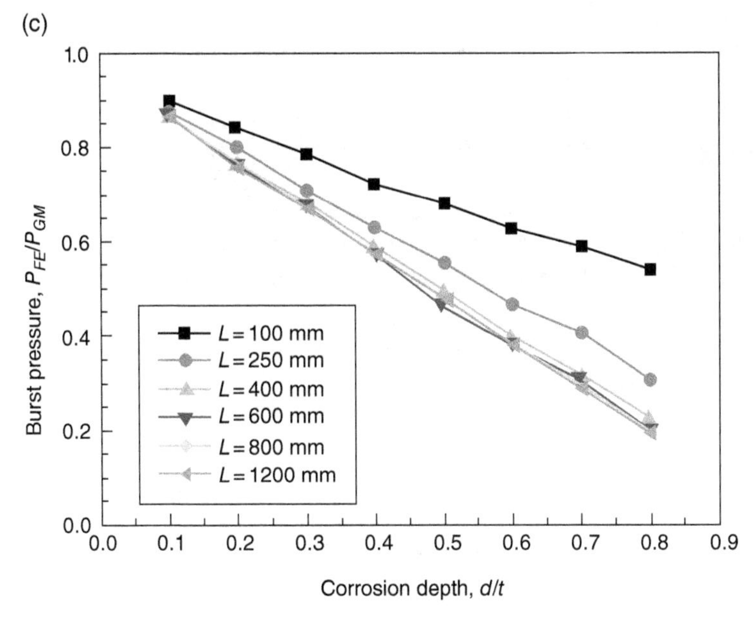

Figure 3.17 (Continued)

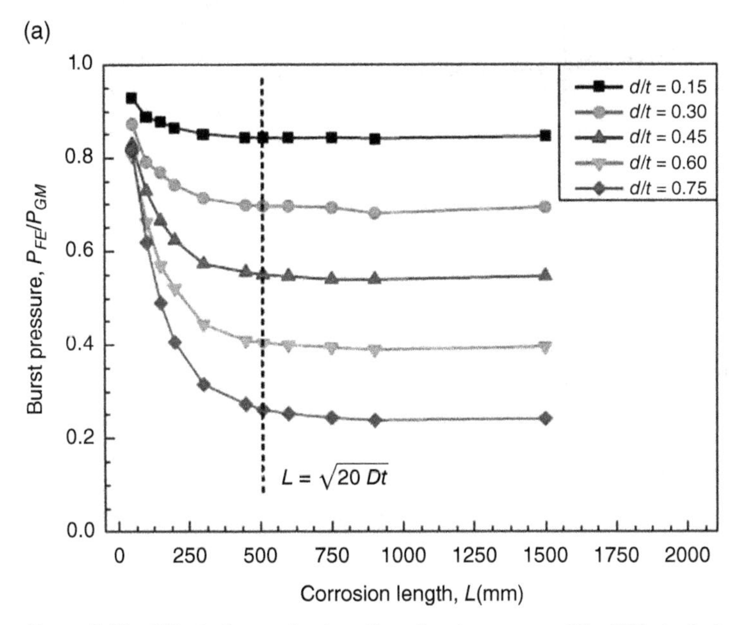

Figure 3.18 Effect of corrosion length on burst pressure of the X80 steel pipe elbow while considering the corrosion defect depth, where the defects are located at (a) intrados, (b) extrados, and (c) crown of the elbow, respectively. *Source:* From Shuai et al. [2022] / with permission from Elsevier.

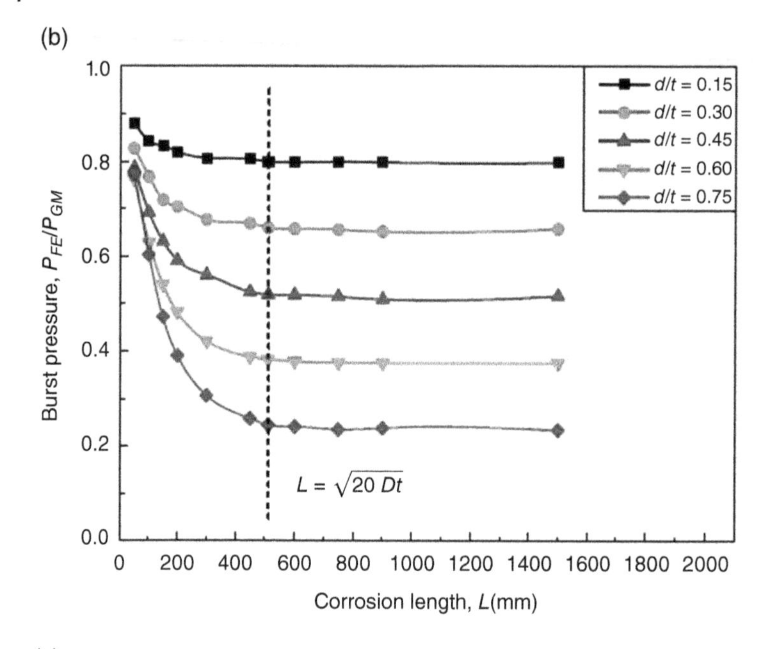

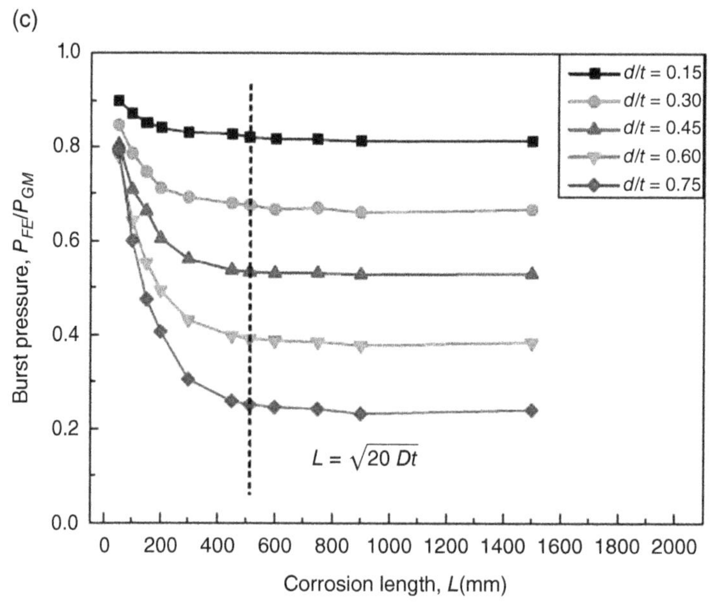

Figure 3.18 (Continued)

defect is located at the elbow are considered. A critical length that affects failure pressure of the elbow is determined as $L/\sqrt{Dt} = \sqrt{20}$. When the corrosion defect length $L < \sqrt{20Dt}$, the burst pressure decreases with increased defect length. However, when $L \geq \sqrt{20Dt}$, the burst pressure is independent of the defect length. In addition, the effect of corrosion length depends on the corrosion depth. For a shallow corrosion defect, the effect of corrosion length on the failure pressure is not obvious. As the corrosion defect becomes deep, the length effect of the defect becomes significant. Thus, both the corrosion defect depth and its length should be considered simultaneously for failure pressure determination at corroded pipe elbow.

Figure 3.19 shows the burst pressure of the corroded elbow as a function of the corrosion defect width at three typical locations (i.e., intrados, extrados, and crown) of the elbow. Generally, the burst pressure is almost independent of the defect width, which is also same as straight pipes [Mondal and Dhar, 2019; Zhu, 2021]. Therefore, the defect width is usually ignored while assessing the residual strength of corroded pipe elbows.

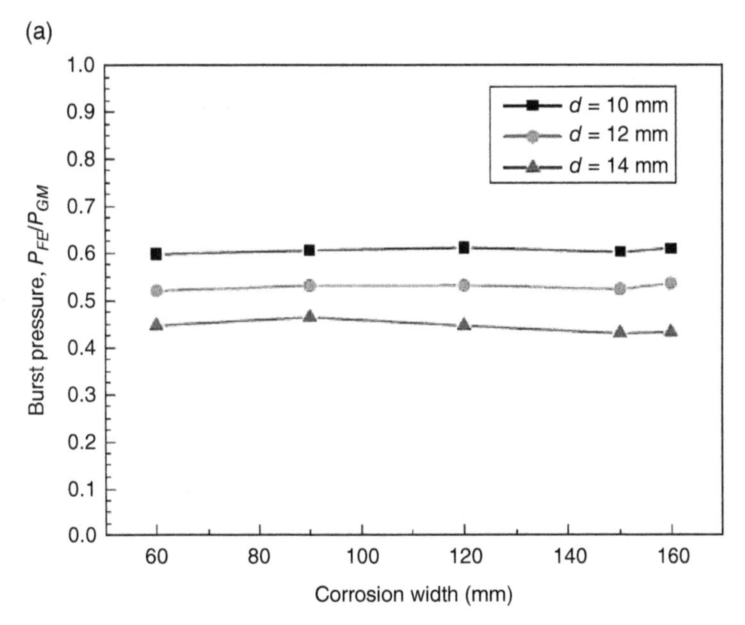

Figure 3.19 Effect of corrosion width on burst pressure of the X80 steel pipe elbow, where the defects are located at (a) intrados, (b) extrados, and (c) crown of the elbow, respectively. *Source:* From Shuai et al. [2022] / with permission from Elsevier.

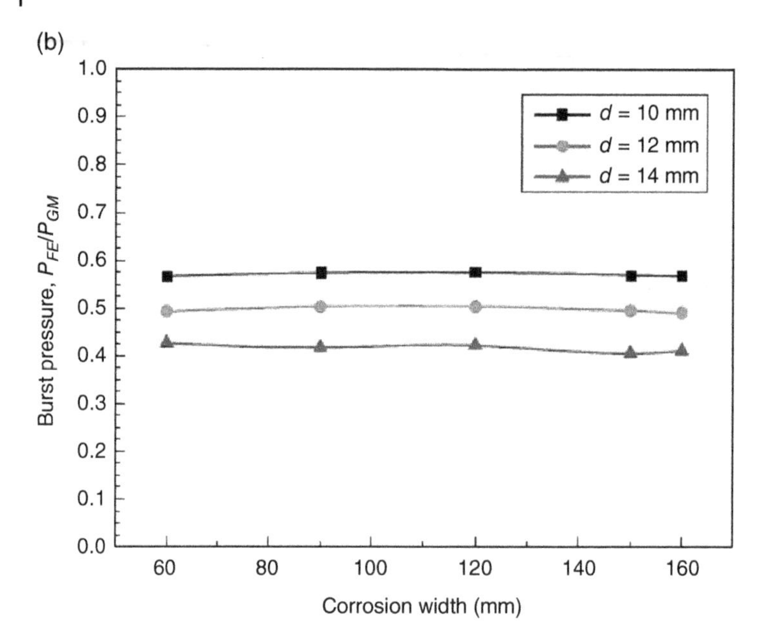

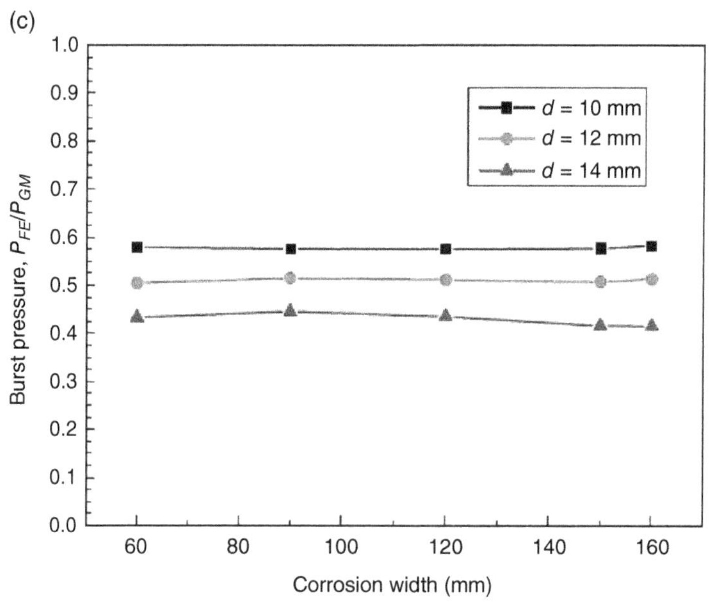

Figure 3.19 (Continued)

3.3.4.4 A New Model for Prediction of Burst Pressure of Corroded Pipe Elbows

For the model developed for determination of the burst capacity of a pipe elbow containing corrosion defect, it is assumed that the corrosion defect is infinitesimal. The upper limit burst pressure, P_{max}, can be determined as the failure pressure of an intact pipe elbow:

$$P_{max} = P_{GM} = \frac{2c}{f(c)} \frac{t}{r} \sigma_u \tag{3.13}$$

$$f(c) = \sqrt{4 - 2c + c^2} \tag{3.14}$$

$$c = \frac{R + r \cdot \sin\theta}{R + \frac{1}{2} r \cdot \sin\theta} \tag{3.15}$$

where c is defined as the curvature coefficient of the elbow. The lower limit burst pressure, P_{min}, is obtained from an intact pipe elbow with a residual wall thickness of $(t - d)$:

$$P_{min} = \frac{2c}{f(c)} \frac{t - d}{r} \sigma_u \tag{3.16}$$

$$\frac{P_{min}}{P_{max}} = \frac{t - d}{t} \tag{3.17}$$

For a pipe elbow containing a corrosion defect, the failure pressure, P_F, is between P_{min} and P_{max}:

$$P_F = P_{min} + \left(P_{max} - P_{min}\right) g\left(\frac{d}{t}, \frac{L}{\sqrt{Dt}}\right) \tag{3.18}$$

where $g\left(\frac{d}{t}, \frac{L}{\sqrt{Dt}}\right)$ is a function that represents the geometric factor of the corrosion defect, d/t is normalized defect depth by pipe wall thickness, and $L/\sqrt{Dt}$ is normalized defect length by the pipe segment length. The corrosion width is not considered in the geometric factor function because it does not apparently affect the burst pressure of the elbow. Eq. (3.18) is then rewritten as:

$$P_F = P_{GM}\left\{1 - \frac{d}{t}\left[1 - g\left(\frac{d}{t}, \frac{L}{\sqrt{Dt}}\right)\right]\right\} \tag{3.19}$$

Two boundary conditions are considered for the geometric factor function g. When corrosion defect does not exist, g = 1. When the corrosion defect has a width of approximately πD and an infinite length or a depth of the pipe wall thickness, g = 0. The expression of the function g can be assumed as:

$$g\left(\frac{d}{t}, \frac{L}{\sqrt{Dt}}\right) = p\exp\left(\frac{qL}{\sqrt{Dt}}\right)\left(1 - \frac{d}{t}\right)^k \tag{3.20}$$

where the parameters p, q, and k are coefficients to be determined based on a given condition. Therefore, Eq. (3.18) is written as:

$$P_F = \frac{2ct\sigma_u}{f(c)r}\left\{1 - \frac{d}{t}\left(1 - p\exp\left(\frac{qL}{\sqrt{Dt}}\right)\left(1 - \frac{d}{t}\right)^k\right)\right\} \tag{3.21}$$

The angle θ in Eq. (3.15) can be considered a variable defining the clock position of corrosion defect in the circumferential direction of the pipe elbow. As a result, the developed FE model enables determination of the failure pressure of corroded elbow under given conditions while considering the effects of pipe dimensions and corrosion geometry.

To verify the modeling results, the burst pressures obtained from the model are compared with the results from FE analysis method, while two reported models [Zhang et al., 2014; Khalajestani et al., 2015], are also used for comparison, as shown in Figure 3.20. The dotted lines represent the error

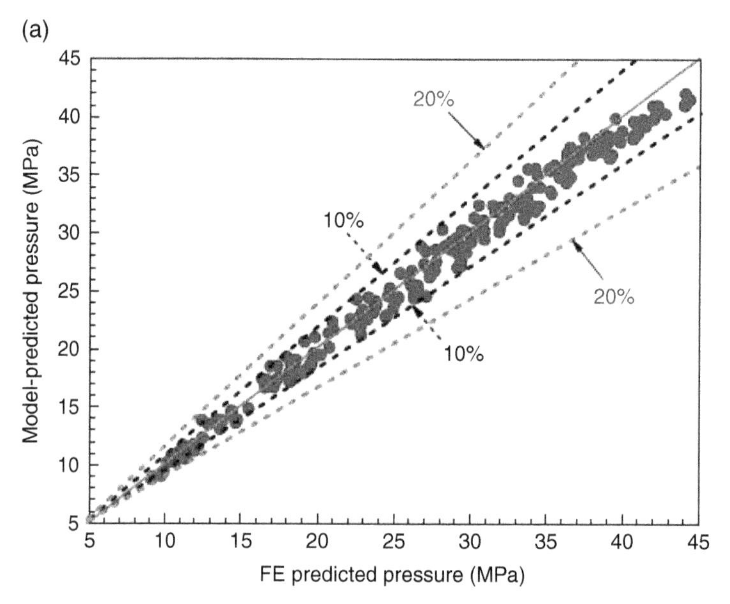

Figure 3.20 Comparison of the burst pressures predicted from the developed model with two reported models: (a) proposed model, (b) Zhang et al. model, P_Z, and (c) Khalajestani et al. model, P_K. *Source:* From Shuai et al. [2022] / with permission from Elsevier.

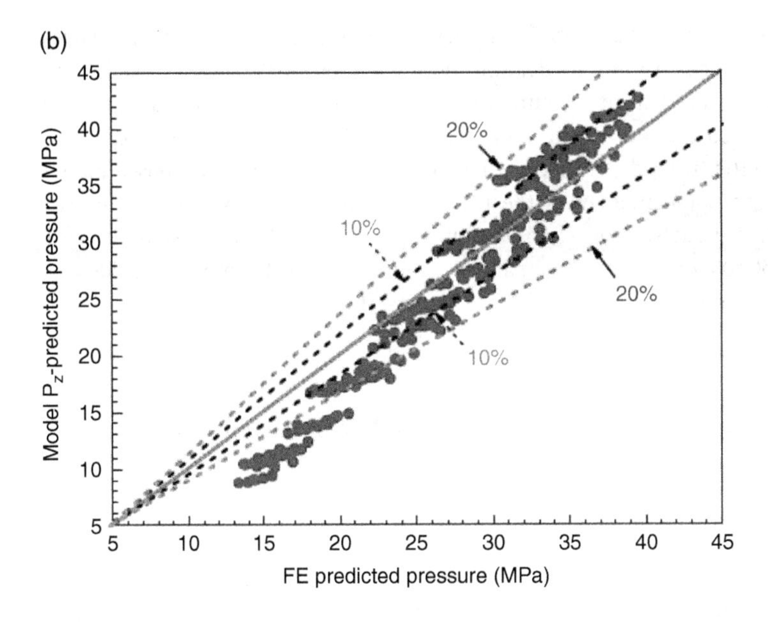

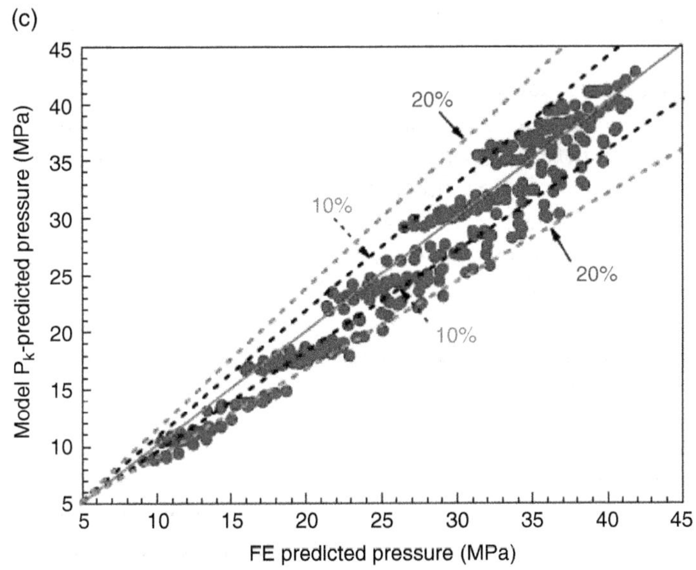

Figure 3.20 (Continued)

lines with 10% and 20% confidence levels, respectively, of the burst pressures predicted from the FE model and those from the two reported models. The two groups of failure pressure are linearly related with an average relative error of 3.63% and a maximum error of 10.35%. Most discrete points fall within the two red ±10% error lines. This indicates that the proposed model is in good agreement with the FE analysis method. However, for the two reported models, a considerable number of data points fall outside the error line of 10% or even 20%. Apparently, the developed FE model provides an accurate method to predict burst pressure of pipe elbows containing corrosion defect.

3.3.5 Interaction Between Internal and External Corrosion Defects on Pipelines

Buried pipelines can suffer from corrosion attacks both externally and internally. External corrosion occurs due to coating degradation and ineffective CP on the pipelines which are exposed to either electrolyte trapped under disbonded coatings or soil environments [Cheng and Norsworthy, 2017]. Internal corrosion of pipelines in corrosive environments generated on the pipe wall surface is usually attributed to the synergism of fluid properties, microorganisms, flow dynamics, and pipe geometry [Han and Cheng, 2013; Liu and Cheng, 2018; Qian and Cheng, 2019]. Thus, both external and internal corrosion defects are detected by ILI tools. Pipeline defect assessment is usually conducted on either external defects or internal defects. When a pipeline contains both external and internal defects which interact with each other, the existing standards and rules do not apply. To date, a method or model assessing the interaction between internal and external corrosion defects and the effect on failure pressure of the pipelines has not been available. Progress has been made in this aspect recently by developing a new Level III assessment method [Zhang et al., 2021].

3.3.5.1 Model Development

A 3D FE model was developed for an X52 steel pipe containing a semi-ellipsoidal corrosion defect, as shown in Figure 3.21. The basic geometrical parameters, including the pipe outer diameter (D), pipe wall thickness (t), the length (L_{P}) of the pipe segment, the primary axial length ($2c$) in z-axis direction (i.e., the longitudinal direction of the pipe), the secondary axial length ($2a$) in x-axis direction, and the depth (d) of the defect are included in Zhang et al. [2021]. The internal pressure (P) is the principal source of stress applied on the pipe. Thus, the stress of the pipe is mainly hoop tension

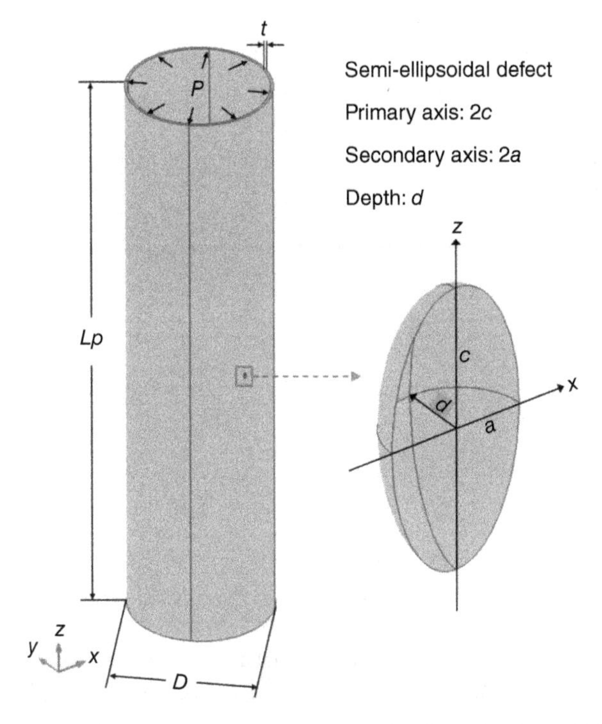

Figure 3.21 Schematic diagram of a 3D FE model for an X52 steel pipe segment containing a semi-ellipsoidal corrosion defect. *Source:* From Zhang et al. [2021] / with permission from Elsevier.

and longitudinal tension caused by the internal pressure. It is assumed that pressure fluctuations are slight so that fatigue failure is not considered.

Three types of distributions between internal and external corrosion defects were modeled, as shown in Figure 3.22, where the subscripts "ext" and "int" refer to external defect and internal defect, respectively. The type I distribution contains two defects, i.e., $D_{21,\text{ext}}$ and $D_{22,\text{ext}}$ (or $D_{22,\text{int}}$), which are aligned in the longitudinal direction of the pipe with varied spacing (S_L). The type II distribution contains three defects, i.e., $D_{31,\text{ext}}$, $D_{32,\text{ext}}$, and $D_{33,\text{ext}}$ (or $D_{33,\text{int}}$), where $D_{31,\text{ext}}$ and $D_{32,\text{ext}}$ are located in the hoop direction of the pipe with a fixed circumferential spacing (S_C) of 10 mm, and $D_{33,\text{ext}}$ (or $D_{33,\text{int}}$) is located in the center between $D_{31,\text{ext}}$ and $D_{32,\text{ext}}$, but with various S_L. The type III distribution contains four defects, i.e., $D_{41,\text{ext}}$, $D_{42,\text{ext}}$, $D_{43,\text{ext}}$, and $D_{44,\text{ext}}$ (or $D_{44,\text{int}}$), where $D_{41,\text{ext}}$ and $D_{42,\text{ext}}$ are located in the hoop direction of the pipe with an S_C of 10 mm, and $D_{43,\text{ext}}$ and $D_{44,\text{ext}}$ (or $D_{44,\text{int}}$) are in the center between $D_{41,\text{ext}}$ and $D_{42,\text{ext}}$. The S_L between $D_{43,\text{ext}}$ and $D_{41,\text{ext}}$ (or $D_{42,\text{ext}}$) was i_0,

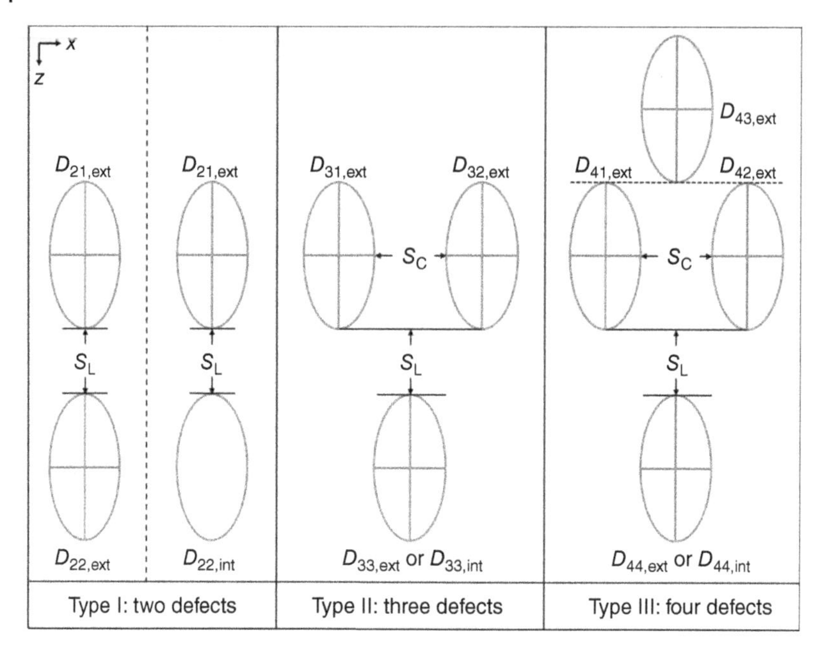

Figure 3.22 Three types of distributions between internal and external defects on the pipe segment to be modeled. *Source:* From Zhang et al. [2021] / with permission from Elsevier.

and the S_L between $D_{44,ext}$ (or $D_{44,int}$) and $D_{41,ext}$ (or $D_{42,ext}$) was varied. The depth of all external defects is equal to 0.5 t, while the depth of internal defects is varied. Like the type I defect distribution, types II and III distributions have two models (i.e., composed of external defects or mixed internal and external defects) distinguished by $D_{33,ext}$ or $D_{33,int}$, and $D_{44,ext}$ or $D_{44,int}$, respectively.

The stress–strain behavior of X52 pipeline steel follows the R-O rule, and the yield and ultimate tensile strengths of the steel are 372 and 607 MPa, respectively. The DNV RPF101 standard is selected as the pipe failure criterion, i.e., a corroded pipeline is assumed to fail when the maximum von Mises stress at the corrosion defect exceeds the ultimate tensile strength of pipe steel under operating pressure [Det Norske Veritas, 2004].

3.3.5.2 Stress Distributions and Failure Pressure of a Steel Pipe Containing Corrosion Defects with Various Distribution Types

The stress distribution of the pipe segment containing corrosion defects with different distribution types under 15 MPa internal pressure is shown in Figure 3.23. All internal and external defects are assumed geometrically identical, with the lengths of $2c$, $2a$, and d of 20 mm, 10 mm, and 0.5 t

Figure 3.23 Stress distribution of the pipe segment containing corrosion defects with different distribution types under 15 MPa internal pressure. *Source:* From Zhang et al. [2021] / with permission from Elsevier.

(4.85 mm), respectively. The values of S_L and S_C for all defect distribution types are 0 and 10 mm, respectively. It is seen that when the pipe contains both internal and external defects, the stress distribution is quite different from the pipe stress when all defects are external. Compared with the external defects, the local stress concentration at the internal defect is mainly located on the interior surface and does not apparently affect the external defects. Moreover, the presence of the internal defect makes the stress distribution nonsymmetrical.

Table 3.7 shows the maximum von Mises stress (σ_{max}) at the defects for different distribution types (S_L is 0 mm) obtained from Figure 3.23, where the first column data refer to the defect distributions composed of mixed internal and external defects, and the second column data correspond to the defect type composed of external defects only. It is seen that although the ratio of the external defect depth (d) to pipe wall thickness (t), i.e., $r_{dt,ext}$ is equal to the ratio of the internal defect depth to d, i.e., $r_{dt,Int}$, the σ_{max} on the pipe composed of external defects only is greater than the stress when both external and internal defects are present. Generally, a corrosion defect with a great σ_{max} would burst first with increased internal pressure. Thus, for type II defect distribution, $D_{33,int}$ or $D_{33,ext}$ is the weakest site on the pipe for burst to happen. For type I, $D_{21,ext}$ and $D_{22,ext}$ (or $D_{22,int}$) have approximately equal σ_{max} due to the same defect depth. Thus, either of them would fail with increased internal pressure.

Table 3.7 Maximum von Mises stress (σ_{max}) at defect for different defect distribution types (S_L is 0) obtained from Figure 3.23.

		Mixed defects: $r_{dt,ext}$ = 0.5, $r_{dt,int}$ varied									
Defect distribution		**0.7**	**0.6**	**0.5**	**0.4**	**0.3**	**0.2**	**0.1**	**0**	**External defects $r_{dt,ext}$ = 0.5**	
Type I	$D_{21,ext}$	491.68	481.41	473.75	467.49	462.58	456.55	450.05	449.79	$D_{21,ext}$	593.92
	$D_{22,int}$	506.97	489.43	472.99	458.29	437.70	416.97	392.67	—	$D_{22,ext}$	593.48
Type II	$D_{31,ext}$	463.51	457.90	453.79	449.52	445.36	441.22	440.95	440.77	$D_{31,ext}$	460.88
	$D_{32,ext}$	463.62	457.67	453.71	450.02	445.87	441.64	440.28	440.69	$D_{32,ext}$	460.01
	$D_{33,int}$	495.99	482.52	468.11	459.16	439.19	425.73	401.07	—	$D_{33,ext}$	481.09
Type III	$D_{41,ext}$	480.42	476.76	471.47	468.14	462.40	460.24	460.32	460.88	$D_{41,ext}$	475.51
	$D_{42,ext}$	480.58	476.69	471.62	468.42	462.57	460.13	460.09	460.01	$D_{42,ext}$	475.09
	$D_{43,ext}$	500.30	494.76	490.76	485.89	483.05	480.42	479.23	479.09	$D_{43,ext}$	493.10
	$D_{44,int}$	510.11	495.70	481.96	468.03	451.22	434.93	408.69	—	$D_{44,ext}$	493.52

Source: From Zhang et al. [2021] / with permission from Elsevier.

For type III, the σ_{max} of $D_{43,ext}$ and $D_{44,ext}$ are equal and greater than that of $D_{41,ext}$ and $D_{42,ext}$, indicating that $D_{43,ext}$ or $D_{44,ext}$ is the weak point to fail. When internal defect $D_{44,int}$ is present, σ_{max} at $D_{44,int}$ is lower than that at $D_{43,ext}$, indicating that $D_{43,ext}$ would fail first than other defects.

For pipelines containing external defects only (such as $D_{21,ext}$ for type I, $D_{31,ext}$ and $D_{32,ext}$ for type II, or $D_{41,ext}$, $D_{42,ext}$, and $D_{43,ext}$ for type III), the failure pressure is defined as P_F. When the pipelines contain additional external or internal defects (for example, $D_{22,ext}$ or $D_{22,int}$ added to type I, $D_{33,ext}$ or $D_{33,int}$ added to type II, or $D_{44,ext}$ or $D_{44,int}$ added to type III), the failure pressure is defined as $P_{F,add}$. To determine if the additional defect affects the failure pressure, a criterion is proposed as:

$$\text{Ratio} = \begin{cases} P_{F,add}/P_F \geq 0.99, \text{ No interaction} \\ P_{F,add}/P_F < 0.99, \text{ Interaction exists} \end{cases} \tag{3.22}$$

When the ratio is equal to or larger than 0.99, the additional defect does not interact with the existing external defects. If the ratio is smaller than 0.99, the interaction between them exists.

According to DNV RPF101, when the longitudinal spacing between two adjacent defects exceeds $2(Dt)^{1/2}$ (i.e., ~140 mm in the modeling work), the interaction between the defects is negligible. Figure 3.24 shows the failure

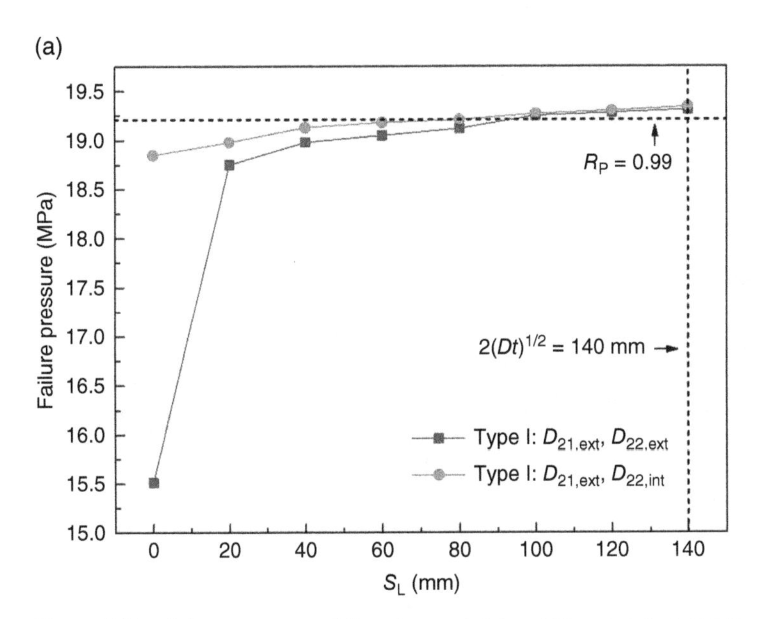

Figure 3.24 Failure pressure of the pipe containing different defect distribution types ($r_{dt,ext} = r_{dt,int} = 0.5$) as a function of S_L: (a) type I, (b) type II, and (c) type III. *Source:* From Zhang et al. [2021] / with permission from Elsevier.

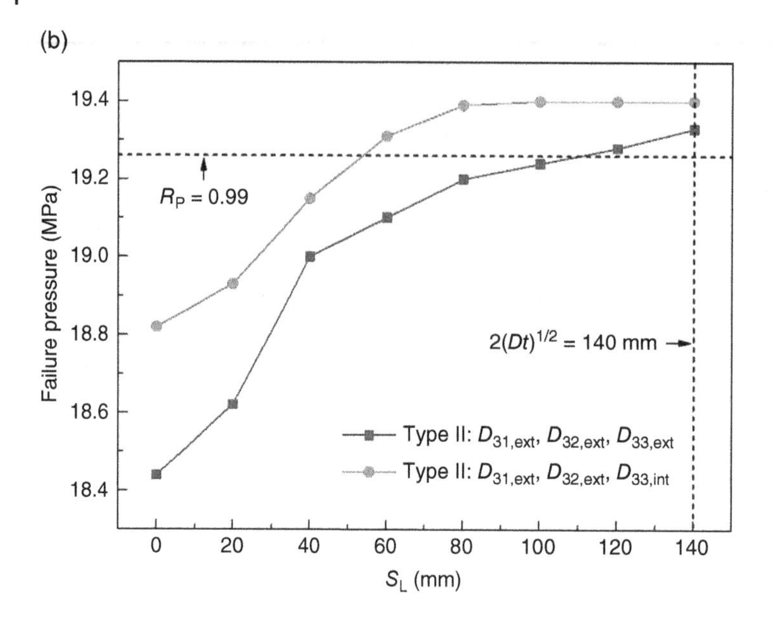

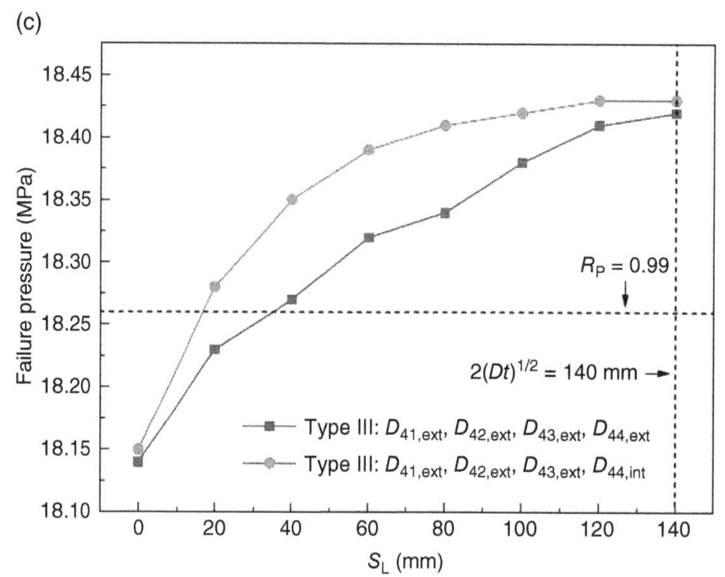

Figure 3.24 (Continued)

pressure of the pipe containing different distribution types of the defects ($r_{dt,ext} = r_{dt,int} = 0.5$) as a function of S_L. It is seen that the ratio is always greater than 0.99 when S_L is more than $2(Dt)^{1/2}$, indicating that there is no interaction between the defects. When the ratio is equal to 0.99, the longitudinal limit spacing for type I defect distribution with all external defects ($D_{21,ext}$ and $D_{22,ext}$), i.e., $S_{L,ext}^{Lim}$, is approximately 94 mm, while the longitudinal limit spacing in the presence of both external and internal defects ($D_{21,ext}$ and $D_{22,int}$), i.e., $S_{L,int}^{Lim}$, is approximately 80 mm. For types II and III defect distributions, the $S_{L,ext}^{Lim}$ are approximately 110 and 35 mm, respectively, and the $S_{L,int}^{Lim}$ are approximately 54 and 17 mm, respectively. Thus, the $S_{L,int}^{Lim}$ for all defect distribution types are smaller than the $S_{L,ext}^{Lim}$ owing to a lower σ_{max} of internal defects as compared with external defects under identical conditions. Particularly, for type I distribution, when S_L is equal to 0, the P_F decreases remarkably because two external defects join together to become one defect with a long longitudinal length, causing an apparent increase of σ_{max}.

3.3.5.3 Assessment of the Interaction Between Internal and External Corrosion Defects and the Implication on Pipeline Integrity Management

Determination of the interaction between internal and external corrosion defects on pipelines is associated with a complex rule where multiple factors affect the assessment. The stress distribution at the defects is related not only to the longitudinal spacing between internal and external defects (S_L), but also to the internal defect depth, As normalized by the ratio of the internal defect depth to pipe wall thickness (i.e., $r_{dt,int}$), the $r_{bB,int}$ has a remarkable impact on the interaction between the internal and external defects, and plays a critical role in defining the $S_{L,int}^{Lim}$. When $r_{dt,int}$ is small, e.g., 0.1, 0.1, and 0.3 for types I, II, and III defect distributions, respectively, the role of the internal defect in determination of failure pressure of the pipelines can be negligible. However, with increased $r_{dt,int}$, the effect of internal defect becomes important, causing an increased $S_{L,int}^{Lim}$. If the $r_{dt,int}$ is too high, e.g., 0.7 for types I and II defect distribution types, the ratio of $P_{F,add}/P_F$ is always smaller than 0.99, resulting in a rapid decrease of the failure pressure.

Compared with the external defects, the effect of internal defects on local stress concentration is not so significant, as indicated by the lower σ_{max} for the same defect distribution type. Thus, the $S_{L,int}^{Lim}$ for internal defects is smaller than the $S_{L,int}^{Lim}$ for external defects. For a given defect distribution type with a fixed S_L, the failure pressure of a pipeline containing external defects only is usually lower than that of the pipeline containing both internal and external defects.

3.4 Commentary Remarks

The development of Level III defect assessment methods has made significant progress in FFS determination and failure prediction of pipelines by solving highly nonlinear problems through FE modeling. However, the methods are not popularly used in industry due to complex background knowledge and engineering mathematics fundamentals, as well as time-consuming computations. Instead, Level III methods have been mainly developed in academic institutions for research purposes. For industry to realize the benefits of improved accuracy in FFS determination and failure pressure prediction by Level III defect assessment methods, a spreadsheet method was developed, attempting to provide an engineering application platform by simplifying the modeling and numerical computation process [Xu and Low, 2006; Low and Tang, 2007].

In principle, the Level III defect assessment methods regard corrosion as 3D volumetric defects. Growth of the corrosion defects is a primary mechanism leading to pipeline perforation and leakage. Electrochemical corrosion reactions under pipeline operating conditions greatly contribute to the defect growth. However, the Level III assessment methods available today rarely consider the effects of electrochemical reactions during corrosion defect growth and the synergism between corrosion reactions with local stress and strain conditions at the defects. As a result, the corrosion defect assessment, especially the defect growth kinetics prediction, is conducted while ignoring the role of electrochemical reactions and the stress-corrosion interaction in defect growth, making the modeling results deviate from the reality. Therefore, an integration of corrosion electrochemistry with mechanical stress field by FE modeling for corrosion defect assessment, while considering the stress-corrosion interaction, is essential at further improvement of the Level III assessment techniques.

References

Al-Owaisi, S.S., Becker, A.A., Sun, W., Al-Shabibi, A., Al-Maharbi, M., Pervez, T., Al-Salmi, H. (2018) An experimental investigation of the effect of defect shape and orientation on the burst pressure of pressurised pipes, *Eng. Fail. Anal.*, 93, 200–213.

American Petroleum Institute (2000) *Fitness-for-Service*, API Recommended Practice 579, API Publishing Services, Ed. 1, Washington DC, USA.

American Society of Civil Engineering (2005) *Guidelines for the Design of Buried Steel Pipe*, Reston, Virginia, USA.

American Society of Mechanical Engineering (1991) *Manual for Determining the Remaining Strength of Corroded Pipelines, A Supplement to ASME B31 Code for Pressure Piping*, ASME B31G-1991 (Revision of ANSI/ASME B31G-1984), ASME, New York, NY, USA.

American Society of Mechanical Engineering (2009) *Manual for Determining the Remaining Strength of Corroded Pipelines – A Supplement to ASME B31 Code for Pressure Piping*, B31G-2009, ASME, New York, NY, USA.

American Society of Mechanical Engineering (2012) *Manual for Determining the Remaining Strength of Corroded Pipelines: A Supplement to ASME B31 Code for Pressure Piping*, ASME, New York, NY, USA.

Amirat, A., Mohamed-Chateauneuf, A., Chaoui, K. (2006) Reliability assessment of underground pipelines under the combined effect of active corrosion and residual stress, *Int. J. Press. Vessel. Pip.*, 83, 107–117.

ANSYS (2013) *ANSYS Fluent Theory Guide*, ANSYS. Inc., Canonsburg, PA, USA

Arumugam, T., Karuppanan, S., Ovinis, M. (2020) Finite element analyses of corroded pipeline with single defect subjected to internal pressure and axial compressive stress, *Mar. Struct.*, 72, 102746.

Benjamin, A.C., Freire, J.L.F., Vieira, R.D., Diniz, J.L., De Andrade, E.Q. (2006) Burst tests on pipeline containing interacting corrosion defects, In: *Proc. Int. Conf. Offshore Mech. Arctic Eng.*, Hamburg, Germany.

Benjamin, A.C., Freire, J.L.F., Vieira, R.D., Cunha, D.J.S. (2016a) Interaction of corrosion defects in pipelines–part 1: fundamentals, *Int. J. Press. Vessel Pip.*, 144, 56–62.

Benjamin, A.C., Freire, J.L.F., Vieira, R.D., Cunha, D.J.S. (2016b) Interaction of corrosion defects in pipelines – part 2: MTI JIP database of corroded pipe tests, *Int. J. Press. Vessel Pip.*, 145, 41–59.

Canadian Standardization Association (2007) *Limit State Equation for Burst of Large Leaks and Rupture for Corrosion Defect: Oil and Gas Pipeline Systems*, CSA-Z662–07, Ottawa, ON, Canada.

Capula Colindres, S., M'endez, G.T., Vel'azquez, J.C., Cabrera-Sierra, R., Angeles-Herrera, D. (2020) Effects of depth in external and internal corrosion defects on failure pressure predictions of oil and gas pipelines using finite element models, *Adv. Struct. Eng.*, 23, 3128–3139.

Chandra B., Mondal B.C. (2016) Burst pressure assessment for pipelines with multiple corrosion defects, *Can. Soc. Civil Eng. Resilient Infrastruct.*, London, ON, Canada.

Chegeni, B., Jayasuriya, S., Das, S. (2019) Effect of corrosion on thin-walled pipes under combined internal pressure and bending, *Thin-Walled Struct.*, 143, 106218.

Chen, H.F., Shu, D. (2001) Simplified limit analysis of pipelines with multi-defects, *Eng. Struct.*, 23, 207–213.

Chen, Y., Zhang, H., Zhang, J., Liu, X., Li, X., Zhou, J. (2015) Failure assessment of X80 pipeline with interacting corrosion defects, *Eng. Fail. Anal.*, 47, 67–76.

Cheng, Y.F. (2020) Technical insights into the long-term integrity and sustainability of China-Russia eastern gas pipeline, *Oi/Gas Storage Transport.*, 39, 1–8.

Cheng, Y.F., Norsworthy, R. (2017) *Pipeline Coatings*, NACE, Houston, TX, USA

Chiodo, M.S.G., Ruggieri, C. (2009) Failure assessments of corroded pipelines with axial defects using stress-based criteria: numerical studies and verification analysis, *Int. J. Press. Vessel Pip.*, 86, 164–176.

Comsol (2018) *Comsol Multiphysics, Structural Mechanics Module*, User's Guide, Stockholm, Sweden.

Cosham, A., Hopkins, P. (2004) An overview of the pipeline defect assessment manual (PDAM), In: *Proc. 4th Int. Pipeline Technol. Conf.*, Oostende, Belgium.

Coulsen, K.E.W., Worthingham, R.W. (1990) New guidelines promise more accurate damage assessment, *Oil Gas J.*, 88, 16.

Cunha, D.J., Benjamin, A.C., Silva, R.C., Guerreiro, J.N., Drach, P.R. (2014) Fatigue analysis of corroded pipelines subjected to pressure and temperature loadings, *Int. J. Press. Vessel Pip.*, 113, 15–24.

De Andrade, E.Q., Benjamin, A.C., Machado Jr., P.R. (2006) Finite element modeling of the failure behavior of pipelines containing interacting corrosion defects, In: *Proc. Int. Conf. Offshore Mech. Arctic Eng.*, Hamburg, Germany.

Det Norske Veritas (2004) *Corroded Pipelines–Recommended Practice, RP-F10*, Norway.

Det Norske Veritas (2007) *Submarine Pipeline Systems, DNV-OS-F101*, Norway.

Det Norske Veritas (2010) *Corroded Pipelines – Recommended Practice*, RP F101, Norway.

Dewanbabee, H. (2009) *Behaviour of Corroded X46 Steel Pipe under Internal Pressure and Axial Load*, Master Thesis University of Windsor, Windsor, ON, Canada.

Enbridge (2014) *Operational Reliability Plan Report, Line 5 and Line 5 Straits of MacKinac Crossing*, Calgary, AB, Canada.

Fu, B., Kirkwood, M.G. (1995) Predicting failure pressure of internally corroded linepipe using the finite element method, In: *Proc. Int. Conf. Offshore Mechanics Arctic Eng.*, Copenhagen, Denmark.

Goodall, I.W. (1978) *Lower Bound Limit Analysis of Curved Tubes Loaded by Combined Internal Pressure and In-Plane Bending Moment*, RD/B/N4360, Center for Electricity Generation Board (CEGB), UK.

Hale, J.R., Lammert, W.F., Allen, D.W. (1991) Pipeline on-bottom stability calculations: comparison of two state-of-the-art methods and pipe-soil model verification, In: *Proc. Offshore Technol. Conf.*, Houston, TX, USA.

Han, D., Cheng, Y.F. (2013) Mechanism of electrochemical corrosion of carbon steel under deoxygenated water drop and sand deposit, *Electrochim. Acta*, 114, 403–408.

Han, C.J., Zhang, H., Zhang, J. (2016) Failure pressure analysis of the pipe with inner corrosion defects by FEM, *Int. J. Electrochem. Sci.*, 11, 5046–5062.

Hopkins, P., Jones, D.G. (1992) A study of the behavior of long and complex-shaped corrosion in transmission pipelines, part A – pipeline technology, In: *Proc. 11th Int. Conf. Offshore Mechanics Arctic Eng.*, Vol. 5, ASME, pp. 211–217.

Jin, Z.J., Qiu, C., Chen, Z.F., Yan, S.T., Shen, X.L. (2020) Integrity assessment of the pipelines containing an isolated corrosion pit, *Eng. Fail. Anal.*, 107, 104539.

Khalajestani, K.M., Bahaari, M.R., Salehi, A., Shahbazi, S. (2015) Predicting the limit pressure capacity of pipe elbows containing single defects, *Appl. Ocean Res.*, 53, 15–22.

Kiefner, J.F., Vieth, P.H. (1990a) Evaluating pipe – 1. New method corrects criterion for evaluating corroded pipe, *Oil Gas J.*, 88, 56–59.

Kiefner, J.F., Vieth, P.H. (1990b) Evaluating pipe conclusion: PC program speeds new criterion for evaluating corroded pipe, *Oil Gas J.*, 88, 91–93.

Kim, S.W., Chang, S.J., Park, D.U. (2020) Failure criteria of a carbon steel pipe elbow for low-cycle fatigue using the damage index, *Thin Wall. Struct.*, 153, 106800.

Kirn, J.W., Min, S.Y., Chi, Y.P. (2013) The effect of load-controlled bending load on the failure pressure of wall-thinned pipe elbows, *Nuclear Eng. Des.*, 265, 174–183.

Kuppusamy, C.S., Karuppanan, S., Patil, S.S. (2016) Buckling strength of corroded pipelines with interacting corrosion defects: numerical analysis, *Int. J. Struct. Stabil. Dynam.*, 16, 1550063.

Lam, C., Zhou, W. (2016) Statistical analysis of incidents on onshore gas transmission pipelines based on PHMSA database, *Int. J. Press. Vessel Pip.*, 145, 29–40.

Lamontagne, M. (2002) Interaction rules – an integral factor, In: *Corrosion'2002*, NACE, Houston, TX, USA.

Lee, Y.K., Kim, Y.P., Moon, M.W., Bang, W.H., Oh, K.H., Kim, W.S. (2005) The prediction of failure pressure of gas pipeline with multi corroded region, *Mater. Sci. Forum*, 475, 3323–3326.

Li, Z., Yinpei, W., Jin, C., Cengdian, L. (2001) Evaluation of local thinned pressurized elbows, *Int. J. Press. Vessel Pip.*, 78, 697–703.

Li, J., Zhou, C.Y., Xue, J.L., He, X.H. (2014) Limit loads for pipe bends under combined pressure and out-of-plane bending moment based on finite element analysis, *Int. J. Mech. Sci.*, 88, 100–109.

Li, X., Bai, Y., Su, C., Li, M. (2016) Effect of interaction between corrosion defects on failure pressure of thin wall steel pipeline, *Int. J. Press. Vessel Pip.*, 138, 8–18.

Liu, H.W., Cheng, Y.F. (2018) Mechanistic aspects of microbially influenced corrosion of X52 pipeline steel in a thin layer of soil solution containing sulphate-reducing bacteria under various gassing conditions, *Corros. Sci.*, 133, 178–189.

Liu, J., Chauhan, V., Ng, P., Wheat, S., Hughes, C. (2009) *Remaining Strength of Corroded Pipe Under Secondary (Biaxial) Loading*, Report no. R9068, GL Industrial Services UK Ltd., UK.

Liu, H., Khan, F., Thodi, P. (2017) Revised burst model for pipeline integrity assessment, *Eng. Fail. Anal.*, 80, 24–38.

Liu, C., Shi, S., Cai, Y. (2018) Ratcheting behavior of pressurized-bending elbow pipe after thermal aging, *Int. J. Press. Vessel Pip.*, 169, 160–169.

Low, B.K., Tang, W.H. (2007) Efficient spreadsheet algorithm for first-order reliability method, *J. Eng. Mech.*, 133, 1378–1387.

Mohammad, D., Amir, F., Ali, N. (2007) Investigation of dynamics and vibration of pig in oil and gas pipelines. ASME, *Int. Mech. Eng. Cong. Exp.*, Orlando, FL, USA.

Mondal, B.C., Dhar, A.S. (2019) Burst pressure of corroded pipelines considering combined axial forces and bending moments, *Eng. Struct.*, 186, 43–51.

Oh, C.S., Kim, Y.J., Park, C.Y. (2007) Plastic loads of elbows with local wall thinning under in-plane bending, *Int. J. Fract.*, 145, 63–79.

Oh, C.K., Kim, Y.J., Park, C.Y. (2009) Effects of local wall thinning on net-section limit loads for pipes under combined pressure and bending, *Nuclear Eng. Des.*, 239, 261–273.

Qian, S., Cheng, Y.F. (2019) Corrosion of X52 steel under thin layers of water condensate in wet gas pipelines, *J. Nat. Gas Sci. Eng.*, 68, 102921.

Qin, G.J., Cheng, Y.F. (2020) Failure pressure prediction by defect assessment and finite element modelling on natural gas pipelines under cyclic loading, *J. Nat. Gas Sci. Eng.*, 81, 103445.

Qin, G.J., Cheng, Y.F. (2021a) A review on defect assessment of pipelines: principles, numerical solutions, and applications, *Int. J. Press. Vessel Pip.*, 191, 104329.

Qin, G.J., Cheng, Y.F. (2021b) Modeling of mechano-electrochemical interaction at a corrosion defect on a suspended gas pipeline and the failure pressure prediction, *Thin-Walled Struct.*, 160, 107404.

Ramberg, W., Osgood, W.R. (1943) *Description of Stress-Strain Curves by Three Parameters*, Report no. NACA-TN-902, NASA Scientific and Technical Information Facility, USA.

Shim, D.J., Choi, J.B., Kim, Y.J., Kim, J.W., Park, C.Y. (2003) Assessment of local wall thinned pipeline under combined bending and pressure, *Int. J. Mod. Phys. B*, 17, 1870–1876

Shuai, Y., Shuai, J., Xu, K. (2017) Probabilistic analysis of corroded pipelines based on a new failure pressure model, *Eng. Fail. Anal.*, 81, 216–233.

Shuai, Y., Wang, X.H., Cheng, Y.F. (2020) Modelling of local buckling of corroded X80 gas pipeline under axial compression loading, *J. Nat. Gas Sci. Eng.*, 81, 103472.

Shuai, Y., Wang, X.H., Cheng, Y.F. (2021) Buckling resistance of an X80 steel pipeline at corrosion defect under bending moment, *J. Nat. Gas Sci. Eng.*, 93, 104016.

Shuai, Y., Zhao, X., Wang, H., Feng, C., Cheng, Y.F. (2022) Development of an empirical model to predict the burst pressure of corroded elbows of pipelines by finite element modelling, *Int. J. Press. Vessel Pip.*, 195, 104602.

Silva, R.C.C., Guerreiro, J.N.C., Loula, A.F.D. (2007) A study of pipe interacting corrosion defects using the FEM and neural networks, *Adv. Eng. Software*, 38, 868–875.

Su, C., Li, X., Zhou, J. (2016) Failure pressure analysis of corroded moderate-to-high strength pipelines, *China Ocean Eng.*, 30, 69–82.

Sun, J.L., Cheng, Y.F. (2018) Assessment by finite element modeling of the interaction of multiple corrosion defects and the effect on failure pressure of corroded pipelines, *Eng. Struct.*, 165, 278–286.

Sun, J.L., Cheng, Y.F. (2019) Modelling of mechano-electrochemical interaction of multiple longitudinally aligned corrosion defects on oil/gas pipelines, *Eng. Struct.*, 190, 9–19.

Sun, J.L., Cheng, Y.F. (2020) Modelling of mechano-electrochemical interaction at overlapped corrosion defects and the implication on pipeline failure prediction, *Eng. Struct.*, 213, 110466.

Sun, J.L., Cheng, Y.F. (2021) Modeling of mechano-electrochemical interaction between circumferentially aligned corrosion defects on pipeline under axial tensile stresses, *J. Petro. Sci. Eng.*, 198, 108160.

Systemes, D. (2015) *Abaqus 6.14 Documentation–Theory Guide*, Providence, RI, USA.

Thesi, I., Kenedi, P., Guimeraes de Souza, L., Pacheco, P. (2010) Modeling of pipe cold bending: a finite element approach, *Proc. 6th National Cong. Mech. Eng.*, Paraíba, Brazil.

Vanaei, H.R., Eslami, A., Egbewande, A. (2017) A review on pipeline corrosion, in-line inspection (ILI), and corrosion growth rate models, *Int. J. Press. Vessel Pip.*, 149, 43–54.

Wang, Q., Zhou, W. (2019) Burst pressure models for thin-walled pipe elbows, *Int. J. Mech. Sci.*, 159, 20–29.

Warman, D.J., Hart, J.D. (2005) *Development of a Pipeline Surface Loading Screening Process and Assessment of Surface Load Dispersing Methods*, Report no. 05-44, Canadian Energy Pipeline Association, Calgary, AB, Canada.

Wu, Y., Li, J. (2019) Finite element analysis on mechanical behavior of semi-exposed pipeline subjected to debris flows, *Eng. Fail. Anal.*, 105, 781–797.

Wu, Y., Zha, S., Jin, P. (2019) Finite element method simulations to study factors affecting buried pipeline subjected to debris flow, *J. Press. Vessel Technol.*, 141, 021701.

Xu, L.Y., Cheng, Y.F. (2012) Reliability and failure pressure prediction of various grades of pipeline steel in the presence of corrosion defects and pre-strain, *Int. J. Press. Vessel Pip.*, 89, 75–84.

Xu, B., Low, B.K. (2006) Probabilistic stability analyses of embankments based on finite element method, *J. Geotech. Geoenviron. Eng.*, 132, 1444–1454.

Yatabe, H., Fukuda, N., Masuda, T., Toyoda, M. (2004) Effect of material stress-strain behavior and pipe geometry on the deformability of high-grade pipelines, *J. Offshore Mech. Arctic Eng.*, 126, 113–119.

Zhang, S., Zhou, W. (2020) Assessment of effects of idealized defect shape and width on the burst capacity of corroded pipelines, *Thin Wall. Struct.*, 154, 106806.

Zhang, S.H., Wang, X.N., Song, B., Zhao, D.W. (2014) Limit analysis based on GM criterion for defect-free pipe elbow under internal pressure, *Int. J. Mech. Sci.*, 78, 91–96.

Zhang, H., Zhang, S., Liu, S., Zhu, X., Tang, B. (2015) Chatter vibration phenomenon of pipeline inspection gauges (PIGs) in natural gas pipeline, *J. Nat. Gas Sci. Eng.*, 27, 1129–1140.

Zhang, H., Zhang, S., Liu, L., Wang, Y. (2017) Collisional vibration of PIGs (pipeline inspection gauges) passing through girth welds in pipelines, *J. Nat. Gas Sci. Eng.*, 37, 15–28.

Zhang, S.Z., Cheng, Y.F., Feng, X.D., Wei, L.J., Wang, R.J., Duo, Y.Q., Han, Y.X. (2019) Performance characteristics and technical challenges of X80 pipeline steel – a review, *Oil Gas Storage Transport.*, 38, 481–495.

Zhang, H., Dong, J., Cui, C., Liu, S. (2020) Stress and strain analysis of spherical sealing cups of fluid-driven pipeline robot in dented oil and gas pipeline, *Eng. Fail. Anal.*, 108, 104294.

Zhang, Z.W., Guo, L.P., Cheng, Y.F. (2021) Interaction between internal and external defects on pipelines and its effect on failure pressure, *Thin-Wall. Struct.*, 159, 107230.

Zhou, H., Liu, M., Ayton, B., Bergman, J., Nanney, S. (2016) Tensile and compressive strain capacity of pipelines with corrosion anomalies, In: *Proc. 2016 Int. Pipeline Conf.*, Calgary, AB, Canada.

Zhu, X.K. (2021) A comparative study of burst failure models for assessing remaining strength of corroded pipelines, *J. Pipeline Sci. Eng.*, 1, 36–50.

4

Mechano-electrochemical Interaction for Level III Assessment of Corrosion Anomalies on Pipelines – A Single Corrosion Defect

4.1 Fundamentals of Mechano-electrochemical Interaction for Pipeline Corrosion

Most engineering structures such as pipelines operate under stress. When corrosion occurs on the structures in service environments, the stress will affect (usually accelerate) corrosion, and even change the corrosion mechanism. Corrosion of stressed metals has been one of the most investigated topics by metallurgists and corrosion researchers/engineers over several decades [Simnad and Evans, 1950; Hoar and West, 1958; Despic et al., 1968; Kim, 2017]. Depending on the materials, an applied stress or even the residual stress due to surface curvature of the metallic structures can shift electrochemical potential of the materials in the environments and amplify or reduce corrosion reaction rates [Swarnavo and Aquino, 2013].

The integrity of a pipeline in operation can be compromised by many factors, of which corrosion is one of the primary mechanisms causing pipeline failures [Cheng, 2016]. Particularly, external corrosion occurs on buried pipelines due to coating degradation and failures (e.g., coating disbondment, inclusion of pinholes and defects in the coatings, etc.), shielding of CP current, and generation of corrosive environments where the pipe steel is exposed under a synergistic effect of soil chemistry, aeration, moisture, microorganism, etc. [Cheng and Norsworthy, 2017]. Pipelines are pressurized infrastructure, and the operating pressure always induces a hoop stress, which serves as the primary stress, on the pipelines. In addition, secondary stresses such as soil stress and bending stress can be generated during ground movement and manufacturing, respectively. In some geotechnically unstable regions, pipe–soil interactions can induce an appreciable soil strain, which can be up to ±3%, on the pipe [Wang and Yeh, 1985; Cheng, 2020]. Therefore, pipeline corrosion is essentially a stress-corrosion process.

Defect Assessment for Integrity Management of Pipelines, First Edition. Y. Frank Cheng.

In 2013, Xu and Cheng [2013] developed a "Mechano-Electrochemical (M–E) interaction" concept to explain pipeline corrosion phenomenon, defining the interaction of mechanical stress (or strain) and electrochemical corrosion of pipeline steels both mechanistically and quantitatively. The concept was then used as a theoretical base to develop a multi-physics field coupling model, which was integrated with the FE-based Level III defect assessment methods for pipeline FFS determination and failure prediction [Xu and Cheng, 2017; Sun and Cheng, 2019a, b, 2020, 2021; Zhang et al., 2020a, 2021a]. Major progress has been made for improved accuracy of defect assessment for pipeline integrity management. Nowadays, the M–E interaction concept has been widely accepted as an accurate and effective methodology in fundamental research of stress-corrosion phenomena of engineering structural materials. Moreover, the concept has been used for FFS determination and failure pressure prediction of stressed structures, primarily pipelines.

4.1.1 The Mechanical–Chemical Interaction of Corrosion of Stressed Metals

The mechanical–chemical (M–C) interaction of corrosion of stressed metals was initially proposed by Gutman [1994, 1998]. The principle of the M–C interaction is that, for an irreversible chemical reaction on a solid proceeding in several consecutive steps (i.e., partial reactions) with formation of certain intermediate compounds which are in equilibrium with the solid, the reaction affinities depend on the difference of chemical potentials (μ) of reactants and products. The relationship between the chemical potential difference, $\Delta\mu$, and pressure (P), which is equivalent to a stress applied on the solid, is approximately calculated by [Eyring and Lin, 1980]:

$$\Delta\mu = \int_{P_1}^{P_2} V(P)\mathrm{d}P = \int_{P_1}^{P_2} V_0 \exp(-\chi P)\mathrm{d}P \approx V_0\Delta P \tag{4.1}$$

where V is volume; P_1 and P_2 are pressures at initial and end stages, respectively; ΔP is pressure difference, which is equal to $P_2 - P_1$; V_o is initial volume; and χ is a compressibility coefficient of the solid, which is approximately 10^{-6}. Generally,
$\chi P \ll 1$.

When the solid experiences a pressure excess (ΔP), the chemical potential can be expressed by [Gutman, 1994]:

$$\mu = \mu_{\Delta P=0} + \Delta P V_m = \mu_0 + RT \ln a + \Delta P V_m = (\mu_0 + \Delta P V_m) + RT \ln a = \mu_0{}' + RT \ln a \tag{4.2}$$

where a is activity, μ_0 is chemical potential of the solid in a standard state (i.e., $a = 1$), V_m is molar volume of the solid, μ_0' is standard chemical potential of the solid affected by the M–C interaction, R is ideal gas constant (8.314 J/mol K), and T is temperature (K). It is seen that an increase in chemical potential is the sum of standard chemical potential (μ_0) and the magnitude of ΔPV_m. Eq. (4.2) is then rewritten as [Gutman, 1994]:

$$\mu = \mu_0 + RT \ln \tilde{a} \tag{4.3}$$

where $\tilde{a}$ is defined as the M–C interaction activity, which is determined by the mechanical effect factor (ΔP):

$$\tilde{a} = a \exp \frac{\Delta PV_m}{RT} \tag{4.4}$$

The chemical potential of the solid is thus affected by the mechanical factor (e.g., stress or pressure difference) through the M–C interaction activity $\tilde{a}$, either accelerating or retarding the involved reaction, such as corrosion.

An M–C interaction exists in corrosion systems where the corroded metal is stressed. It was observed that a plastic deformation of 316L stainless steel increased the anodic passive current density [Gutman et al., 1996]. Characteristic changes in corrosion potential, polarization resistance, and Tafel constants of steel in seawater were observed when deformation of the steel transited from elastic to plastic regimes [Akid and Dmytrakh, 1998]. For active corrosion of steels in neutral solutions, both the anodic and cathodic reactions could be accelerated by external stress [Kuwazuru et al., 2018]. In acidic solutions, an applied stress favored both hydrogen evolution and anodic dissolution reactions [Kim et al., 2012]. For a passivated stainless steel in a chloride solution, it was found that the pit initiation and growth kinetics were affected by local strain [Wang and Han, 2016].

4.1.2 The M–E Interaction for Pipeline Corrosion

Corrosion of metals is electrochemical in nature, and so is pipeline corrosion. Replacement of the M–C interaction, as initially proposed by Gutman [1994, 1998], with the M–E interaction concept for pipeline corrosion, which was developed by Xu and Cheng [2013], is mainly attributed to the presence of multiple physics fields and their synergism affecting the pipeline corrosion process. For example, during corrosion of steels in aqueous environments, an electric field exists in both the solution phase and the steels, affecting the interfacial charge-transfer reactions during corrosion processes. However, the multiple-field coupling effect was not considered in the M–C interaction concept. Without inclusion of the electrochemical factors in pipeline corrosion models, the modeling results will not be representative of reality, making the models incapable of predicting the corrosion growth rate of defects.

The effect of an electrical action on chemical potential of a solid can be thermodynamically calculated by [Gutman, 1994]:

$$\mu = \mu_0 + RT \ln a + zF\varphi = \mu_0 + RT \ln \bar{a} \tag{4.5}$$

where z is chemical valence or charge number of the solid (metal), F is Faraday's constant (96,500 C/mol), φ is electrical potential, and $\bar{a}$ is defined as the electrochemical activity, as determined by:

$$\bar{a} = a \exp \frac{zF\varphi}{RT} \tag{4.6}$$

When both the mechanical and electrical actions are considered simultaneously, the electrochemical potential can be obtained by:

$$\mu = \mu_0 + RT \ln a + \Delta PV_m + zF\varphi = \mu_0 + RT \ln \bar{\bar{a}} \tag{4.7}$$

where $\bar{\bar{a}}$ is defined as the M–E activity:

$$\bar{\bar{a}} = a \exp \frac{zF\varphi + \Delta PV_m}{RT} \tag{4.8}$$

Buried pipelines are usually under complex stress conditions, including both primary hoop stress resulting from internal operating pressure and secondary stresses, such as soil stress, due to ground movement. Moreover, various local stress raisers, e.g., corrosion defects, dents, scratches, and microcracks, frequently exist on the pipe surface, especially for those which have been in service for a long time. These surface anomalies are associated with high stress concentrations. For corrosion defect, the electrochemical reactions occur at the steel/solution interface in the local area, and the electrical field in the solution phase affects mass transfer of corrosive species toward the interface through an electro-migration mechanism, in addition to diffusion and, sometimes, convection. The effect induced by stress (usually an enhanced effect) on the local electrochemical activity of the steels at the corrosion defect changes both corrosion thermodynamics and kinetics.

4.1.2.1 Corrosion Thermodynamics and Kinetics Under an Elastic Stress

Under an elastic stress, the stress-induced effect on thermodynamic activity of steels, as indicated by a change of electrochemical anodic equilibrium potential, $\Delta\varphi^{e}_{a,eq}$, is [Gutman, 1998]:

$$\Delta\varphi^{e}_{a,eq} = -\frac{\Delta PV_m}{zF} \tag{4.9}$$

where the superscript *e* refers to elastic stress, V_m is molar volume (for steels, 7.13×10^{-6} m^3/mol), z is charge number (two for steels), and ΔP is excessive pressure, which is approximately one-third of the uniaxial tensile stress while a pipe is under internal pressure [Xu and Cheng, 2013]:

$$\Delta P = (\sigma_\theta + \sigma_z)/3 \tag{4.10}$$

$$\sigma_\theta = \frac{PD}{2t} \tag{4.11}$$

$$\sigma_z = \frac{PD}{4t} \tag{4.12}$$

where σ_θ is hoop stress, and σ_z is axial stress. It is seen from Eqs. (4.9) to (4.12) that an increase in pipeline operating pressure P causes a negative shift of the electrochemical anodic equilibrium potential, resulting in an increased electrochemical activity of the steels.

Figure 4.1 shows the time dependence of stress and corrosion potential of a pipeline steel during tensile testing in a near-neutral pH 0.01 M $NaHCO_3$ solution [Xu and Cheng, 2012a]. It is seen that, in the elastic range, the corrosion potential shifts negatively by 2.6 mV (vs. saturated calomel electrode,

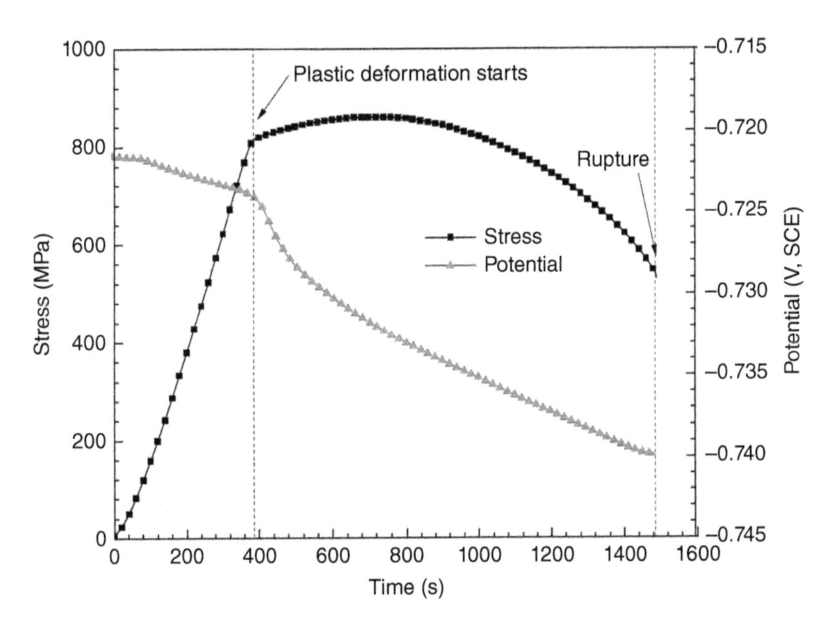

Figure 4.1 Time dependence of stress and corrosion potential of a pipeline steel during tensile testing in a near-neutral pH 0.01 M $NaHCO_3$ solution. *Source:* From Xu and Cheng [2012a] / with permission from Elsevier.

SCE) only. Thus, although an elastic stress can increase the electrochemical corrosion activity of steels, as indicated by the negative shift of corrosion potential, the effect is marginal and, sometimes, negligible.

Kinetically, an applied stress affects the corrosion rate by changing anodic current density of the steels in corrosive environments. The effect of an elastic stress on anodic current density of a stressed steel is given by [Gutman, 1998]:

$$i_a^{\ e} = i_a \exp\left(\frac{\Delta P V_m}{RT}\right) \tag{4.13}$$

where i_a and $i_a^{\ e}$ are anodic current densities of a nonstressed steel and an elastically stressed steel, respectively. It is expected that the elastic stress increases the anodic current density, accelerating the steel corrosion. Since the thermodynamic corrosion potential drops by several millivolts only in the elastic range, the marginal shift in potential is not expected to cause an obvious change in corrosion reaction kinetics. Figure 4.2 shows the electrochemical impedance spectroscopy (EIS) measured on a pipeline steel

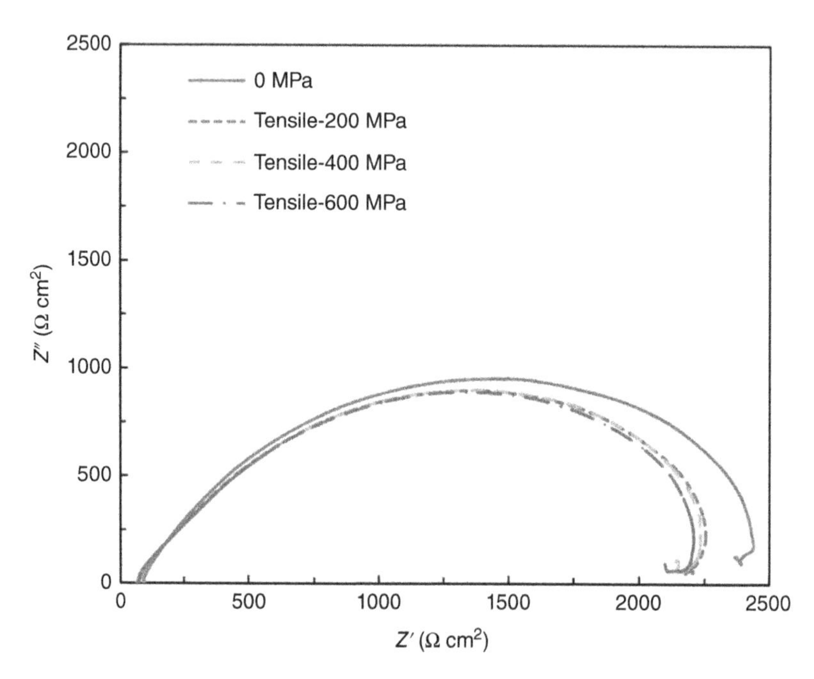

Figure 4.2 EIS plots measured on a pipeline steel under various elastic tensile stresses increasing from 0 to 600 MPa in several seconds in 0.01 M $NaHCO_3$ solution. *Source:* From Xu and Cheng [2012b] / with permission from Elsevier.

under various elastic tensile stresses increasing from 0 to 600 MPa in several seconds in 0.01 M $NaHCO_3$ solution [Xu and Cheng, 2012b]. Generally, the size of the semicircle in EIS plots is proportional to charge-transfer resistance, and thus, inversely proportional to corrosion rate. It is seen from the EIS measurements that the semicircle size decreases slightly with the increased stress. Thus, an elastic tensile stress can increase corrosion rate of the steel, but slightly.

4.1.2.2 Corrosion Thermodynamics and Kinetics Under a Plastic Stress

When a stress applied on pipelines is plastic, resulting in a significant plastic strain on the pipe, both corrosion reaction thermodynamics and kinetics are enhanced at an obvious level. Quantitatively, the change of electrochemical anodic equilibrium potential of the steel under a plastic stress, $\Delta\varphi^{p}_{a,eq}$, is [Gutman, 1998]:

$$\Delta\varphi^{p}_{a,eq} = -\frac{TR}{zF}\ln\left(\frac{\upsilon\alpha}{N_0}\varepsilon_p + 1\right) \quad (4.14)$$

where the superscript p refers to plastic deformation, υ is an orientation-dependent factor ($\upsilon = 0.45$) [Gutman, 1994], α is a coefficient (1.67×10^{11} cm^{-2}) [Gutman, 1994], N_0 is initial density of dislocations prior to plastic deformation (approximately 1×10^8 cm^{-2}) [Gutman, 1994], and ε_p is plastic strain. Combined with both elastic and plastic deformations in Eqs. (4.9) and (4.14), the equilibrium potential of anodic reaction under the M–E interaction is expressed as:

$$\varphi_{a,eq} = \varphi^{0}_{a,eq} - \frac{\Delta PV_m}{zF} - \frac{RT}{zF}\ln\left(\frac{\upsilon\alpha}{N_0}\varepsilon_p + 1\right) \quad (4.15)$$

where $\varphi_{a,\ eq}$ and $\varphi^{0}_{a,eq}$ are equilibrium potential and standard equilibrium potential of the anodic reaction, respectively. As the plastic strain increases, the anodic reaction equilibrium potential shifts negatively, indicating an increased reaction activity. As shown in Figure 4.1, the potential drop increases to 15.8 mV (SCE) at the plastic deformation stage, approximately six times the potential drop in the elastic stage. Therefore, the plastic strain apparently increases the corrosion activity of pipeline steels in corrosive environments.

Kinetically, the effect of plastic strain on anodic current density, i.e., corrosion rate, of a steel in corrosive environments is written as [Gutman, 1998]:

$$i_a{}^{p} = i_a\left(\frac{\upsilon\alpha}{N_0}\varepsilon_p + 1\right)\exp\left(\frac{\Delta PV_m}{RT}\right) \quad (4.16)$$

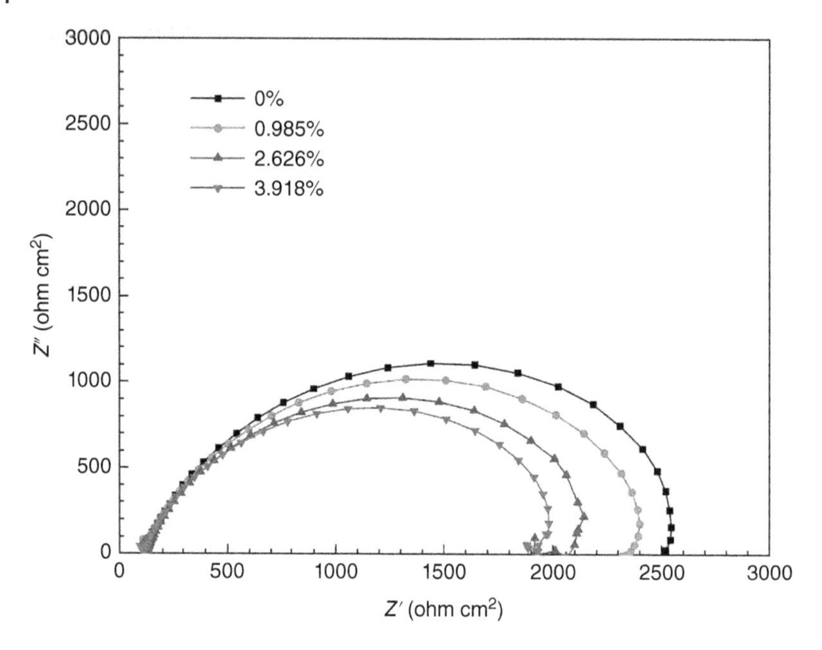

Figure 4.3 EIS plots measured on a pipeline steel under various plastic strains in 0.01 M $NaHCO_3$ solution. *Source:* From Xu and Cheng [2012a] / with permission from Elsevier.

where i_a^p is anodic current density of a plastically stressed steel in an aqueous solution. An increase in plastic strain will increase the anodic current density, i.e., the corrosion rate, of the steel. Figure 4.3 shows the EIS plots of a pipeline steel under various plastic strains in a near-neutral pH 0.01 M $NaHCO_3$ solution [Xu and Cheng, 2012a]. The semicircle size, which is inversely proportional to the corrosion rate, decreases with increased strain. Thus, the plastic deformation increases the corrosion rate of the steel obviously.

In summary, the M–E interaction for pipeline corrosion is unapparent, and, sometimes, undetectable in elastic stage. However, the M–E interaction becomes significant in plastic stage and must be considered in corrosion defect assessment on pipelines. The equilibrium potential and current density of anodic dissolution reaction of pipeline steels during continuous elasto-plastic deformation are defined by Eqs. (5.15) and (5.16), respectively.

4.2 Multi-Physics Field Coupling at a Corrosion Defect on Pipelines

According to the M–E interaction for pipeline corrosion, multiple physics fields exist at a corrosion defect. These fields and their coupling effect affect performance and failure pressure of the pipelines. In 2013, Cheng and coworkers developed a FE-based multi-physics field coupling model for corrosion defect assessment, determining the distributions of stress and strain, corrosion potential, and anodic/cathodic reaction current densities (i.e., anodic/cathodic reaction rates) at corrosion defect, and further, predicting the pipeline failure pressure and the defect growth rate [Xu and Cheng, 2013; Qin and Cheng, 2021a; Wang et al., 2016; Wang et al., 2021]. The model integrates a mechanical stress field, an electrical field, an electrochemical corrosion field, and the resulting M–E interaction at the corrosion defect, accurately defining various mechanical and electrochemical corrosion parameters. The modeling results can evaluate the FFS of a pipeline containing corrosion defects, and further predict the defect growth rate and remaining service life of the pipeline under the multi-physics field coupling effect [Xu and Cheng, 2017]. Cheng group further expands the model assessment from a single corrosion defect to multiple, adjacent defects with various orientations [Sun and Cheng, 2019a, b, 2020], from low-grade steel pipelines to the pipelines made of high grades of steel [Xu and Cheng, 2013, 2017; Zhang et al., 2020a], from external defects to both external and internal defects [Zhang et al., 2020b], from static stress to cyclic stress [Qin and Cheng, 2020], and from fully supported pipelines to pipelines in suspension [Qin and Cheng, 2021b], covering a wide variety of pipeline operation conditions in the field.

4.2.1 Electrochemical Anodic and Cathodic Reactions and Relevant Parameters

To establish an electrochemical corrosion model for a corrosion defect on pipelines, mechanistic aspects of the corrosion process should be defined. These include electrochemical anodic and cathodic reactions, corrosive environments, solution chemistry, pipeline steels, coating failure modes, and CP performance. For example, for a steel pipe containing a corrosion defect exposed to an electrolyte generated under disbonded coatings where the CP current is shielded, the electrolyte usually features an anaerobic, near-neutral pH (about 6.5), diluted bicarbonate solution [Cheng and

Norsworthy, 2017]. The electrochemical anodic and cathodic reactions during corrosion of the steels are mainly iron oxidation and hydrogen evolution, respectively [King et al., 2000; Parkins, 2000; Cheng, 2013]. The pipeline steels are in an active dissolution state in the solution [Parkins, 2000; Yang and Cheng, 2016]. A description of the steel dissolution reaction kinetics is reasonable to model the steel corrosion.

For an activation-controlled corrosion process (i.e., the charge transfer is the rate-determining step of the entire corrosion process), the corrosion reaction kinetics are expressed by anodic and cathodic Tafel laws, respectively:

$$i_a = i_{0,a} \exp\left(\frac{\eta_a}{b_a}\right) \tag{4.17}$$

$$i_c = i_{0,c} \exp\left(\frac{\eta_c}{b_c}\right) \tag{4.18}$$

$$\eta = \varphi - \varphi_{eq} \tag{4.19}$$

where i_a and i_c are anodic and cathodic reaction current densities, respectively; $i_{0,a}$ and $i_{0,c}$ are anodic and cathodic exchange current densities, respectively; η_a and η_c are anodic and cathodic activation overpotentials, respectively; b_a and b_c are anodic and cathodic Tafel slopes, respectively; φ is electrode potential; and φ_{eq} is equilibrium electrode potential. According to Nernst equation [Bagotsky, 2006], the anodic and cathodic equilibrium potentials are written as:

$$\varphi_{a,eq} = \varphi^0_{a,eq} + \frac{0.0592}{2} \log\left[\mathrm{Fe}^{2+}\right] \tag{4.20}$$

$$\varphi_{c,eq} = \varphi^0_{c,eq} + 0.0592 \log[\mathrm{H}^+] = -0.0592\mathrm{pH} \tag{4.21}$$

The standard equilibrium potentials of the anodic and cathodic reactions are −0.409 V (vs. standard hydrogen electrode, SHE) and 0 V (SHE), respectively. Assume that the concentration of ferrous ions in the solution is 10^{-6} M. At a near-neutral solution pH of 6.8, the equilibrium potentials of the anodic and cathodic reactions calculated by Nernst equations are −0.618 V (SHE) and −0.403 V (SHE), respectively. The exchange current densities of the anodic and cathodic reactions can be derived from potentiodynamic polarization measurements on the steel in the solution through Tafel extrapolation.

4.2.2 Electrical Field in the Solution Phase

According to the electrical field theory [Zahn, 1979], the distribution of current field in a solution during electrochemical reactions is defined by:

$$\nabla i_k = Q_k \tag{4.22}$$

$$i_k = -\sigma_k \nabla \varphi_k \tag{4.23}$$

where Q_k is a general source term, k is an index for liquid (l), e.g., the corrosive solution the pipeline steel is exposed to, or for solid (s), e.g., the pipeline steel, σ_k is conductivity, and φ_k is potential of liquid or solid phase.

4.2.3 Mechanical Stress Field on Pipelines

Energy pipelines are pressurized infrastructure. The primary stress results from the internal operating pressure. An elasto-plastic solid stress simulation is performed on pipe steel, where an isotropic hardening model is usually selected, as described in Eq. (3.1). A von Mises yielding criterion is used for the elasto-plastic simulation. To simulate the soil strain induced by ground movement, various tensile strains are applied on the steel pipe along the longitudinal direction. It is noted that the applied tensile strain is the total strain, which is equal to the total deformation divided by the pipe length. Local strain is varied due to the presence of corrosion defect.

4.3 The M–E Interaction at a Single Corrosion Defect on Pipelines

4.3.1 A Single Corrosion Defect with a Regular Geometrical Shape

Initially, the M–E interaction was integrated with Level III assessment methods for either a single corrosion defect or multiple corrosion defects but without mutual interaction, determining the FFS and failure pressure of pipelines [Xu and Cheng, 2013]. As stated, the pipelines are under complex stress conditions, including hoop stress resulting from the internal operating pressure and a longitudinal soil stress (or strain) due to pipe–soil interactions. Moreover, a local stress concentration is generated at the corrosion defect. For modeling convenience, the corrosion defect is usually simplified as either an elliptical or a rectangular shape so that the defect dimension can be defined by the combination of either longitudinal diameter/circumferential diameter/depth or length/width/depth, respectively [American Society of Mechanical Engineering, 2012].

Figure 4.4 shows the distributions of von Mises stress, corrosion potential, and anodic/cathodic current densities at the bottom of an elliptical corrosion defect with a depth of 11.46 mm (i.e., 60% of pipe wall thickness) on an X100 steel pipe under various axial tensile strains in a near-neutral pH bicarbonate solution. All the modeled parameters follow a symmetrical distribution

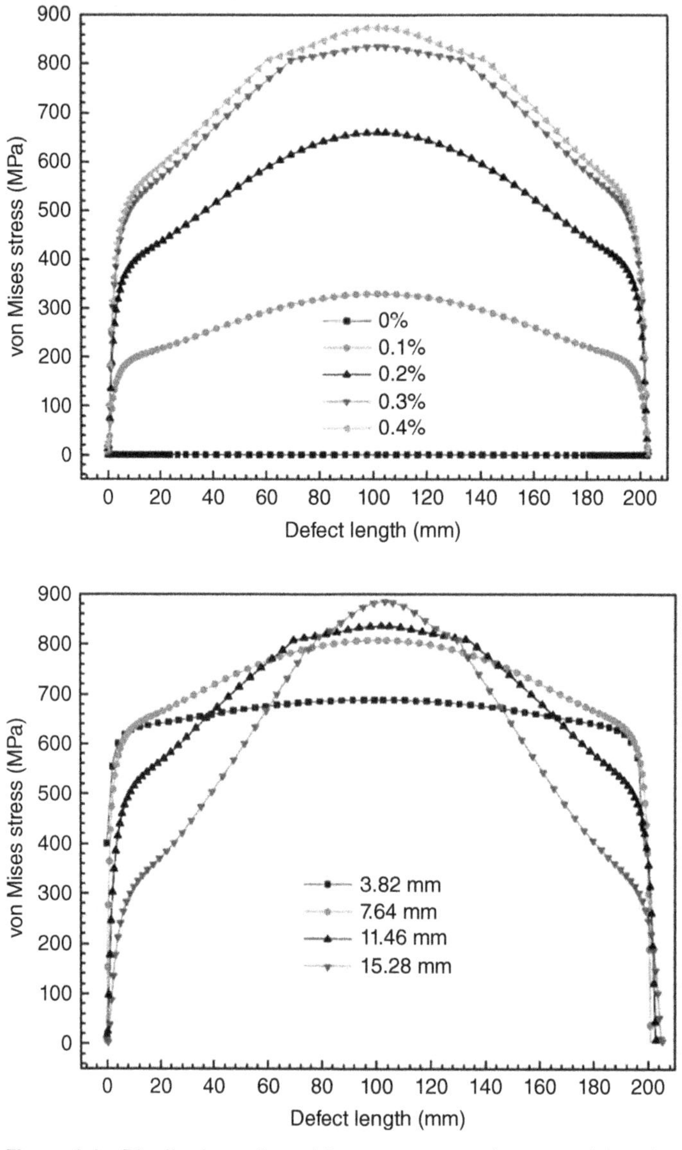

Figure 4.4 Distributions of von Mises stress, corrosion potential, and anodic/cathodic current densities at the bottom of an elliptical corrosion defect with a depth of 11.46 mm (i.e., 60% of pipe wall thickness) on an X100 steel pipe under various axial tensile strains in a near-neutral pH bicarbonate solution. *Source:* From Xu and Cheng [2013] / with permission from Elsevier.

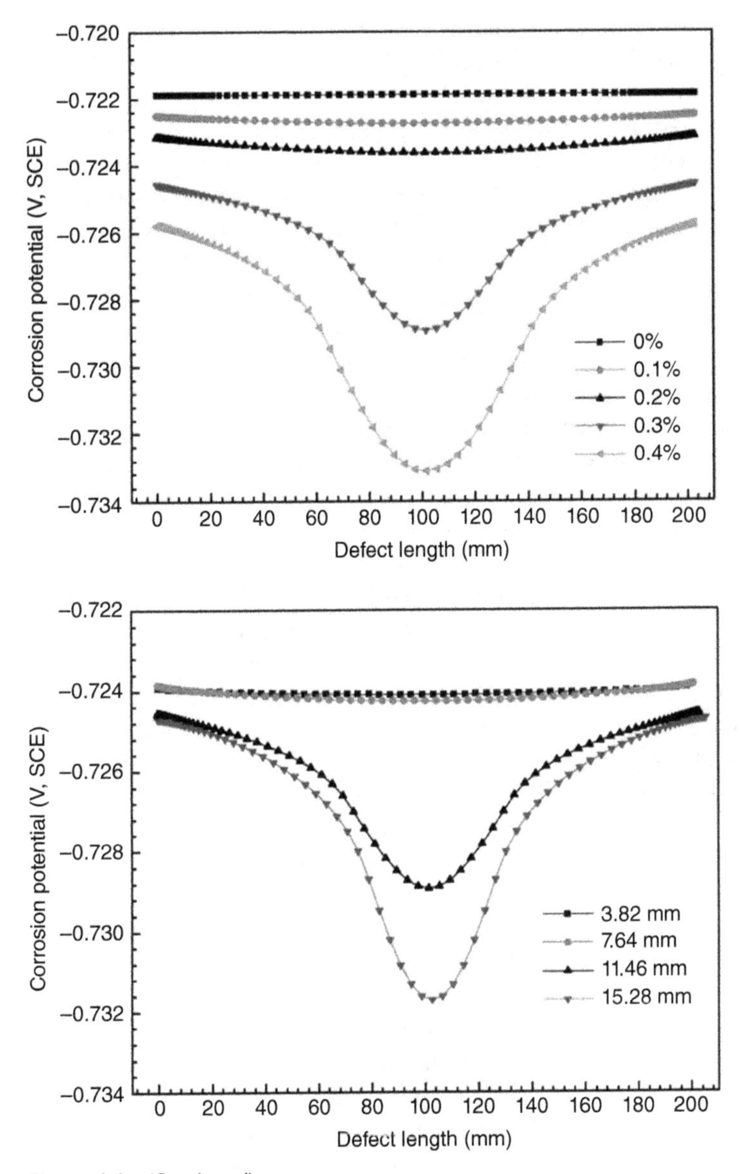

Figure 4.4 (Continued)

relative to the defect center. Obviously, there are the greatest stress, the most negative corrosion potential, and the greatest absolute values of both anodic and cathodic current densities at the defect center. With the increase in tensile strain, these parameters further increase. When the tensile strain reaches 0.3% and 0.4%, the stress concentration at the corrosion defect exceeds yield strength of the steel (i.e., 806 MPa for X100 steel), indicating that a plastic deformation occurs at the central area of the defect. As a result, the corrosion potential shifts negatively, and the absolute values of both anodic and cathodic current densities increase at the defect center. The results are attributed to the remarkable effect of plastic deformation on increased corrosion due to the M–E interaction. As a comparison, an elastic deformation such as 0.1% of tensile strain does not apparently affect the corrosion parameters. Generally, the applied tensile strain causes an overall elevation of the stress level, corrosion activity (i.e., corrosion potential) and corrosion kinetics (i.e., anodic/cathodic current densities) at the defect.

The von Mises stress, corrosion potential, and anodic/cathodic current densities at the corrosion defect are also dependent on the defect depth, which is usually believed as the most important parameter affecting the local stress distribution, as compared with the defect length and width. The modeling results on the elliptical corrosion defect with various depths on the X100 steel pipe under a fixed 0.3% axial tensile strain in the near-neutral pH bicarbonate solution are shown in Figure 4.5. Same as the results in Figure 4.4, there are the greatest von Mise stress, the most negative corrosion potential, and the greatest absolute values of anodic/cathodic current densities at the center of the defect. It is further seen that, as the defect depth increases, these parameters become more concentrated at the defect center with increased magnitude. Thus, the deeper the corrosion defect, the greater the local stress concentration and the more remarkable the resulting M–E interaction at the defect center. As a result, the defect suffers from accelerated corrosion at its center.

The modeling results confirm that a difference in electrical potential exists between the bottom of the corrosion defect and the defect sides due to a non-uniform stress distribution. Thus, each corrosion defect can be regarded as a galvanic cell. The area with a higher stress and a more negative potential, such as the defect center, serves as the anode, while the area under a lower stress and a less negative potential, such as the defect sides, serves as the cathode. Electrons generated at the anode during corrosion reaction flow to the cathode for cathodic reduction such as electrochemical reduction of dissolved oxygen. The ferrous ions as corrosion products transport to the cathode due to concentration gradient. Thus, the M–E interaction at the corrosion defect accelerates localized corrosion at the defect center, where a much higher corrosion rate is recorded compared with the steel corrosion in the absence of the M–E interaction. It is expected that corrosion progression will cause pipeline leakage at the defect.

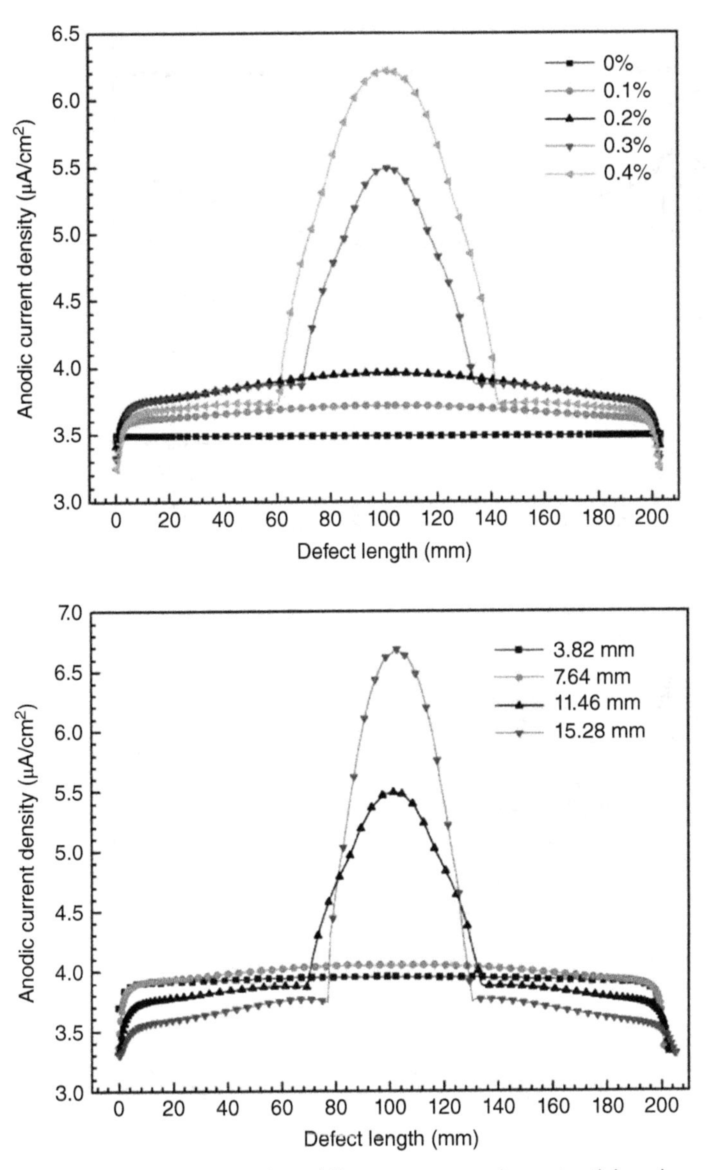

Figure 4.5 Distributions of von Mises stress, corrosion potential, and anodic/cathodic current densities at an elliptical corrosion defect with various depths on the X100 steel pipe under a fixed 0.3% axial tensile strain in the near-neutral pH bicarbonate solution. *Source:* From Xu and Cheng [2013] / with permission from Elsevier.

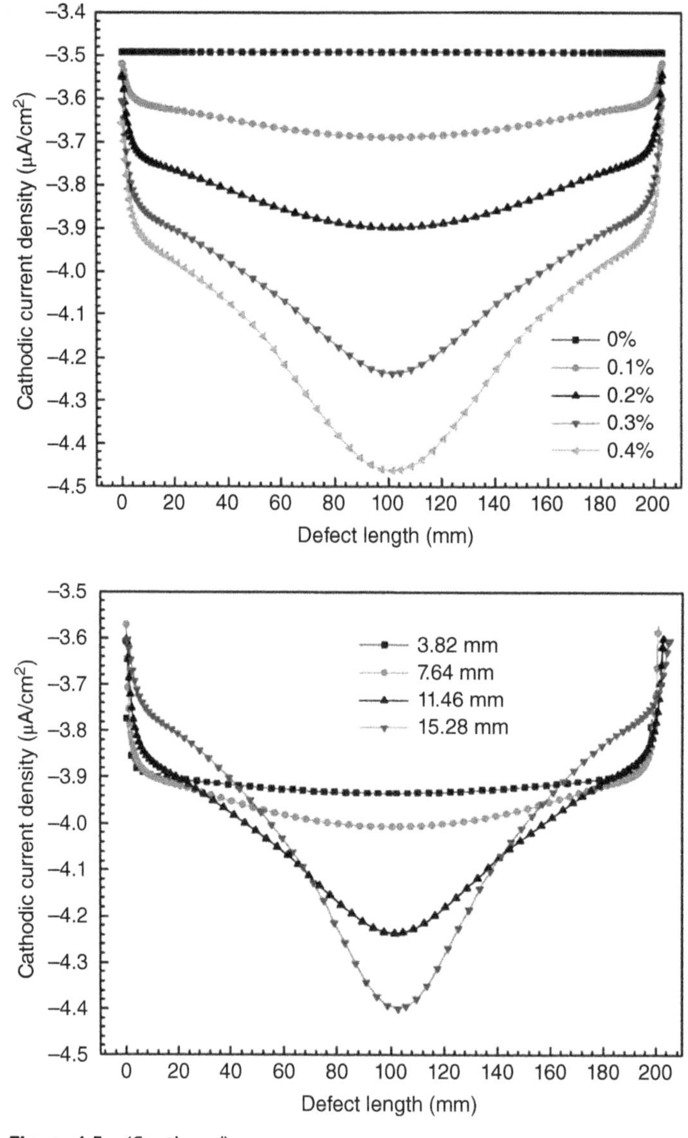

Figure 4.5 (Continued)

4.3.1.1 Corrosion Defect with Various Inclinations on Pipelines

As stated, the corrosion defect on pipelines is usually treated as a semi-ellipsoid for assessment purpose [Yi et al., 2012; Xu and Cheng, 2013; Xie and Tian, 2018], where the defect depth and the longitudinal/circumferential lengths are the main geometrical parameters affecting the local stress distribution [Noda and Hayashida, 2000; Huang et al., 2014]. It is usually thought that the main axes of the ellipsoidal defect are either perpendicular or parallel to the longitudinal direction of the pipeline [Xu and Cheng, 2013]. As a result, the stress distribution at the defect is symmetric relative to the axes, as shown in Figures 4.4 and 4.5. However, the actual corrosion defects on a pipeline are never ideal semi-ellipsoids [Kim and Park, 2017]. The orientation of the defects is often random, with various inclinations relative to the axial direction of the pipeline. This causes the local stress distribution and M–E interaction at the corrosion defect to depend on the defect's inclination angle (θ_{incl}), i.e., the angle between the primary axis of the corrosion defect and the longitudinal axis of the pipe.

An FE model developed for an X100 steel pipe containing a semi-ellipsoidal corrosion defect exposed to a near-neutral pH solution is shown in Figure 4.6 [Zhang et al., 2021b]. For the corrosion defect, $2c$ and $2a$ are lengths of the primary and secondary axes, respectively. When the inclination angle, θ_{incl}, is 0° and 90°, the primary axis ($2c$) of the corrosion defect is perpendicular and parallel to the longitudinal direction of the pipe, respectively.

The distribution of von Mises stress at the corrosion defect with varied θ_{incl} degrees is shown in Figure 4.7, under an internal pressure of 20 MPa. When the inclination angle is 0° or 90°, a stress concentration occurs at the center and the tips of the defect, while its sides are under low stress, as indicated by the color bar. With the angle changing from 0° to 90°, the stress concentration decreases. Moreover, only when θ_{incl} is equal to 0° or 90°, the stress distribution at the defect is symmetric. The most asymmetrical stress distribution occurs at 45°. Figure 4.8 shows the distribution of corrosion potential at the defect with various θ_{incl} degrees. When the angle changes from 0° to 90°, the corrosion potential shifts less negatively. At 0°, the most negative corrosion potential is recorded at the defect center. When the angle is 90°, there is a uniform distribution of the corrosion potential at the defect. The most asymmetrical distribution of corrosion potential is also observed at 45°. By comparison of the stress and corrosion potential distributions at the corrosion defect, the sites with a higher stress concentration are always associated with a more negative corrosion potential, and thus a greater corrosion activity in the solution.

The distributions of anodic and cathodic current densities (i.e., i_a and i_c) along the z-axis of the corrosion defect with various θ_{incl} angles are shown in Figure 4.9. The absolute values of both i_a and i_c are the greatest when θ_{incl} is equal to 0°. When the θ_{incl} changes from 0° to 90°, the i_a and i_c decrease. The results show that the M–E interaction accelerates both the anodic and cathodic reaction rates of the corrosion defect in the solution. The difference

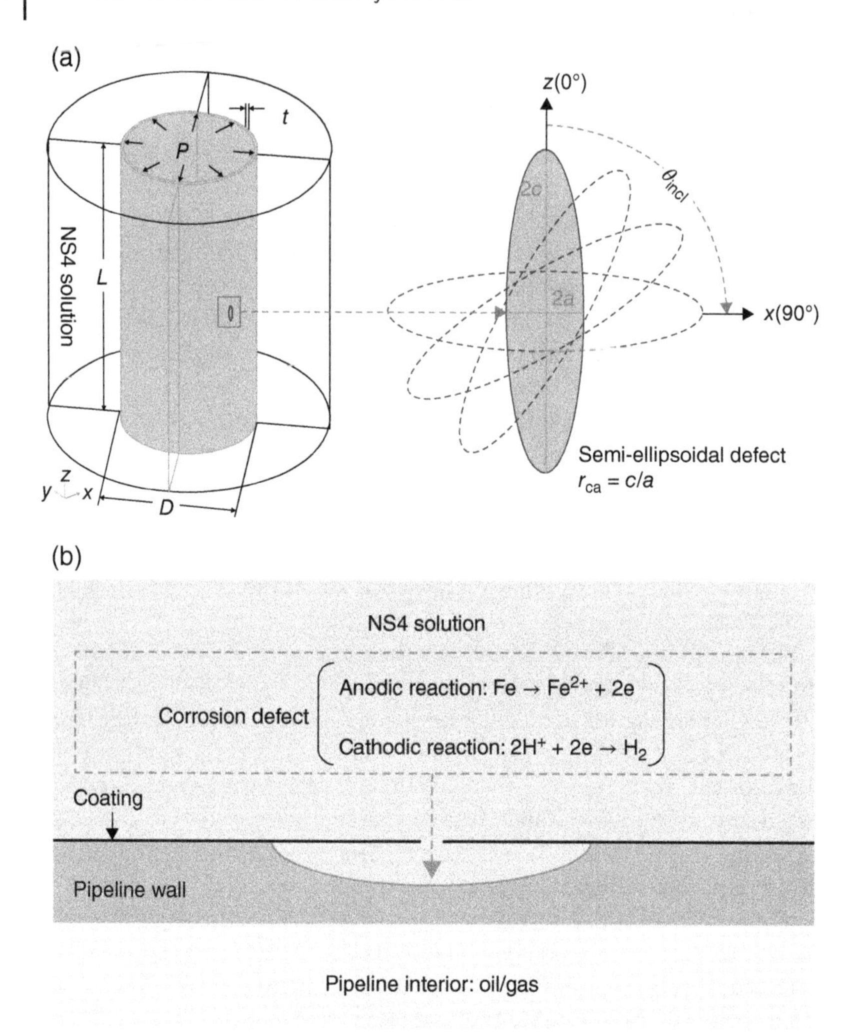

Figure 4.6 (a) Schematic diagram of a 3D FE model for an X100 steel pipe segment containing a semi-ellipsoidal corrosion defect exposed to a near-neutral pH NS4 solution. (b) Schematic diagram of the boundary condition for electrochemical corrosion occurring at the defect under disbonded coating. *Source:* From Zhang et al. [2021b] / with permission from Elsevier.

of i_a values between the defect center and the sides reaches a maximum at 45°, indicating the greatest nonuniformity of the anodic reaction rate (i.e., corrosion rate) at this inclination angle.

The M–E interaction at corrosion defect on pipelines highly depends on the defect orientation, i.e., the inclination angle. Since the hoop stress

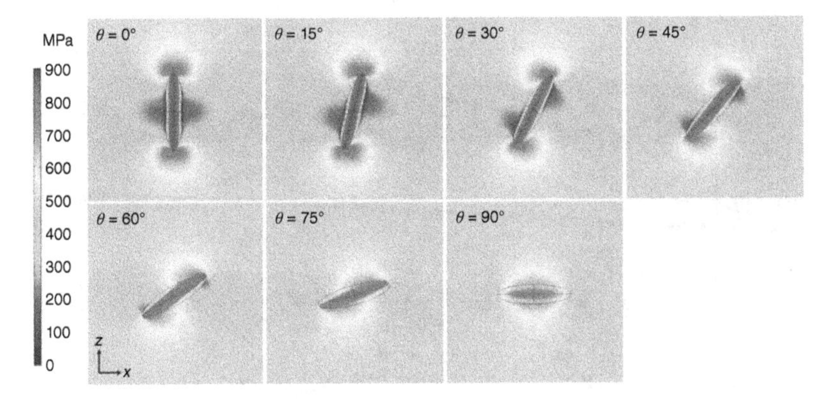

Figure 4.7 Distribution of von Mises stress at the corrosion defect with varied inclination angles between the defect's primary axis and the longitudinal direction of the pipe. *Source:* From Zhang et al. [2021b] / with permission from Elsevier.

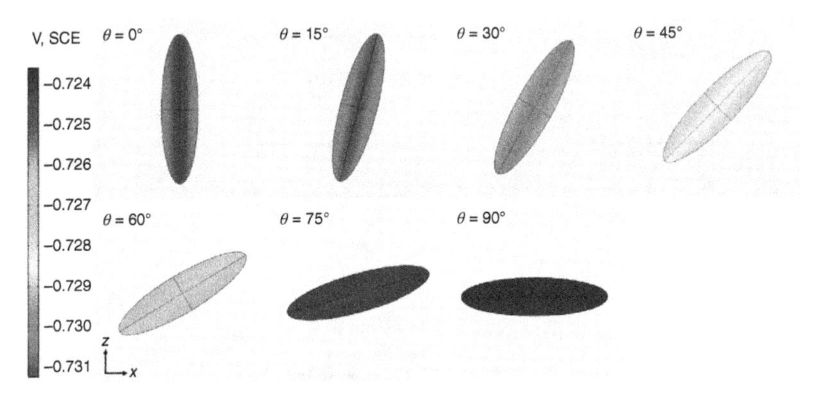

Figure 4.8 Distribution of corrosion potential at the defect with varied inclination angles. *Source:* From Zhang et al. [2021b] / with permission from Elsevier.

resulting from the internal pressure is about three times the axial stress on the pipelines, the stress concentration at the corrosion defect is not uniform when it is inclined to the pipe's longitudinal direction with an angle. When the inclination angle is zero, i.e., the primary axis of the corrosion defect is perpendicular to the longitudinal direction of the pipeline, there is the greatest stress concentration at the defect, resulting in the most negative corrosion potential and the greatest anodic current density, i.e., corrosion rate. Thus, the corrosion defect experiences accelerated corrosion at this inclination. As a comparison, the corrosion defect with the primary axis parallel to

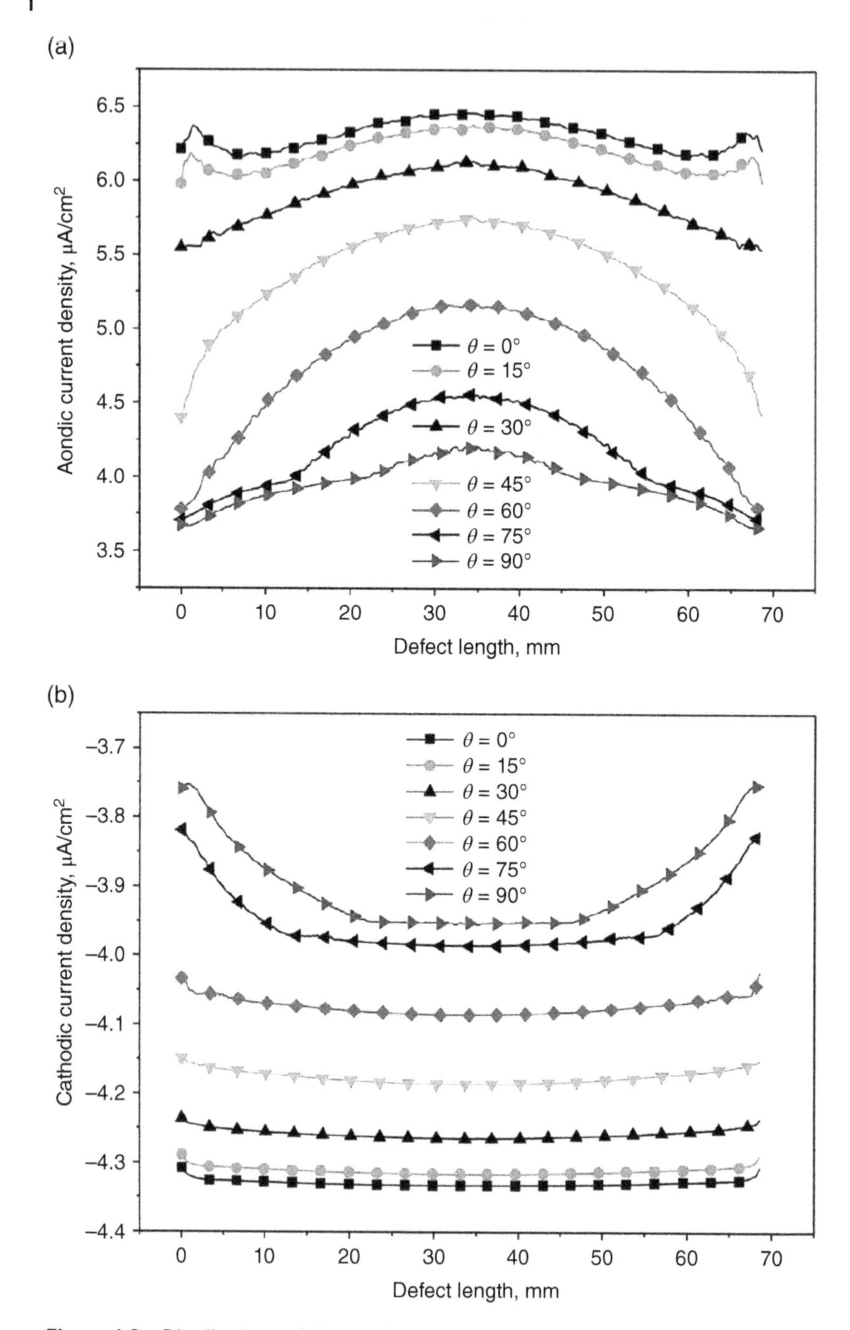

Figure 4.9 Distributions of (a) anodic and (b) cathodic current densities along the primary axial direction of the defect with varied inclination angles. *Source:* From Zhang et al. [2021b] / with permission from Elsevier.

the longitudinal direction of the pipe, i.e., $\theta_{incl} = 90°$, the stress concentration decreases at the defect, accompanied with a decreased corrosion rate. In addition, the dependence of the stress concentration level and M–E effect on the inclination angle is also affected by the internal pressure and the defect size, i.e., the relative lengths of the primary and secondary axes of the defect.

4.3.1.2 Corrosion Defect at an Elbow of Pipelines

As stated previously, elbows are common and key constituents of a pipeline system, playing an important role in changing the direction of a pipe's route. Owing to the bending curvature, a nonuniform stress distribution is generated at the pipe elbows. As a result, the external coating on the elbows gets peeled off more easily compared to the coating applied on straight pipes, thereby exposing the pipe steel to soil solution to cause corrosion [Khalaj et al., 2015; Lam and Zhou, 2016]. Nowadays, the standards and methods used for defect assessment mostly consider straight pipes, and are not suitable for assessment of pipe elbows as the intrados and extrados in pipe elbows are under nonuniform circumferential and axial stresses under the internal pressure. There are some empirical methods to determine burst pressure of corroded elbows. For example, Goodall [1978] proposed a model to predict the burst pressure of intact elbows subject to a pipe's internal pressure. Based on this model, modifications were made to calculate the failure pressure of pressurized pipe elbows using full-scale burst tests or FE numerical methods [Kim et al., 2009, 2013; Wang and Zhou, 2019a, b]. However, these models mainly focused on mechanical strength of the pipe elbows in the absence or presence of a corrosion defect, ignoring the effect of electrochemical corrosion reaction, especially the M–E interaction, on degradation and failure of the elbows. Generally, the stress of a straight pipe is uniformly distributed on the pipe body under internal pressure. The bending curvature of pipe elbows generates different pressurized areas on the inner surface of intrados and extrados, thereby resulting in a nonuniform stress distribution at the elbows. This will change the electrochemical corrosion rate and local residual strength.

A new FE model was developed by Shuai et al. [2021] to determine the M–E interaction, primarily corrosion rate quantified by anodic reaction current density, at a corrosion defect on an elbow of X100 steel pipe in a near-neutral pH bicarbonate solution by integrating the local stress with electrochemical corrosion reaction. A 3D model for an elbow of an X100 steel pipe containing an external corrosion defect and the toroidal coordinate system illustrating the structure of the pipe elbow and the location of the defect are shown in Figure 3.15. Generally, long-distance transmission pipelines do not include high-angle elbows for route change. Instead, the upstream gathering pipelines often have this type of elbow. Different from a corrosion defect on a straight pipe, the location of the defect at pipe elbow, e.g., intrados ($\theta = 0°$), central line ($\theta = 90°$), and extrados ($\theta = 180°$), would experience different stress conditions and thus the M–E interaction effects. Figure 4.10 shows the burst pressure of the pipe elbow containing a corrosion defect (10, 100, and 60 mm in depth, length, and width, respectively) at the three typical locations with various bending radii. It is seen that, with increased bending radius, the failure pressure of the pipe elbow containing a defect on the intrados increases gradually. When the defect is located on the extrados surface, the failure pressure decreases with the bending radius. If the corrosion defect is at the central line, the bending radius does not obviously affect the burst pressure. Moreover, the elbow with the corrosion defect on the intrados has the failure pressure which is always lower than the failure pressure when the defect is on the extrados and the central line. Therefore, the

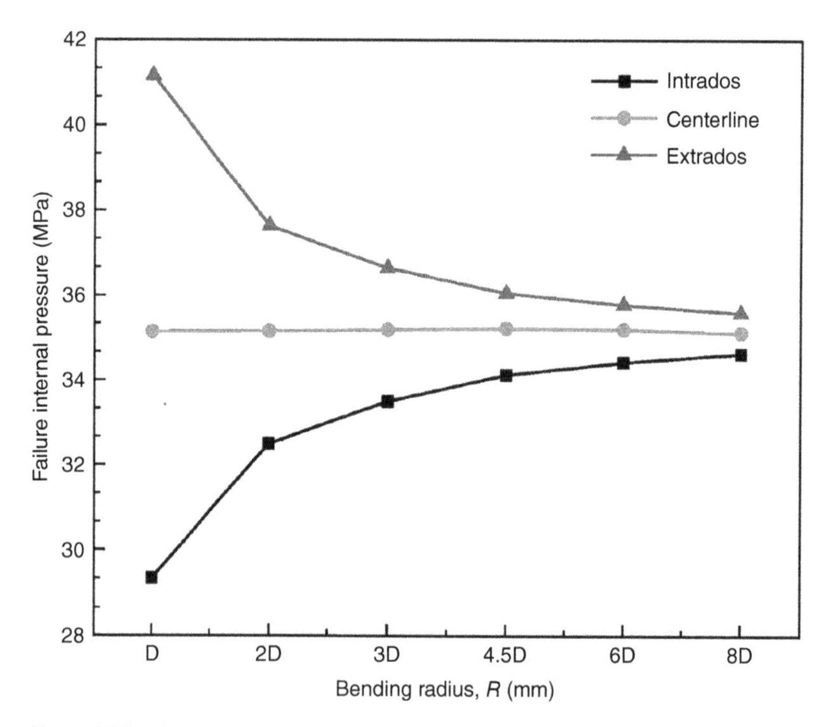

Figure 4.10 Burst pressure of the pipe elbow containing a corrosion defect (10, 100, and 60 mm in depth, length, and width, respectively) on intrados, central line, and extrados with various bending radii. *Source:* From Shuai et al. [2021] / with permission from Elsevier.

elbow burst is dominated by the corrosion defect located on the intrados of the pipe elbow.

Two important parameters defining a pipe elbow and affecting the local stress distribution are bending radius, as labeled in Figure 3.15b, and internal pressure. Figure 4.11 shows the maximum anodic current density (i.e., corrosion rate) at the corrosion defect as a function of bending radius under various internal pressures. While the influence of bending radius on the anodic current density is not obvious, the maximum anodic current density increases with the internal pressure. The obtained results are attributed to the local stress level and its effect on corrosion. It is normally expected that a change in bending radius affects local bending and the state of stress. However, when the stress concentration at the defect does not exceed the yield strength of the steel [Shuai et al., 2021], the resulting M–E interaction is not apparent. With the modeled bending radii, the corrosion defect is still in an elastic range. Thus, a change in bending radius does not affect the corrosion rate due to the negligible M–E interaction. Although the internal pressure can increase corrosion at the defect, the corrosion enhancement is not obvious when the internal pressure is lower than 10 MPa, under which the stress

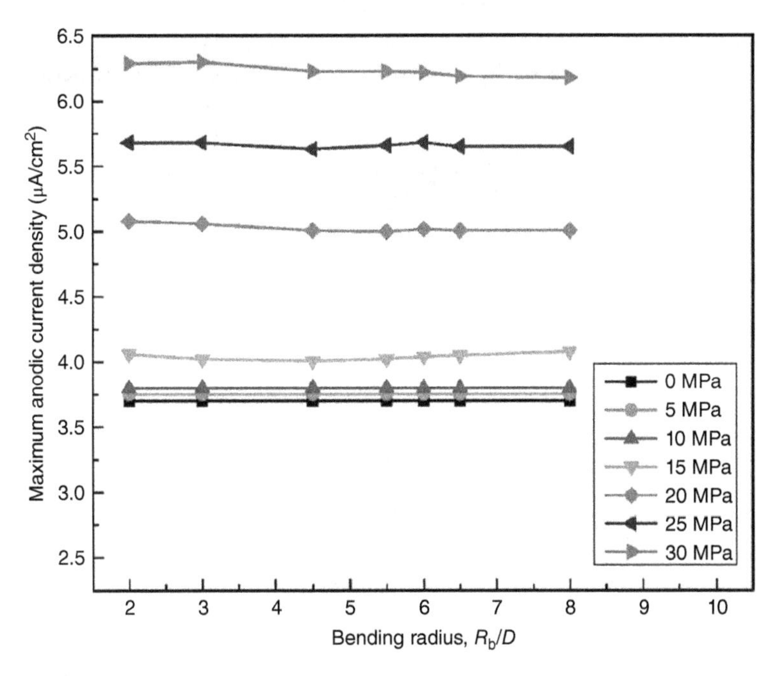

Figure 4.11 The relationship between the maximum anodic current density (i.e., corrosion rate) at a corrosion defect and the bending radius under various internal pressures. *Source:* From Shuai et al. [2021] / with permission from Elsevier.

level does not reach the yield strength of the steel. As a result, the M–E interaction is ignorable. Only when the internal pressure is over 15 MPa, the maximum von Mises stress at the defect exceeds the yield strength of the steel, making the defect area under plastic deformation. The maximum corrosion rate increases significantly with the internal pressure, showing an appreciable M–E interaction effect.

In summary, when a corrosion defect is present on the pipe elbow, the existing standards are not applicable for defect assessment. New assessment methods should be developed and used for determination of failure pressure of the elbow. In addition, when the corrosion defect is located on the intrados, the failure pressure of the elbow is always the lowest, compared with the elbows containing a corrosion defect on the extrados and the central line, due to different M–E interactions. This is quite different from straight pipes.

4.3.2 A Single Corrosion Defect with Complex Shape

As stated, most defect assessment techniques treat corrosion defect as an ellipsoid with defined depth, width, and length, where the deepest site is located at the bottom center of the defect [Adib-Ramezani et al., 2006; American Society of Mechanical Engineering, 2012; Zelmati et al., 2017; Amaya-G'omez et al., 2019]. The corrosion defect is generally approximated in the way as marked by the dashed curve in Figure 4.12a. However, actual

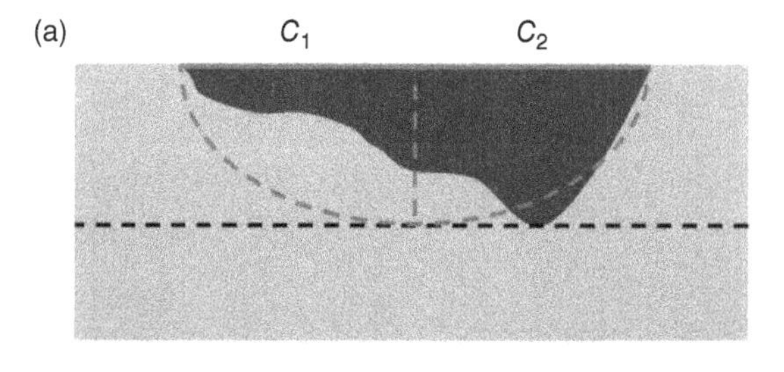

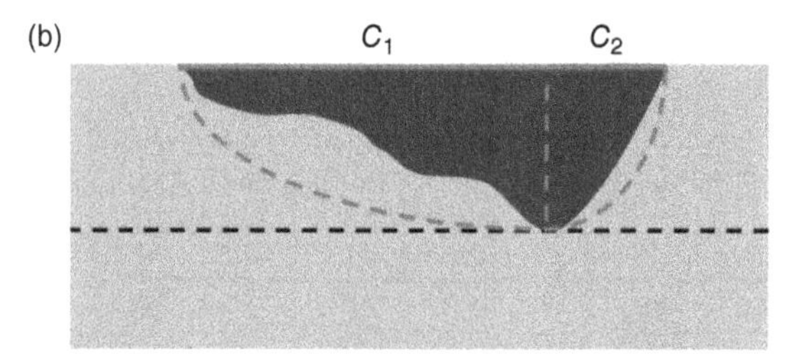

Figure 4.12 Schematic diagram of the approximation for a corrosion defect on pipelines (a) The standard ellipsoid approach. *Source:* Adapted from American Society of Mechanical Engineering [2012] and (b) the double ellipsoid approach. *Source:* From Zhang et al. [2020a] / with permission from Elsevier.

corrosion defects rarely shape an ellipsoid. Moreover, the deepest site is not necessarily at the bottom center. Many corrosion defects tend to be a double ellipsoid which is composed of two overlapped, incomplete ellipsoids, as shown in Figure 4.12b, where the deepest site is close to the ellipsoid with a shorter axial length, i.e., c_2 in the figure. Although the current ILI techniques such as MFL and UT can detect defects with an improved resolution, providing an accurate definition of the defect geometry, the assessment models and codes used today usually simplify the defect shape for modeling and computational convenience. However, accurate information on the defect geometry is critical to determination of the local stress concentration and thus the assessment of pipeline FFS.

A 3D FE-based model for an X100 steel pipe containing a double-ellipsoidal corrosion defect exposed to a near-neutral pH, diluted bicarbonate solution, i.e., the so-called NS4 solution [Cheng and Norsworthy, 2017], was developed, as shown in Figure 4.13, where $2a$ and $2c$ refer to the lengths of the secondary and the primary axes of the defect, and d is the defect depth. The double-ellipsoidal defect is composed of two

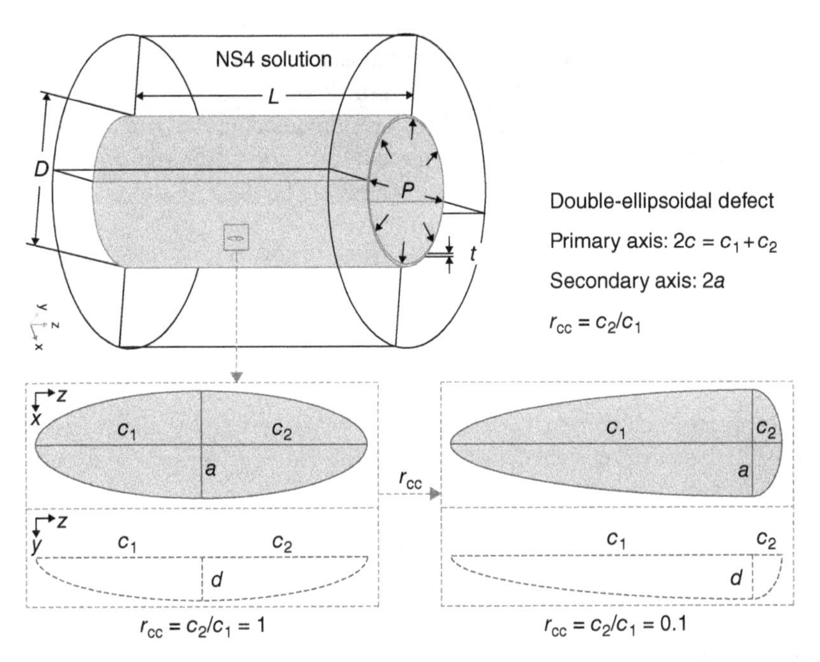

Figure 4.13 Schematic diagram of a 3D FE-based model for an X100 steel pipe containing a double-ellipsoidal corrosion defect exposed to a near-neutral pH, diluted bicarbonate solution (i.e., NS4 solution). *Source:* From Zhang et al. [2020a] / with permission from Elsevier.

overlapped semi-ellipsoidal defects, where one defect is bigger with a longer primary semi-axis (c_1) and the other one is smaller with a shorter primary semi-axis (c_2). If the deepest site of the defect is at its bottom center, the double-ellipsoidal defect is assumed to consist of two identical semi-ellipsoids (i.e., $c_1 = c_2$), where the ratio of c_2 to c_1, i.e., r_{cc}, is equal to 1. If the deepest site is not at the bottom center of the defect, r_{cc} is smaller than 1 due to $c_2 < c_1$. Furthermore, when the primary axis ($2c$) of the corrosion defect is parallel to the axial direction of the pipe, the inclination angle (θ_{incl}) of the defect is defined as 0°. When they become perpendicular to each other, the angle θ_{incl} is 90°.

4.3.2.1 At the Inclination Angle of 90°

The distribution of von Mises stress at a corrosion defect of $\theta_{incl} = 90°$ and various r_{cc} ratios on the pipe under an internal pressure of 15 MPa is shown in Figure 4.14a. It is seen that, when r_{cc} is equal to 1, the stress distribution is symmetric relative to both the x- and z-axes, and a maximum stress is recorded at the center of the defect. The results reflect the symmetry of stress relative to the defect center. With the r_{cc} ratio changing from 1 to 0.1, the stress symmetry is kept in the z-axis direction only. The maximum stress is always located at the secondary axis (i.e., the deepest site of the defect), and moves to the smaller semi-ellipsoid defect which has a shorter axial length, i.e., the c_2 axis. Figure 4.14b shows the distribution of von Mises stress along the bottom of the corrosion defect in the primary axial direction. With the r_{cc} changing from 1 to 0.1, the stress at the defect center decreases slightly, but the maximum stress at the deepest site increases obviously. For example, when the r_{cc} ratios are 1, 0.7, 0.4, and 0.1, the maximum stresses are 587.13, 590.80, 612.03, and 808.48 MPa, respectively. When r_{cc} is equal to 0.1, the maximum stress exceeds the yield strength of the steel (i.e., 806 MPa).

Figure 4.15 shows the distribution of anodic current density (i.e., corrosion rate) along the corrosion defect at $\theta_{incl} = 90°$ and various r_{cc} ratios on the steel pipe under an internal pressure of 15 MPa. As the r_{cc} ratio changes from 1 to 0.1, the anodic current density at the defect center decreases slightly, but the maximum anodic current density at the deepest site of the defect increases from 3.84 to 4.19 $\mu A/cm^2$. The result shows that corrosion is accelerated with the decreased c_2 axis length, causing the maximum stress and corrosion rate at the deepest site of the defect.

4.3.2.2 At the Inclination Angle of 0°

Similarly, the distribution of von Mises stress of a corrosion defect at $\theta_{incl} = 0°$ and various r_{cc} ratios and the nonlinear distribution of stress along the defect

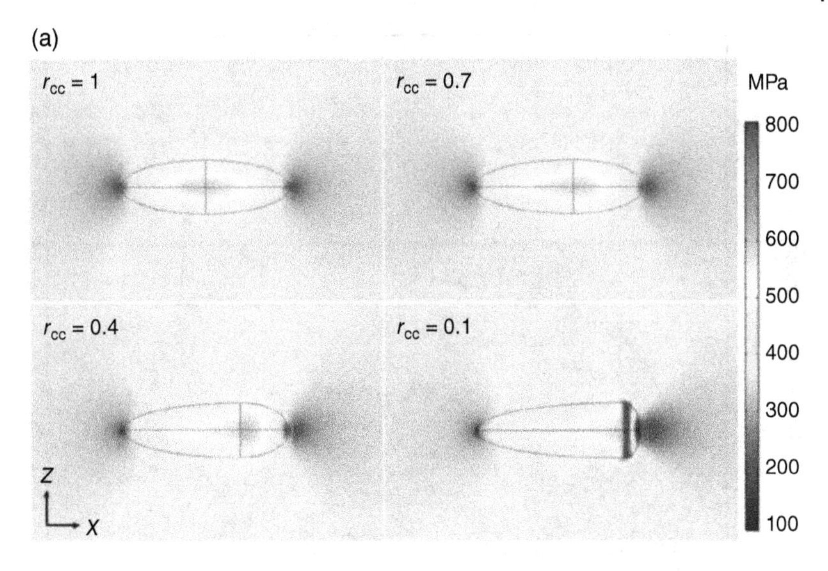

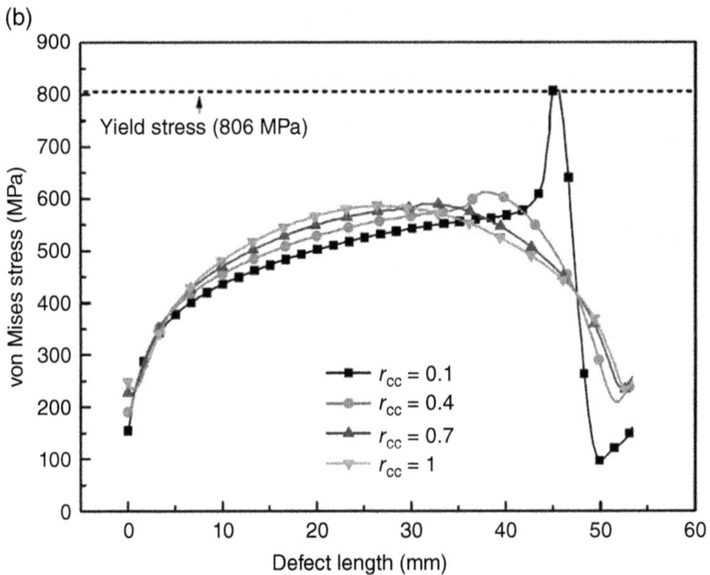

Figure 4.14 (a) Distribution of von Mises stress of a corrosion defect at θ_{incl} of 90° and various r_{cc} ratios on an X100 steel pipe under an internal pressure of 15 MPa. (b) Distribution of von Mises stress along the bottom of the corrosion defect in the primary axial direction. *Source:* From Zhang et al. [2020a] / with permission from Elsevier.

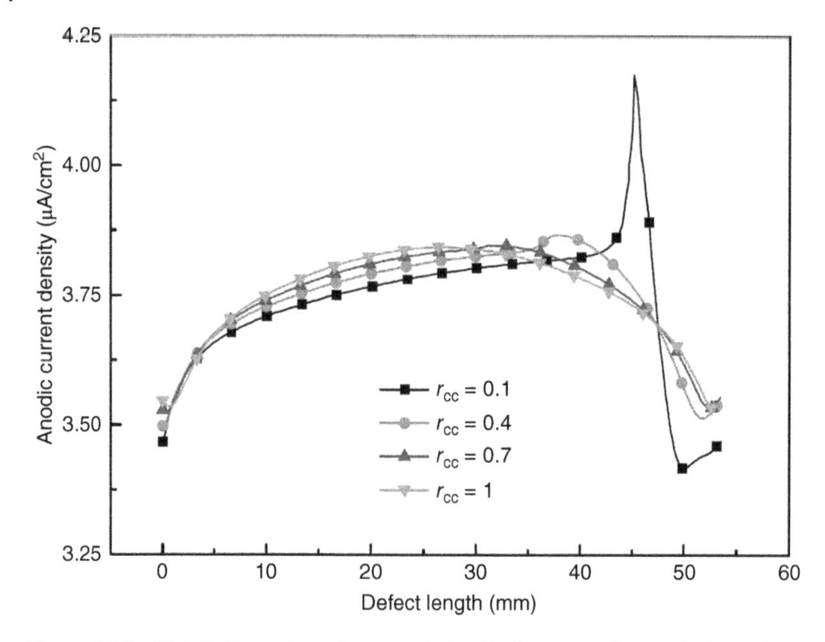

Figure 4.15 Distribution of anodic current density (i.e., corrosion rate) along the corrosion defect at θ_{incl} of 90° with various r_{cc} ratios on the steel pipe. *Source:* From Zhang et al. [2020a] / with permission from Elsevier.

are shown in Figure 4.16. Identical to the results in Figure 4.15, when r_{cc} is equal to 1, the stress distribution is symmetric relative to the x- and z-axes. When r_{cc} decreases from 1 to 0.1, the symmetry of the stress distribution is maintained in the z-axis direction only. However, as r_{cc} decreases, the maximum stress increases very slightly from 826.58 to 827.89 MPa. As a comparison, the stress at the deepest point of the defect decreases from 826.58 to 804.42 MPa. As compared with the results obtained at the inclination angle of 90°, it is concluded that the inclination of the corrosion defect remarkably affects the local stress condition.

For the corrosion defect at $\theta_{incl} = 0°$, the greatest anodic current density is observed at the center of the defect, as shown in Figure 4.17. With the r_{cc} decreasing from 1 to 0.1, the maximum anodic current density increases slightly from 5.13 to 5.22 $\mu A/cm^2$, but the anodic current density at the deepest site decreases rapidly from 4.84 to 3.95 $\mu A/cm^2$. The results are opposite to those obtained at the defect with $\theta_{incl} = 90°$. The change in anodic current density is consistent with the stress level as the inclination angle of the defect changes on the pipe.

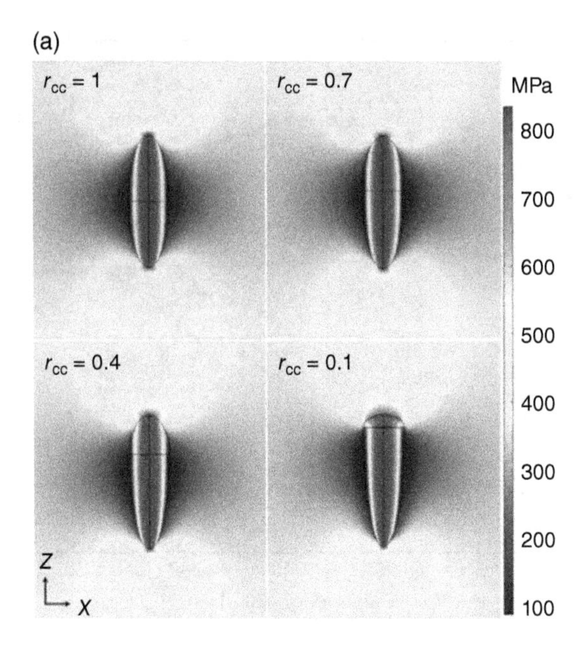

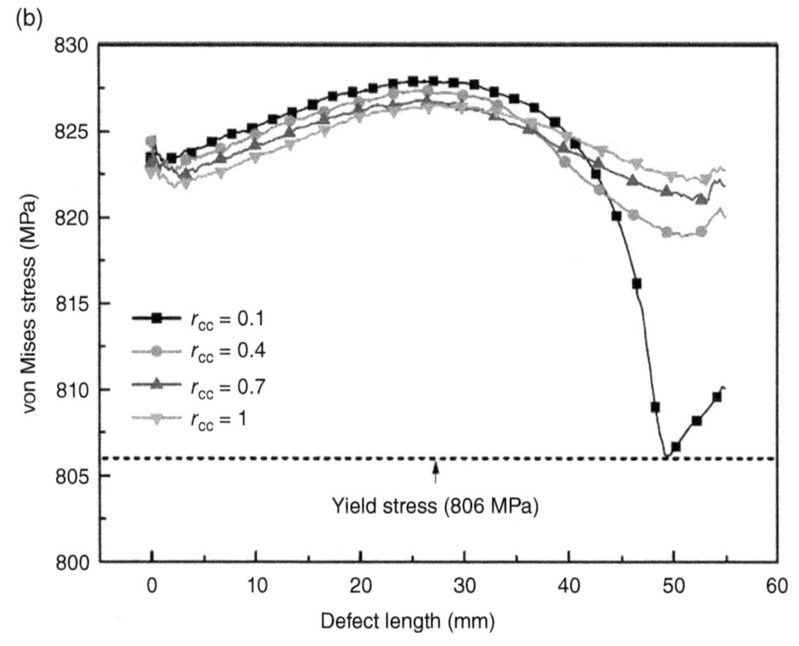

Figure 4.16 (a) Distribution of von Mises stress of a corrosion defect at θ_{incl} of 0° and various r_{cc} ratios on the pipe under an internal pressure of 15 MPa. (b) Distribution of von Mises stress along the bottom of the defect in the primary axial direction. *Source:* From Zhang et al. [2020a] / with permission from Elsevier.

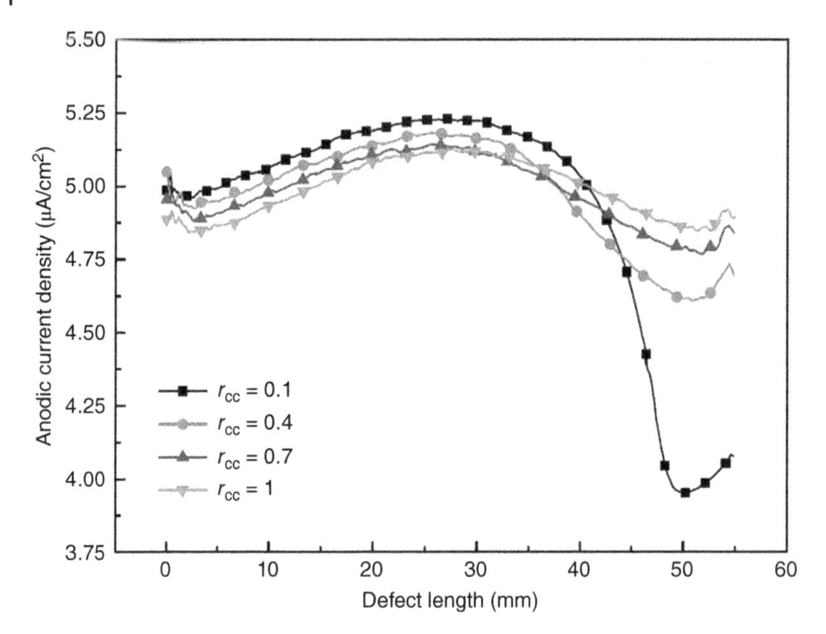

Figure 4.17 Distribution of anodic current density (i.e., corrosion rate) along the bottom of a corrosion defect at θ_{incl} of 0° with various r_{cc} ratios on the steel pipe. *Source:* From Zhang et al. [2020a] / with permission from Elsevier.

4.3.2.3 Effect of the Corrosion Defect Geometry on Assessment Accuracy

The modeling results obtained on the double-ellipsoidal corrosion defect, which can better simulate the shape of actual corrosion defects on a pipeline than a regular ellipsoidal defect, show that both the stress and anodic current density (i.e., corrosion rate) distributions at the defect depend heavily on its geometry. Apparently, an accurate definition of corrosion defect in terms of its geometrical shape and dimension is essential at the reliability of pipeline FFS determination. Particularly, the r_{cc} ratio, i.e., the ratio of axial lengths of the two overlapped ellipsoidal defects, is a critical parameter to define the defect geometry. Another important parameter is the inclination angle of the defect on pipelines. Generally, the corrosion defect with a small r_{cc} ratio such as 0.1 is more likely to fail owing to an elevated local stress concentration at the defect bottom. When the local stress exceeds the yield stress or even ultimate tensile stress, cracks will initiate at the defect base. Moreover, the greatest anodic current density is obtained at defects with a small r_{cc} ratio. The local M–E interaction mutually “catalyzes” the stress concentration and corrosion reaction at the defect, which will accelerate pipeline failure.

Furthermore, the inclination of corrosion defect relative to the axial direction of pipelines is critical to accurate defect assessment. When the primary axis (i.e., $c_1 + c_2$ in Figure 4.13) of the corrosion defect is parallel to the axial direction of the pipe, i.e., the inclination angle is 0°, a reduction in r_{cc} ratio does not affect obviously the stress concentration and anodic current density. However, with the increase of the inclination angle up to 90°, the reduced r_{cc} ratio causes a remarkable increase in local stress concentration and anodic current density, especially at the deepest site of the defect. This is expected to result in pipeline failure preferentially occurring at the defect bottom due to the local stress exceeding yield strength of the steel or a rapid pipeline leaking due to accelerated corrosion growth. Thus, an increase in inclination angle of the corrosion defect increases the threat to pipeline integrity. The exact orientation of a corrosion defect on pipelines must be defined for defect assessment. Presently, ILI tools, especially UT, are developed with an improved resolution to provide accurate shape and geometrical information on corrosion defects, as well as their orientation on the pipelines. The developed model will base this information on accurate defect assessment.

Finally, for a corrosion defect with complex geometrical parameters, the center of the defect is not always the deepest site of the defect. Ignorance of the realistic geometry will result in a mis-assessment of the corrosion defect in terms of both the maximum stress concentration and the greatest anodic current density. Thus, the available defect assessment methods where the corrosion defect is treated as a single ellipsoidal shape should be improved by defining a more accurate defect geometry such as double ellipsoid for numerical modeling.

4.3.3 Corrosion Defect Growth on Pipelines Under the M–E Interaction

Corrosion progression is time dependent. As a result, corrosion-induced pipeline degradation is a time-dependent process [Miran et al., 2016]. Generally, defect assessment focuses on determination of pipeline FFS and failure pressure at given instants. The available methods for corrosion growth prediction are mainly based on historical ILI data and further modeling and calculations [Vanaei et al., 2017]. Some commonly used methods include inverse Gaussian processes [Zhou et al., 2017], Gamma process [Wang et al., 2020], Poisson square wave process [Bazan and Beck, 2013], and Brownian motion with drift model [Marc, 2013]. A major limitation of these methods is that it is difficult, if not impossible, to use statistical ILI data to obtain accurate corrosion defect growth rate on pipelines while ignoring the mechanistic aspects such as the M–E interaction occurring at the corrosion defect. Moreover, the reliability of the data-based methods is highly dependent on the availability and quality of ILI data.

As stated, defect assessment including Level III methods generally focuses on pipeline FFS determination at a specific moment under given conditions. They are not capable of providing time-dependent corrosion defect growth on the pipelines. In 2017, Xu and Cheng [2017] developed a 2D FE-based multi-physical field coupling model to predict corrosion defect growth rate under the synergism of stress and electrochemical corrosion reactions, while considering the effects of internal pressure, corrosive environments, defect geometry, and steel grade. This was probably the first attempt to derive the defect growth rate as a function of time from the mechanistic aspect. However, the work suffered from a limitation in that the model considered the corrosion defect growth at circumferential (i.e., the defect width) and radial (i.e., the defect depth) directions only, while ignoring the defect growth along the longitudinal direction (i.e., the defect length). Moreover, the model was not capable of predicting failure pressure of the corroded pipeline as a function of service time. In a recent work, Cheng group [Qin et al., 2021] furthered the effort to develop a 3D FE model, enabling prediction of corrosion defect growth at circumferential, radial, and longitudinal directions and, simultaneously, determination of the failure pressure of corroded pipeline as a function time. It is expected that, upon validation by substantial data collected from the field and laboratories, the developed method can provide accurate and reliable prediction of long-term performance of pipelines containing corrosion defect and their remaining service life.

4.3.3.1 The Model and Modeling Process

An FE-based 3D multi-physics field coupling model was developed to simulate and define the corrosion defect growth on an X100 steel pipe in a simulated near-neutral pH NS4 solution using a COMSOL Multiphysics 5.4 software package. The physical model and a half model are shown in Figure 4.18a,b. The pipe segment had an outer diameter of 812.8 mm and a wall thickness of 19.1 mm. According to Saint Venart's principle, the length of the modeled pipe (i.e., 2000 mm) was sufficient to avoid the influence of boundary conditions on corrosion defect on the outer surface of the pipe segment [Xu and Cheng, 2012a, b]. The geometric characteristics of the corrosion defect are shown in Figure 4.18c,d, where the defect length, width, and maximum depth were modeled in the longitudinal, circumferential, and radial directions of the pipe, respectively.

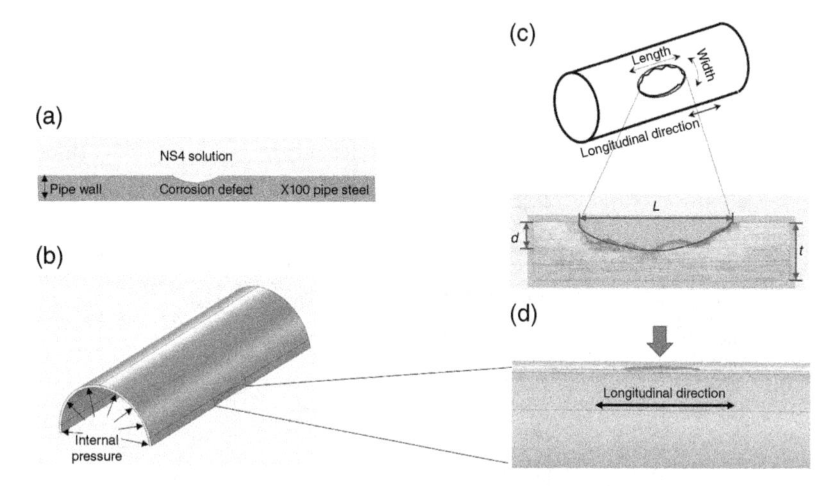

Figure 4.18 Schematic diagrams of (a) a physical model showing a corrosion defect on the outer surface of an X100 steel pipe exposed to a simulated near-neutral pH NS4 solution, (b) a half model of the corroded pipe, (c) geometry of the corrosion defect on the pipe segment, and (d) a cross-sectional view of the corrosion defect on the pipe. *Source:* From Qin et al. [2021] / with permission from Elsevier.

A flowchart showing the modeling process for 3D corrosion defect growth on the steel pipe and the failure pressure prediction is shown in Figure 4.19. Three steps are included. Step 1: Define geometries of the pipe and corrosion defect and determine corrosion reactions occurring in the solution and the stress condition of the pipe. Step 2: Develop an FE-based multi-physics field coupling model integrating the mechanical stress field, electrochemical corrosion field, and the M–E interaction. The corrosion growth rate of the defect is expressed as the anodic reaction current density. Step 3: Use the widely accepted stress-based criterion as the pipeline failure criterion, where the ultimate tensile strength of the steel is selected as the failure stress [Ahammed and Melchers, 1997]. According to ASME B31G [American Society of Mechanical Engineering, 2012], the maximum defect depth is used for defect assessment to predict the failure pressure of the corroded pipeline as a function of time.

4.3.3.2 Corrosion Defect Growth and Failure Pressure Prediction Under Various Internal Pressures

Modeling was conducted on the corroded pipe under a fixed axial stress of 375 MPa and various internal pressures, and the time dependences of the

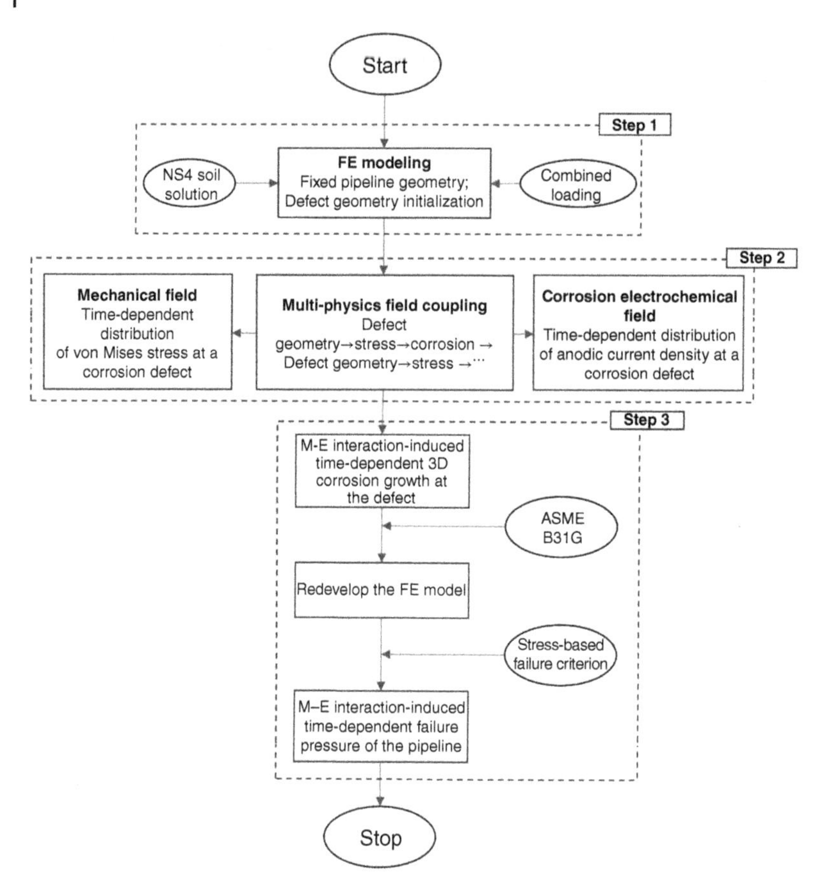

Figure 4.19 Flowchart for modeling of corrosion defect growth on an X100 steel pipe and prediction of the time-dependent failure pressure of the corroded pipeline. *Source:* From Qin et al. [2021] / with permission from Elsevier.

defect length, maximum depth and width are shown in Figure 4.20. It is seen that both the defect length and the maximum depth increase with internal pressure, while the defect width changes slightly. For example, the defect length increases from 8.0 to 8.9 mm (i.e., an increase by 11%) after 10 years at the internal pressure of 18 MPa, while the defect length increases by 16% after 10 years when the internal pressure is 26 MPa. Similarly, the maximum defect depth increases by 27% and 34% after 10 years at the internal pressures of 18 and 26 MPa, respectively. However, the corrosion width increase maintains at about 22% when the internal pressure increases from 18 to 26 MPa. The M–E interaction contributes to increased defect length, maximum depth, and width with time.

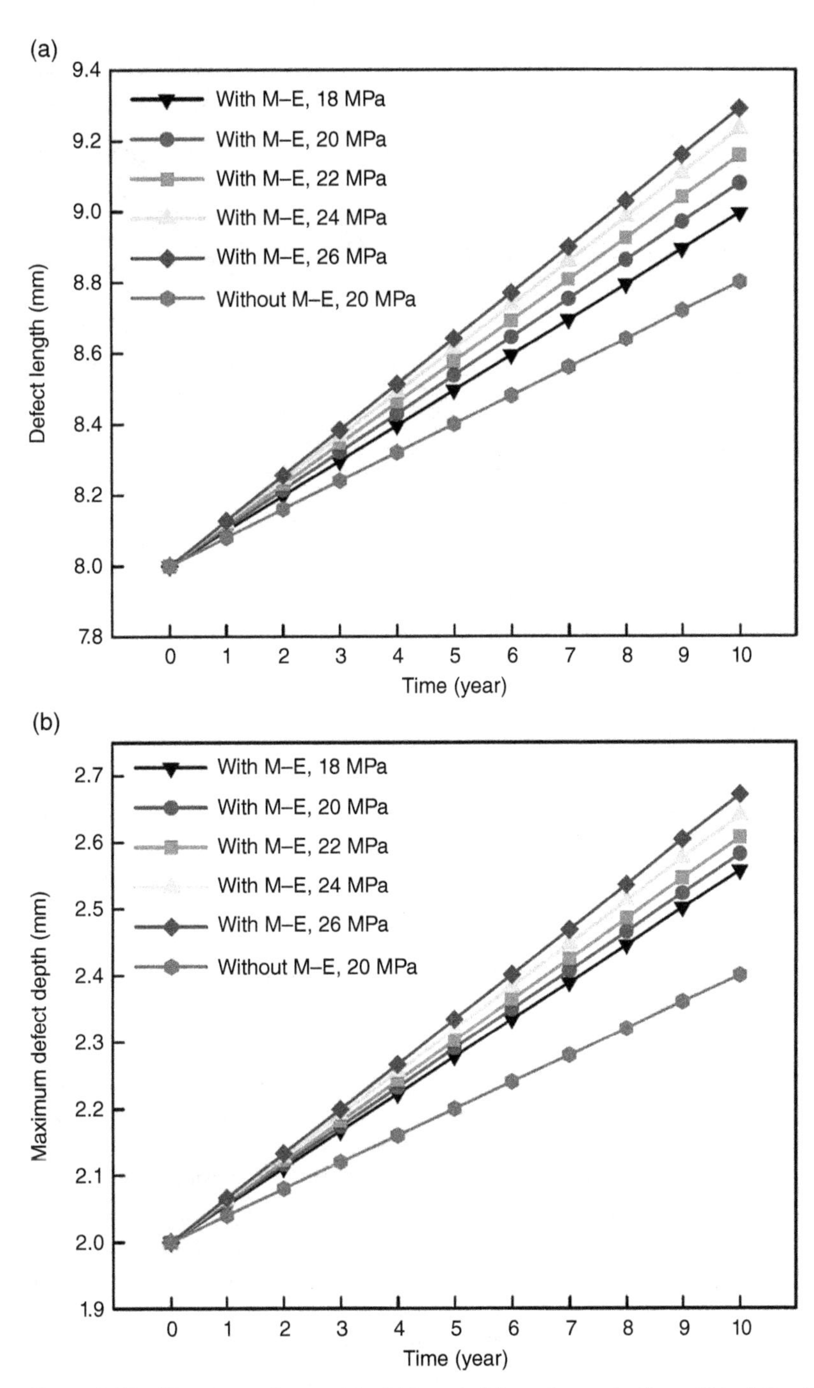

Figure 4.20 Time dependence of the (a) length, (b) maximum depth, and (c) width of a corrosion defect on X100 steel pipe under a fixed axial stress of 375 MPa and various internal pressures in the absence and presence of the M–E interaction. *Source:* From Qin et al. [2021] / with permission from Elsevier.

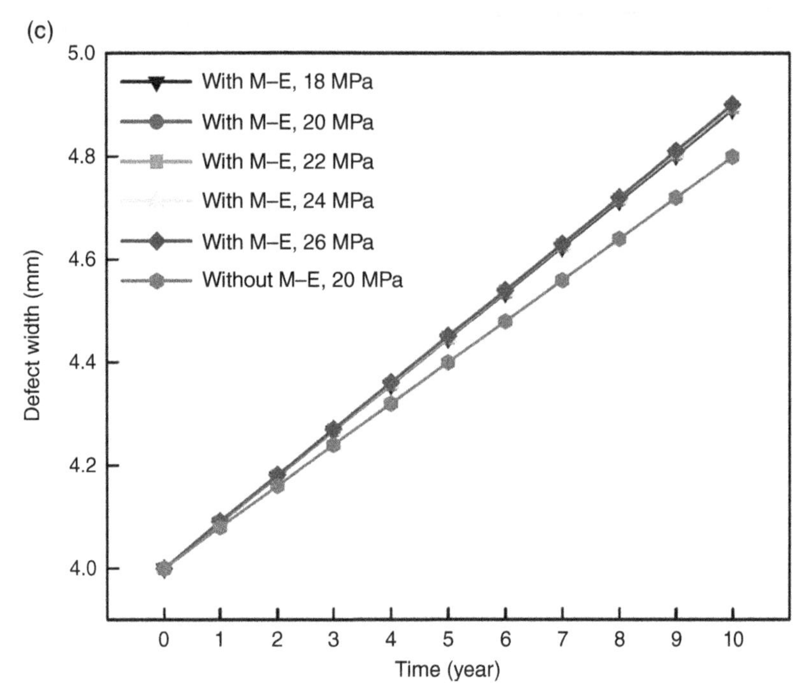

Figure 4.20 (Continued)

The distributions of von Mises stress and anodic current density (i.e., corrosion rate) at the corrosion defect initially and after 10 years under a fixed axial tensile stress of 375 MPa and various internal pressures are shown in Figure 4.21. If the axial tensile stress is absent, an internal pressure of 18 MPa does not cause an apparent stress concentration at the defect. However, a combined axial tensile stress of 375 and 18 MPa of internal pressure causes a plastic stress at the defect initially. After 10 years, the plastic area expands. With the increase in internal pressure, the stress level at the defect further increases. Similarly, the anodic current density at the corrosion defect also increases with the internal pressure. Thus, an increased internal pressure promotes corrosion defect growth on pipelines.

The failure pressures of the steel pipe containing a corrosion defect under different internal pressures and a fixed axial tensile stress of 375 MPa are shown in Table 4.1. The failure pressure decreases as the internal pressure increases after 10 years of service. This is attributed to the increased growth rate of the corrosion defect under high internal pressures. However, the effect of internal pressure on failure pressure of the corroded pipe (with a fixed axial tensile stress) is not as significant as the effect of varied axial

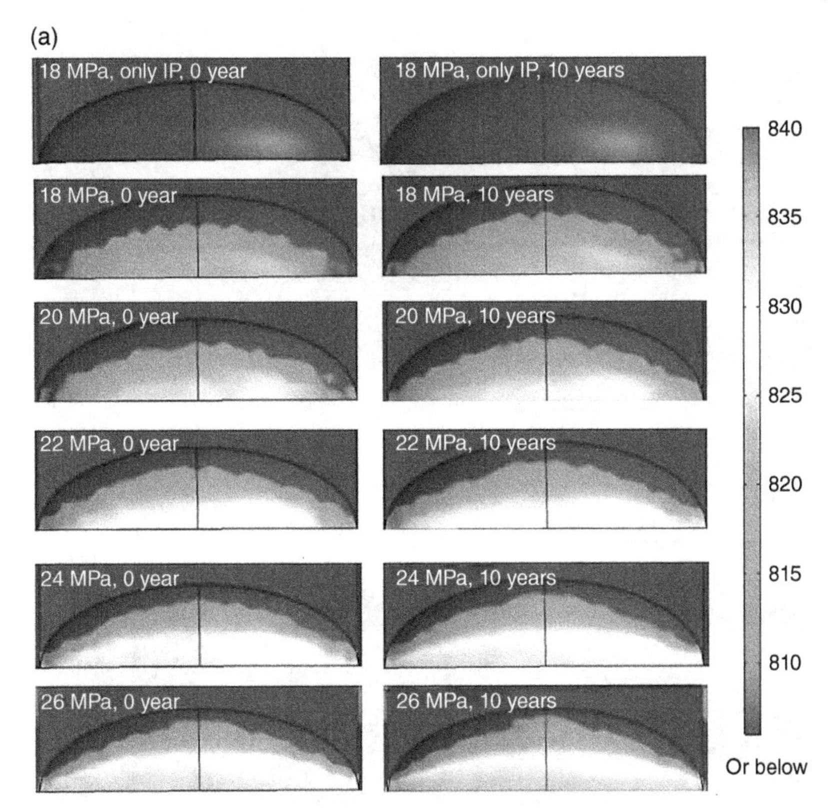

Figure 4.21 Distributions of (a) von Mises stress (MPa) and (b) anodic current density (μA/cm^2) at the corrosion defect on an X100 steel pipe initially and after 10 years under different internal pressures and a fixed axial tensile stress of 375 MPa. *Source:* From Qin et al. [2021] / with permission from Elsevier.

tensile stress on the failure pressure [Qin et al., 2021]. It is thus implied that a slight change in internal pressure within several MPa, as modeled in this work, does not affect the failure pressure of the pipe at an apparent level. The main reason is that a change in internal pressure does not change the defect size remarkably, as shown in Figure 4.21.

4.3.3.3 Implications on Long-Term Performance of Corroded Pipelines

The developed model predicts that the growth rate of the corrosion defect follows the order by maximum defect depth > length > width. This is mainly attributed to two factors, i.e., a stress concentration generated at the defect bottom and a unique corrosive environment inside the defect. Generally, the

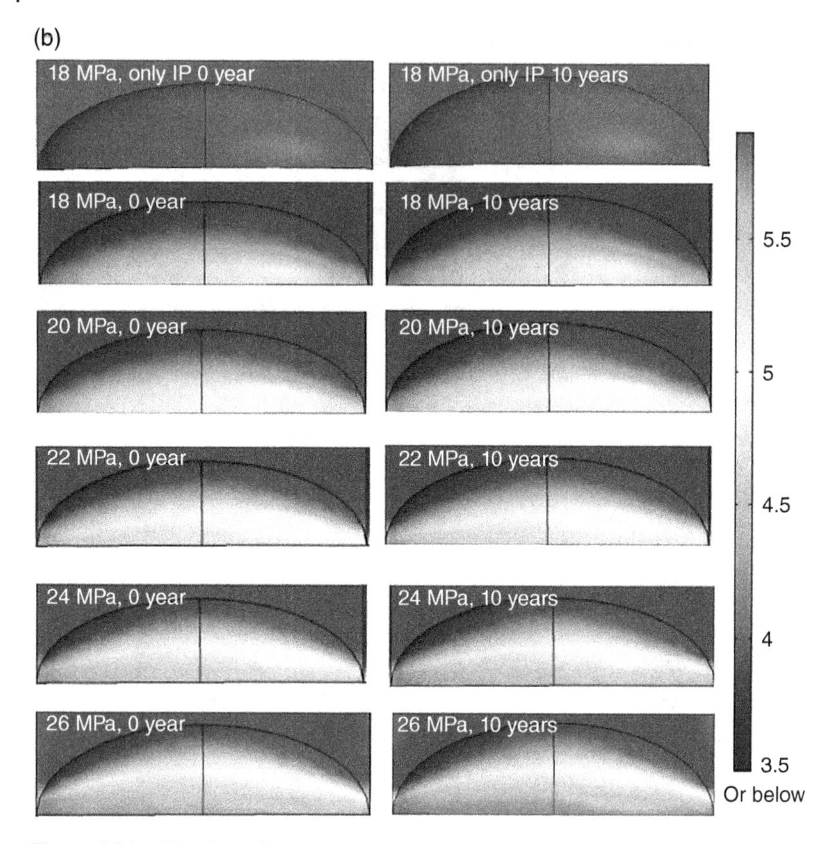

Figure 4.21 (Continued)

Table 4.1 Failure pressures of X100 steel pipe containing a corrosion defect (initially 8 mm in length, 2 mm in maximum depth, and 4 mm in width) initially and after 10 years under various internal pressures and a fixed axial tensile stress of 375 MPa.

Internal pressure	**18 MPa**	**20 MPa**	**22 MPa**	**24 MPa**	**26 MPa**
Initially	29.25 MPa	29.25 MPa	29.25 MPa	29.25 MPa	29.25 MPa
10 years	29.11 MPa	29.09 MPa	29.05 MPa	29.01 MPa	28.95 MPa

Source: From Qin et al. [2021] / with permission from Elsevier.

sharper the defect bottom, the greater the local stress concentration. Moreover, the environmental chemistry in a defect is usually more aggressive (e.g., a low solution pH and a high chloride concentration) than the corrosive environment outside the defect. Both factors contribute to an increased corrosion defect growth rate along the depth direction (i.e., the radial direction of the pipe). Thus, for pipeline leaking accidents caused by corrosion, an accelerated corrosion resulting in pipeline perforation is usually the key reason. The order of the corrosion defect growth in different directions provides a recommendation to the industry to rank the defect depth analysis as a top priority in the defect assessment program.

With the increase in stress, the maximum defect depth and length increase apparently, but the defect width changes slightly. Thus, the stress-enhanced corrosion defect growth mainly along with the depth and length directions. For defect assessment on pipelines, the depth and length dimensions of the corrosion defect should be focused, while its width is not as important as the other two dimensions.

An important factor affecting the corrosion defect growth on pipelines is the M–E interaction, which can accelerate 3D growth of the corrosion defect. As stated, the M–E interaction results from a synergism of mechanical force (i.e., stress) and electrochemical force (i.e., corrosion reactions), which coexist at the corrosion defect on pipelines. The interaction can accelerate local corrosion and facilitate the corrosion defect growth. Thus, the corrosion defect assessment program cannot ignore the M–E interaction. Ignorance of the interaction will cause an underestimation of the defect growth rate and the failure pressure of corroded pipelines as a function of service time.

It is noted that the effect of corrosion products on defect growth is not considered in the model. Generally, corrosion products can affect further corrosion of the steels by reducing the corrosion rate, especially when the corrosion products deposit on the steel surface [Waseda and Suzuki, 2006]. Further work should be conducted in two aspects. One is to characterize the structure (e.g., porosity), thickness, and protectiveness of corrosion products generated under service environmental conditions. The other is to modify the developed model by including the protective effect of corrosion products in the defect assessment program. It is expected that this will improve the accuracy of corrosion defect assessment and the failure pressure prediction on corroded pipelines.

4.3.4 The M–E Interaction at a Corrosion Defect on Pipelines in Suspension and the Failure Pressure Prediction

Buried pipelines can become suspended over soil supports in geologically unstable regions such as mountains and riversides as environments, weather, and human factors, such as heavy rainfalls and mining subsidence, can cause the soil to erode and collapse [Moncada and Aguirre-Pe, 1999; Wang et al., 2008]. Depending on soil erosion scenarios, the pipelines can be either completely suspended (i.e., a free span) when surrounding soil is totally lost, or partially suspended (i.e., a hidden suspension) when the supporting soil under the pipelines is eroded. Generally, the hidden suspension effect is more difficult to identify compared with the complete suspension. Although the soil under the pipelines is gone, the covering soil provides an environment for corrosion to occur on the top surface of the pipelines when the external coating is degraded, and the CP is not properly functional. As a result, corrosion occurs on the suspended pipelines under an M–E interaction. To date, none of available models is able to assess corrosion defect on pipelines in hidden suspension. For the purpose, Cheng group developed an FE-based, multi-physics field coupling model to determine the M–E interaction at an external corrosion defect on an X100 steel pipe under the hidden suspension and predict the failure pressure of the pipeline [Qin and Cheng, 2021b].

4.3.4.1 The Model and Modeling Process

A COMSOL Multiphysics 5.4 software was used to model by FE analysis the M–E interaction at a corrosion defect on a hidden-suspended X100 steel pipe. Figure 4.22 shows the physical, mechanical, and FE models of a steel pipe segment in hidden suspension. The pipe is subject to a gravity load applied by covering soil (Figure 4.22a), which generates a uniform load on the pipe segment, as labeled in Figure 4.22b. A corrosion defect on the external surface of the pipe is exposed to a near-neutral pH NS4 solution. The internal pressure is assumed static, although it is realized that pressure fluctuations can occur on pipelines and reduce the failure pressure. Under simultaneous stress and corrosion conditions, a local M–E interaction occurs at the corrosion defect. As the pipe was axially symmetric, a half of the pipe along the axial direction is selected for modeling to save computational time (Figure 4.22c). For the pipe segment, an axially symmetric constraint (i.e., Z-axis) is adopted at the suspended end of the pipe (i.e., section 1 in Figure 4.22). Rotation of the pipe end (i.e., section 2 in Figure 4.22) in three primary directions (i.e., X-, Y-, and Z-axis) is constrained. A longitudinal displacement (i.e., Z-axis) of section 2 is also constrained. The other pipe segment is set free. For soil, the top surface is set free, and other surfaces are completely constrained. The corrosion defect is elliptical in shape, as shown in Figure 4.22d. For numerical modeling, the mesh type is tetrahedral. The pipe body is in coarse meshing, and the mesh at the corrosion defect is refined. A mesh sensitivity analysis shows that, when the number of

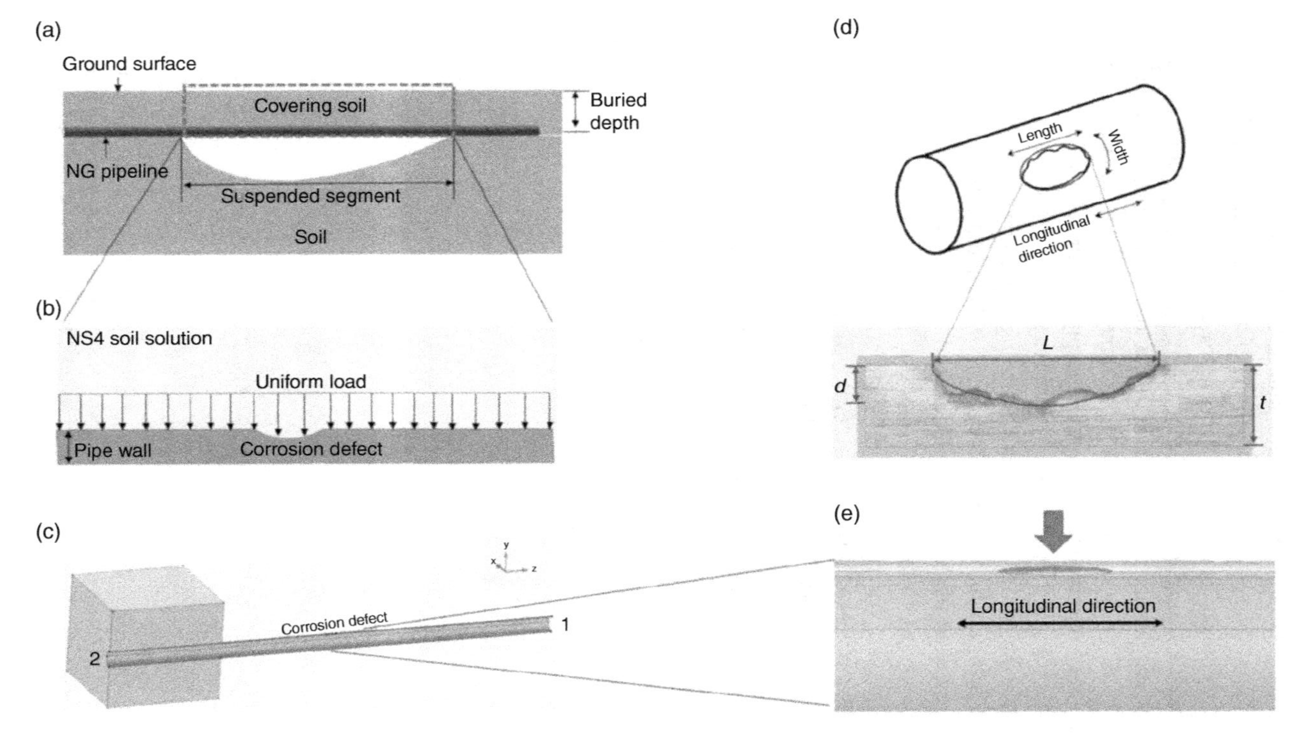

Figure 4.22 Schematic diagram showing (a) the physical model, (b) the mechanical model, and (c) the FE model of an X100 steel pipe segment in hidden suspension, as well as (d) the geometry of a corrosion defect on the pipe surface, and (e) an FE model of the corrosion defect, where section 1 in (c) represents the suspended end of the pipe and section 2 refers to the buried end. *Source:* From Qin and Cheng [2021b] / with permission from Elsevier.

elements exceeds 200,000, the maximum von Mises stress and anodic current density (i.e., corrosion rate) are approximately independent of the number of elements.

An elastic–plastic solid mechanics simulation is carried out on the steel for stress modeling. The criterion for pipeline failure determination is based on whether the equivalent von Mises stress at the corrosion defect exceeds the ultimate tensile strength of the steel. A suspended pipe segment is usually not in a stress balance condition due to the loss of supporting soil below the pipe. Thus, the gravities of the pipe segment and the carried medium cannot be ignored in modeling. A suspended pipe segment is subject to multiple stresses resulting from internal pressure, gravities of the pipe segment, and the carried medium, and a uniform loading applied by the covering soil. The steel corrosion in the NS4 solution (pH 6.8) includes the anodic and cathodic reactions of the iron oxidation and hydrogen evolution, respectively. All relevant electrochemical corrosion parameters are contained in previous chapter. The coupling of mechanical stress and electrochemical corrosion is conducted through the developed multi-physics field coupling model based on the M–E interaction principle [Wang et al., 2021].

4.3.4.2 Modeling of von Mises Stress and Anodic Current Density at a Corrosion Defect on a Suspended Pipe

The distributions of displacement of a suspended X100 steel pipe (40 m in length and 1 m in burial depth) and the von Mises stress at a corrosion defect located at the middle of the pipe segment are shown in Figure 4.23. Due to

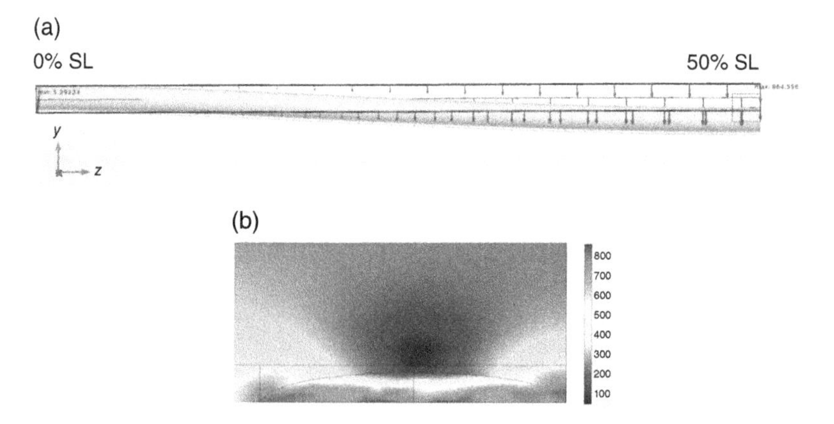

Figure 4.23 Distributions of (a) displacement of a suspended X100 steel pipe (40 m in length and 1 m in burial depth) and (b) von Mises stress at a corrosion defect located at the middle of the pipe segment. *Source:* From Qin and Cheng [2021b] / with permission from Elsevier.

the axial symmetry, a half of the pipe along the axial direction is selected for modeling to save computational time. As expected, the largest displacement occurs at the center of the suspended pipe (i.e., 50% of suspension length, SL), with a direction downward as indicated by the red arrow in the figure. It is attributed to the gravities of both the covering soil and the pipe segment that is not supported by soil underneath. A stress concentration occurs at the corrosion defect owing to the locally reduced wall thickness. The maximum von Mises stress is located at the bottom of the corrosion defect, where the stress level in the length direction is greater than the stress in the width direction of the defect.

Figure 4.24 shows the distributions of von Mises stress and anodic current density at the corrosion defect with different locations on the suspended pipe segment. When the corrosion defect is located at the end of the pipe (i.e., 0% SL), a stress concentration occurs in both the defect length and width directions. However, when the corrosion defect is away from the pipe end, the stress concentration occurs along the defect length direction only. When the corrosion defect is located at the pipe end, it is subject to tension.

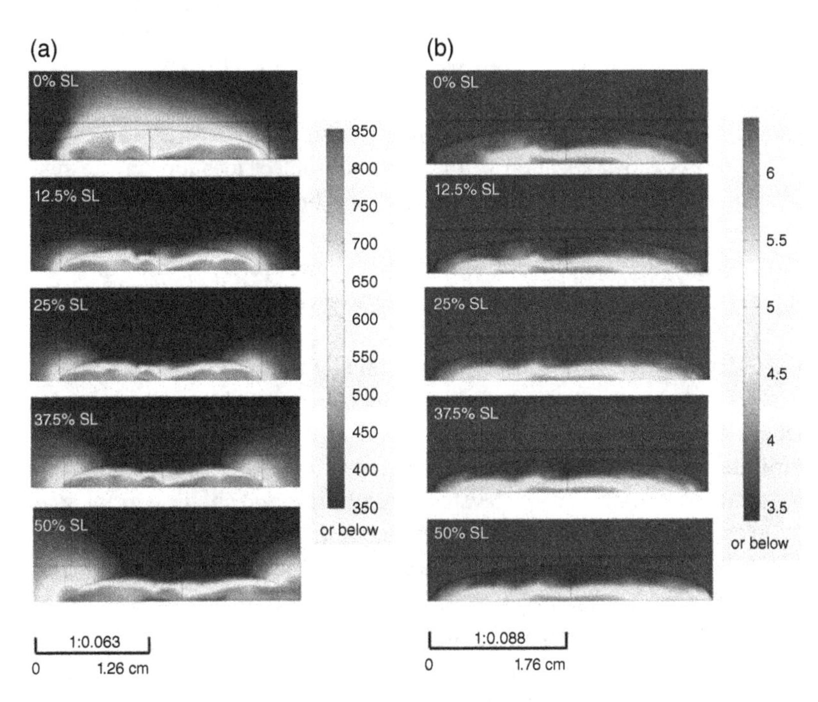

Figure 4.24 Distributions of (a) von Mises stress (MPa) and (b) anodic current density ($\mu A/cm^2$) at a corrosion defect at different locations on the suspended pipe segment. *Source:* From Qin and Cheng [2021b] / with permission from Elsevier.

Table 4.2 Failure pressure of the suspended X100 steel pipe containing a corrosion defect with different locations on the pipe.

Location of corrosion defect on the pipe (% of the SL)	0	12.5	25	37.5	50
Failure pressure (MPa)	24.5	24.1	24.2	24.6	24.3

Source: From Qin and Cheng [2021b] / with permission from Elsevier.

However, when the defect is located at other locations of the suspended pipe, it is under compression due to the suspension-induced bending effect on the pipe. Obviously, the von Mises stress distribution at the corrosion defect located at 0% SL is different from the stress when the defect is located at other locations on the pipe. However, the maximum stress level does not change obviously when the corrosion defect is located differently. Furthermore, the maximum anodic current density is distributed along the defect length direction. There is little effect of the defect location on the anodic current density, i.e., corrosion rate. The failure pressure of the suspended X100 steel pipe containing the corrosion defect located differently on the pipe is calculated, and the results are shown in Table 4.2. It is seen that the location of the corrosion defect does not affect failure pressure of the suspended pipe at an appreciable level. Specifically, the difference in failure pressures is smaller than 0.5 MPa from the modeling results.

In addition to the location of the corrosion defect location on the suspended pipe and the suspension length, the burial depth of the pipe is another important factor affecting the stress distribution, M–E interaction at the corrosion defect, and thus, the failure pressure of the pipe. Figure 4.25 shows the maximum von Mises stress and anodic current density at the corrosion defect on suspended X100 steel pipe segment with different burial depths. When the burial depth is smaller than 2.5 m, the maximum von Mises stress and anodic current density increase slightly with the burial depth. Above 2.5 m, both values increase rapidly. This phenomenon is attributed to the loading effect from covering soil, which becomes significant at a critical depth of 2.5 m to elevate the local stress and anodic current density at the corrosion defect. The interesting finding is the existence of a critical burial depth, i.e., 2.5 m under this condition, and its effect on stress and anodic current density distributions.

The failure pressure of the suspended pipe containing a corrosion defect as a function of the pipe burial depth is shown in Table 4.3. Generally, the failure pressure decreases as the pipe burial depth increases. When the burial depth exceeds 2.5 m, the failure pressure drops rapidly. Therefore, a deep pipe burial will adversely affect the integrity of suspended pipelines.

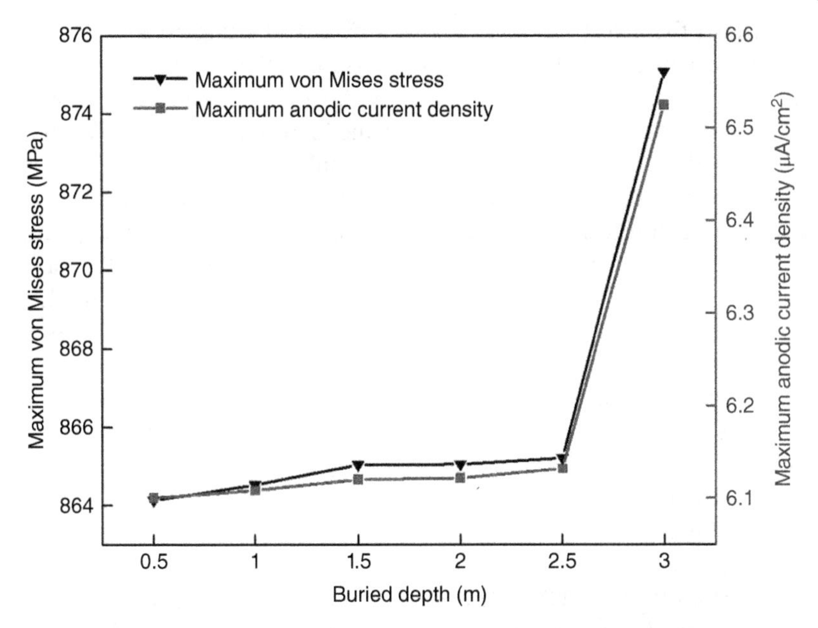

Figure 4.25 Maximum von Mises stress and maximum anodic current density at a corrosion defect on a suspended X100 steel pipe segment with different burial depths. *Source:* From Qin and Cheng [2021b] / with permission from Elsevier.

Table 4.3 Failure pressure of the suspended X100 steel pipe containing a corrosion defect with different pipe burial depths.

Buried depth (m)	1	1.5	2	2.5	3
Failure pressure (MPa)	24.3	23.8	23.3	22.6	19.5

Source: From Qin and Cheng [2021b] / with permission from Elsevier.

4.3.4.3 Failure Prediction of Suspended Pipelines Containing a Corrosion Defect

The model developed for assessment of corrosion defect on a suspended pipeline is critical to determination of the FFS and failure pressure when the pipeline is in suspension. In general, the M–E effect at the corrosion defect increases the local corrosion rate and decreases failure pressure of the corroded pipeline. The effect becomes more apparent when the pipe is in suspension.

Of various affecting parameters, the von Mises stress and anodic current density at the corrosion defect depend on the defect location on the suspended pipe, suspension length, and the pipe burial depth.

Generally, the defect location does not affect the local stress and anodic current density, as well as the failure pressure of the suspended pipeline at an appreciable level. However, the local stress and anodic current density at the corrosion defect increase with the increased suspension length, causing a decreased failure pressure. As a result, the failure risk increases when the suspension length increases. Furthermore, a critical pipe burial depth exists in terms of its effect on local stress and anodic current density at the corrosion defect, as well as the failure pressure of the suspended pipeline. When the burial depth is smaller than 2.5 m, the effect is marginal. After the 2.5 m depth is exceeded, the failure pressure decreases rapidly.

The hidden suspension makes the pipeline produce a downward (i.e., gravity-direction) displacement under the effect of pipe and fluid weights and the load imposed by covering soil. The displacement is the greatest at the middle of the suspension (i.e., 50% SL). The M–E interaction at the corrosion defect under the suspension effect can result in a stress concentration and accelerated corrosion. This potentially leads to stress corrosion crack (SCC) [Cheng, 2013]. The SCCs initiate and propagate under the synergism of elevated stress and corrosion reaction, causing pipeline failures. This kind of accident has been reported on pipelines buried in mountain areas, where geotechnical hazards usually cause soil erosion and pipe suspension. The failure pressure of the suspended pipeline can decrease rapidly with changes in defect geometry, suspension length, and soil condition, constituting a threat to the integrity of the pipelines. Therefore, frequent inspections of geotechnical conditions where the pipelines are buried should be integral to the pipeline integrity management program. Moreover, the suspension effect, once it is identified, must be considered in the defect assessment program.

References

Adib-Ramezani, H., Jeong, J., Pluvinage, G. (2006) Structural integrity evaluation of X52 gas pipes subjected to external corrosion defects using the SINTAP procedure, *Int. J. Press. Vessel Pip.*, 83, 420–432.

Ahammed, M., Melchers, R.E. (1997) Probabilistic analysis of underground pipelines subject to combined stresses and corrosion, *Eng. Struct.*, 19, 988–994.

Akid, R., Dmytrakh, I.M. (1998) Influence of surface deformation and electrochemical variables on corrosion and corrosion fatigue crack development, *Fatigue Frac. Eng. Mater. Struct.*, 21, 903–911.

Amaya-G'omez, R., S'anchez-Silva, M., Bastidas-Arteaga, E., Schoefs, F., Munoz, F. (2019) Reliability assessments of corroded pipelines based on internal pressure – a review, *Eng. Fail. Anal.*, 98, 190–214.

American Society of Mechanical Engineering (2012) *Manual for Determining the Remaining Strength of Corroded Pipelines: A Supplement to ASME B31 Code for Pressure Piping*, ASME, New York, NY, USA

Bagotsky, V.S. (2006) *Fundamentals of Electrochemistry*, Ed. 2, Wiley, New York, NY, USA.

Bazan, F.A.V., Beck, A.T. (2013) Stochastic process corrosion growth models for pipeline reliability, *Corros. Sci.*, 74, 50–58.

Cheng, Y.F. (2013) *Stress Corrosion Cracking of Pipelines*, John Wiley, Hoboken, NJ, USA.

Cheng, Y.F. (2016) Environmental hazard: monitor safety of aged fuel pipelines, *Nature*, 529, 156.

Cheng, Y.F. (2020) Technical insights into the long-term integrity and sustainability of China-Russia eastern gas pipeline, *Oi/Gas Storage Transport.*, 39, 1–8.

Cheng, Y.F., Norsworthy, R. (2017) *Pipeline Coatings*, NACE, Houston, TX, USA.

Despic, A.R., Raicheff, R.G., Bockris, J.O'M. (1968) Mechanism of the acceleration of the electrode dissolution of metals during yielding under stress, *J. Chem. Phys.*, 49, 926–938.

Eyring, H., Lin, S.H. (1980) *Basic Chemical Kinetics*, Wiley Publishing, New York, NY, USA.

Goodall, I.W. (1978) *Lower Bound Limit Analysis of Curved Tubes Loaded by Combined Internal Pressure and In-Plane Bending Moment*, Central Electricity Generation Board (CEGB), RD/B/N4360, UK.

Gutman, E.M. (1994) *Mechanochemistry of Solid Surfaces*, World Scientific Publishing, Singapore.

Gutman, E.M. (1998) *Mechanochemistry of Materials*, Cambridge International Science Publishing, London, UK.

Gutman, E.M., Solovioff, G., Eliezer, D. (1996) The mechanochemical behavior of type 316L stainless steel, *Corros. Sci.*, 38, 1141–1145.

Hoar, T.P., West, J.M. (1958) Mechano–chemical anodic dissolution, *Nature*, 181, 835.

Huang, Y.F., Wei, C., Chen, L.J., Li, P.F. (2014) Quantitative correlation between geometric parameters and stress concentration of corrosion pits, *Eng. Fail. Anal.*, 44, 168–178.

Khalaj, K.M., Bahaari, M.R., Salehi, A., Shahbazi, S. (2015) Predicting the limit pressure capacity of pipe elbows containing single defects, *Appl. Ocean Res.*, 53, 15–22.

Kim, S.J. (2017) Effect of elastic tensile load on the electrochemical corrosion behavior and diffusible hydrogen content of ferritic steel in acidic environment, *Int. J. Hydrogen Energy*, 42, 19367–19375.

Kim, Y., Park, K.J. (2017) Characterization of axial and oblique defects in pipes using fundamental torsional guided modes, *NDT Int.*, 92, 149–158.

Kim, J.W., Lee, S.H., Park, C.Y. (2009) Experimental evaluation of the effect of local wall thinning on the failure pressure of elbows, *Nuclear Eng. Des.*, 239, 2737–2746.

Kim, S.J., Jung, H.G., Kim, K.Y. (2012) Effect of tensile stress in elastic and plastic range on hydrogen permeation of high-strength steel in sour environment, *Electrochim. Acta*, 78, 139–146.

Kim, J.W., Yoon, M.S., Park, C.Y. (2013) The effect of load-controlled bending load on the failure pressure of wall-thinned pipe elbows, *Nuclear Eng. Des.*, 265, 174–183.

King, F., Jack, T., Chen, W., Wilmott, M., Fessler, R.R., Krist, K. (2000) Mechanistic studies of initiation and early-stage crack growth for near-neutral pH SCC on pipelines, *Corrosion'2000*, paper no. 361, NACE, Houston, TX, USA.

Kuwazuru, O., Ode, K., Yamada, M., Kassab, A.J., Divo, E. (2018) Experimental and boundary element method study on the effect of stress on the polarization curve of cast aluminum alloy in sodium chloride solution, *Corros. Sci.*, 132, 136–145.

Lam, C., Zhou, W. (2016) Statistical analysis of incidents on onshore gas transmission pipelines based on PHMSA database, *Int. J. Press. Vessel Pip.*, 145, 29–40.

Marc, A.M. (2013) *Corrosion Growth Models and ILI-Based Estimation Procedures for Reliability-based and Deterministic Pipeline Integrity Assessments*, Pipeline Research Council International (PRCI), Chantilly, VA, USA

Miran, S.A., Huang, Q., Castaneda, H. (2016) Time-dependent reliability analysis of corroded buried pipelines considering external defects, *J. Infrastructure Syst.*, 22, 4016019.

Moncada, M.A.T., Aguirre-Pe, J. (1999) Scour below pipeline in river crossings, *J. Hydraul. Eng.*, 125, 145–151.

Noda, N.A., Hayashida, H. (2000) Interaction between elliptical and ellipsoidal inclusions under bending stress fields, *Arch. Appl. Mech.*, 70, 612–624.

Parkins, R.N. (2000) A review of stress corrosion cracking of high pressure gas pipelines, *Corrosion'2000*, paper no. 363, NACE, Houston, TX, USA.

Qin, G.J., Cheng, Y.F. (2020) Failure pressure prediction by defect assessment and finite element modelling on natural gas pipelines under cyclic loading, *J. Nat. Gas Sci. Eng.*, 81, 103445.

Qin, G.J., Cheng, Y.F. (2021a) A review on defect assessment of pipelines: principles, numerical solutions, and applications, *Int. J. Press. Vessel Pip.*, 191, 104329.

Qin, G.J., Cheng, Y.F. (2021b) Modeling of mechano-electrochemical interaction at a corrosion defect on a suspended gas pipeline and the failure pressure prediction, *Thin Wall. Struct.*, 160, 107404.

Qin, G.J., Cheng, Y.F., Zhang, P. (2021) Finite element modeling of corrosion defect growth and failure pressure prediction of pipelines, *Int. J. Press. Vessel Pip.*, 194, 104509.

Shuai, Y., Wang, X.H., Li, J., Wang, J.Q., Wang, T.T., Han, J.Y., Cheng, Y.F. (2021) Assessment by finite element modelling of the mechano-electrochemical interaction at corrosion defect on elbows of oil/gas pipelines, *Ocean Eng.*, 234, 109228.

Simnad, M.T., Evans, U.R. (1950) The influence of stress upon the electrode potential and polarization of iron and steel in acid solution, *Trans. Faraday Soc.*, 46, 175–186.

Sun, J.L., Cheng, Y.F. (2019a) Modelling of mechano-electrochemical interaction of multiple longitudinally aligned corrosion defects on oil/gas pipelines, *Eng. Struct.*, 190, 9–19.

Sun, J.L., Cheng, Y.F. (2019b) Investigation by numerical modeling of the mechano–electrochemical interaction of circumferentially aligned corrosion defects on pipelines, *Thin Wall. Struct.*, 144, 106314.

Sun, J.L., Cheng, Y.F. (2020) Modelling of mechano-electrochemical interaction at overlapped corrosion defects and the implication on pipeline failure prediction, *Eng. Struct.*, 213, 110466.

Sun, J.L., Cheng, Y.F. (2021) Modeling of mechano-electrochemical interaction between circumferentially aligned corrosion defects on pipeline under axial tensile stresses, *J. Petro. Sci. Eng.*, 198, 108160.

Swarnavo, S., Aquino, W. (2013) Changes in electrodic reaction rates due to elastic stress and stress-induced surface patterns, *Electrochim. Acta*, 111, 814–822.

Vanaei, H.R., Eslami, A., Egbewande, A. (2017) A review on pipeline corrosion, in-line inspection (ILI), and corrosion growth rate models, *Int. J. Press. Vessel Pip.*, 149, 43–54.

Wang, H.T., Han, E.H. (2016) Computational simulation of corrosion pit interactions under mechanochemical effects using a cellular automaton/finite element model, *Corros. Sci.*, 103, 305–311.

Wang, L.R., Yeh, Y.H. (1985) A refined seismic analysis and design of buried pipeline for fault movement, *Earthquake Eng. Struct. Dynam.*, 13, 75–96.

Wang, Q., Zhou, W. (2019a) Burst pressure models for thin-walled pipe elbows, *Int. J. Mech. Sci.*, 159, 20–29.

Wang, Q., Zhou, W. (2019b) A new burst pressure model for thin-walled pipe elbows containing metal-loss corrosion defects, *Eng. Struct.*, 200, 10972.
Wang, X., Shuai, J., Ye, Y., Zuo, S. (2008) Investigating the effects of mining subsidence on buried pipeline using finite element modeling, *7th Int. Pipeline Conf.*, ASME, Calgary, AB, Canada.
Wang, Y., Wharton, J.A., Shenoi, R.A. (2016) Mechano-electrochemical modelling of corroded steel structures, *Eng. Struct.*, 128, 1–14.
Wang, Y.H., Dann, M.R., Zhang, P. (2020) Reliability analysis of corroded pipelines considering 3D defect growth, *Thin Wall. Struct.*, 157, 107028.
Wang, Y.C., Xu, L.Y., Sun, J.L., Cheng, Y.F. (2021) Mechano-electrochemical interaction for pipeline corrosion: a review, *J. Pipeline Sci. Eng.*, 1, 1–16.
Waseda, Y., Suzuki, S. (2006) *Characterization of Corrosion Products on Steel Surfaces*, Springer, Berlin, Germany.
Xie, M.J., Tian, Z.G., 2018. A review on pipeline integrity management utilizing inline inspection data, *Eng. Fail. Anal.*, 92, 222–239.
Xu, L.Y., Cheng, Y.F. (2012a) Corrosion of X100 pipeline steel under plastic strain in a neutral pH bicarbonate solution, *Corros. Sci.*, 64, 145–152.
Xu, L.Y., Cheng, Y.F. (2012b) An experimental investigation of corrosion of X100 pipeline steel under uniaxial elastic stress in a near-neutral pH solution, *Corros. Sci.*, 59, 103–109.
Xu, L.Y., Cheng, Y.F. (2013) Development of a finite element model for simulation and prediction of mechano-electrochemical effect of pipeline corrosion, *Corros. Sci.*, 73, 150–160.
Xu, L.Y., Cheng, Y.F. (2017) A finite element-based model for prediction of corrosion defect growth on pipelines, *Int. J. Press. Vessel Pip.*, 152, 70–79.
Yang, Y., Cheng, Y.F. (2016) Stress enhanced corrosion at the tip of near-neutral pH stress corrosion cracks on pipelines, *Corrosion*, 72, 1035–1043.
Yi, D., Idapalapati, S., Xiao, Z.M., Kumar, S.B. (2012) Fracture capacity of girth welded pipelines with 3D surface cracks subjected to biaxial loading conditions, *Int. J. Press. Vessel Pip.*, 92, 115–126.
Zahn, M. (1979) *Electromagnetic Field Theory: A Problem Solving Approach*, Wiley, New York, NY, USA
Zelmati, D., Ghelloudj, O., Amirat, A. (2017) Correlation between defect depth and defect length through a reliability index when evaluating of the remaining life of steel pipeline under corrosion and crack defects, *Eng. Fail. Anal.*, 79, 171–185.
Zhang, Z.W., Ni, X., Cheng, Y.F. (2020a) Assessment by finite element modelling of the mechanoelectrochemical interaction at double-ellipsoidal corrosion defect with varied inclinations on pipelines, *Construct. Build. Mater.*, 260, 120459.

Zhang, Z.W., Ni, X., Cheng, Y.F. (2020b) Interaction between internal and external defects on pipelines and its effect on failure pressure, *Thin Wall. Struct.*, 159, 107230.

Zhang, Z.W., Guo, L.P., Cheng, Y.F. (2021a) Interaction between internal and external defects on pipelines and its effect on failure pressure, *Thin Wall. Struct.*, 159, 107239.

Zhang, Z.W., Wang, J.C., Zhang, J.H., Cheng, Y.F. (2021b) Modeling of the mechano-electrochemical effect at corrosion defect with varied inclinations on oil/gas pipelines, *Petro. Sci.*, 18, 2520–2529.

Zhou, W.X., Xiang, W., Hong, H.P. (2017) Sensitivity of system reliability of corroding pipelines to modeling of stochastic growth of corrosion defects, *Reliability Eng. Syst. Safety*, 167, 428–438.

5

Mechano-electrochemical Interaction for Level III Assessment of Corrosion Anomalies on Pipelines – Multiple Corrosion Defects

5.1 Introduction

As stated, most corrosion defects on pipelines are in proximity and mutual interactions exist between them. The interaction affects not only the stress distribution but also electrochemical corrosion rate [Liu et al., 2011; Wang et al., 2016]. At the same time, an M–E effect always exists at corrosion defects, affecting their mutual interaction if the interaction does exist. However, the available rules and codes have not considered electrochemical corrosion reactions in the interaction rules for assessment of multiple corrosion defects on pipelines. In recent years, progress has been made to improve the assessment of interacting corrosion defects while including the M–E interaction effect. The developed models and methods can assess the pipeline FFS and determine failure pressure in the presence of multiple corrosion defects with mutual interaction. Moreover, the models are capable of predicting the remaining service life by modeling the corrosion defect growth on pipelines.

5.2 Assessment of Multiple Corrosion Defects on Pipelines and Development of Interaction Rules

5.2.1 Longitudinally Aligned Corrosion Defects Under the M–E Interaction

An accurate assessment of multiple corrosion defects on pipelines depends on both integration of multi-physics fields at the defects, specifically, the M–E interaction, and the interaction between the adjacent defects. To date, the interaction rules developed for defining the mutual interaction of

Defect Assessment for Integrity Management of Pipelines, First Edition. Y. Frank Cheng.

corrosion defects are mostly based on evaluation of mechanical factors (i.e., stress and strain) [Benjamin et al., 2005, 2006, 2016a, b; Benjamin and Cunha, 2007; DNV, 2004], and the role of the M–E interaction has rarely been investigated. At the same time, the interaction of multiple corrosion defects is dependent on their orientation on pipelines. However, the defects can have varied orientations. The most typical orientations between adjacent corrosion defects include longitudinal and circumferential orientations. This section introduces the progress in development of numerical models based on FE analysis for assessment of multiple, longitudinally oriented corrosion defects, by considering the mechanical and electrochemical corrosion factors, as well as their interaction, on an X46 steel pipe exposed to a near-neutral pH soil solution.

5.2.1.1 The Model

Modeling of the M–E interaction of two longitudinally adjacent corrosion defects on an X46 steel pipe was conducted using a COMSOL software package. While the stress over the entire pipe segment was simulated by a 3D model, only a half of the pipe was modeled considering the symmetry of the pipe to save computational time. For comparison, a 2D model was used to simulate the stress, corrosion potential, and anodic current density (i.e., corrosion rate) of the pipe under axial tensile stresses. Figure 5.1 shows the 2D and 3D FE models, where two corrosion defects are present on the pipe surface longitudinally [Sun and Cheng, 2019a]. For the 2D model, the left end of the pipe was fixed, and the right end was applied with a pre-determined tensile load. The bottom of the pipe was set as electric grounding, and the steel/solution interface was set as a free boundary. The mesh type used was triangular. The pipe body was under a coarse meshing to reduce computation, and a mesh refinement was conducted at the corrosion defects to ensure modeling accuracy. The maximum and minimum element sizes were 5 and 0.1 mm, respectively. For the 3D model, the pipe was symmetric relative to the middle of the two defects. Symmetrical constraints were applied on the planes to be modeled. Displacement in z-direction of both ends was constrained. The pipe length was sufficient to avoid the effect of boundary conditions on modeling of the defects. The outer surface of the pipe was set as a free boundary, while other surfaces were set as electrically isolated. The mesh type for the 3D model was tetrahedral. Similarly, the pipe body was under a coarse mesh, and the meshing refinement was applied at the defect area. The maximum and minimum element sizes were 10 and 1 mm, respectively, and the total number of elements was 42,612. The FE modeling was conducted

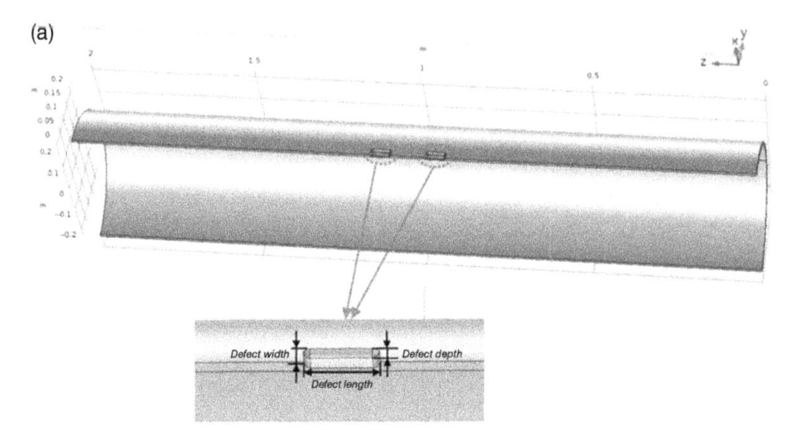

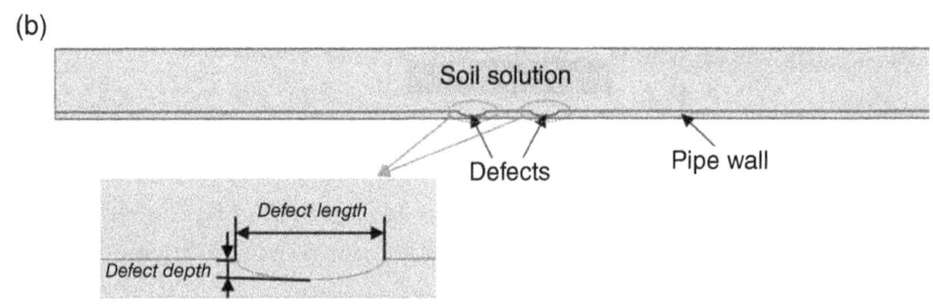

Figure 5.1 Schematic diagram of a steel pipe containing two longitudinally oriented corrosion defects (a) 3D model and (b) 2D model. *Source:* From Sun and Cheng [2019a] / with permission from Elsevier.

considering three physics fields, i.e., stress, electrochemical corrosion, and the M–E coupling effect, existing at the corrosion defects.

The mechanical properties of X46 pipeline steel were determined previously [Yang and Cheng, 2016]. The stress–strain behavior follows a nonlinear isotropic hardening model at the plastic stage [Xu and Cheng, 2012a]. The electrochemical corrosion properties of the steel in deoxygenated, near-neutral pH soil solutions were included in Yang and Cheng [2016].

5.2.1.2 Distributions of Stress and Anodic Current Density of the Pipe Containing Two Corrosion Defects Under Axial Tensile Stresses

Figure 5.2 shows the distribution of von Mises stress at the corrosion defects with various longitudinal spacings from 5 to 150 mm and a fixed length of 60 mm under an axial stress of 196 MPa in a soil solution. Stress concentration occurs at the defect, especially at its center bottom.

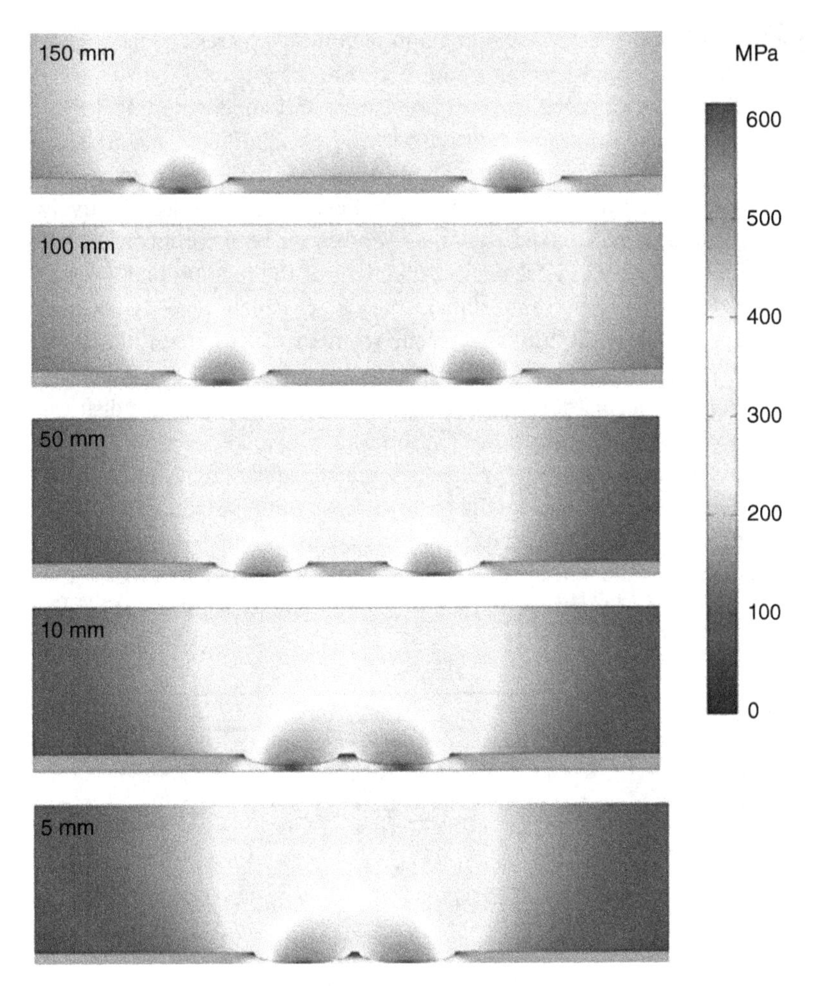

Figure 5.2 Distributions of von Mises stress at two adjacent corrosion defects with various longitudinal spacings from 5 to 150 mm and a fixed length of 60 mm under an axial stress of 196 MPa in a soil solution. *Source:* From Sun and Cheng [2019a] / with permission from Elsevier.

When the corrosion defects become closer from 150 to 50 mm, the stress level changes slightly. The stress level between the defects drops with decreased spacing, as indicated by the color change from light blue to dark blue. Thus, a smaller longitudinal spacing between the corrosion defects results in a lower stress in the adjacent area. It is noted that the color change is hard to detect with respect to the defect spacing. This is attributed to the fact that the maximum von Mises stress varies slightly with the longitudinal spacing under the axial loading.

The linear distributions of von Mises stress and anodic current density along the steel/solution interface with various longitudinal spacings from 5 to 150 mm and a fixed defect length of 60 mm under an axial tensile stress of 196 MPa in the soil solution are shown in Figures 5.3 and 5.4, respectively. It is seen that there are identical stress profiles for both corrosion defects. The stress level at the defects is independent of the longitudinal spacing. Although the stress concentration at the defect center exceeds 600 MPa (i.e., approximately the ultimate tensile strength of X46 steel), the stress in the adjacent area between the defects is below 200 MPa (i.e., the elastic state). Further, the distribution of anodic current density is like the distribution of von Mises stress. As the longitudinal spacing increases from 5 to 150 mm, the anodic current density does not change apparently. As the corrosion defects get closer, the anodic current density drops slightly.

The modeling results of von Mises stress and anodic current density show that, under the axial tensile stress of 196 MPa, the stress and corrosion rate distributions are independent of the longitudinal spacing between the

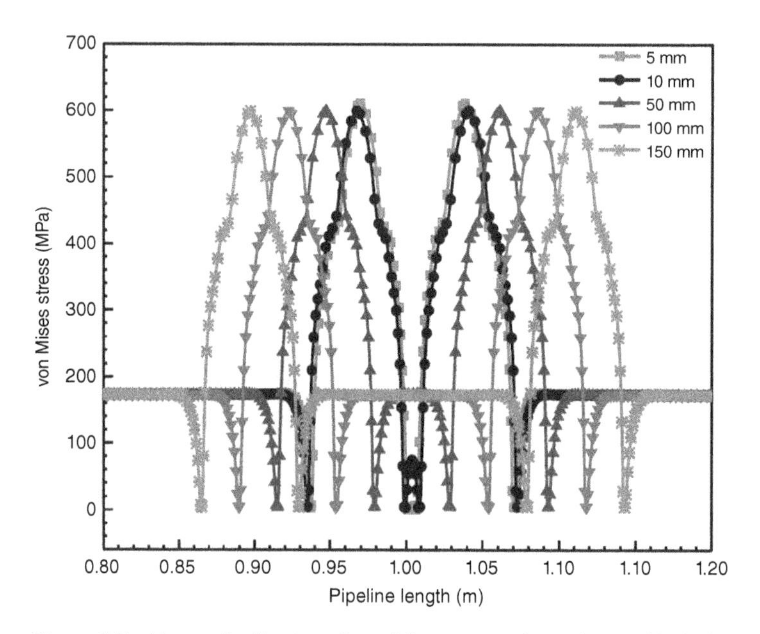

Figure 5.3 Linear distribution of von Mises stress along the steel/solution interface with various longitudinal spacings from 5 to 150 mm and a fixed defect length of 60 mm under an axial tensile stress of 196 MPa in the soil solution. *Source:* From Sun and Cheng [2019a] / with permission from Elsevier.

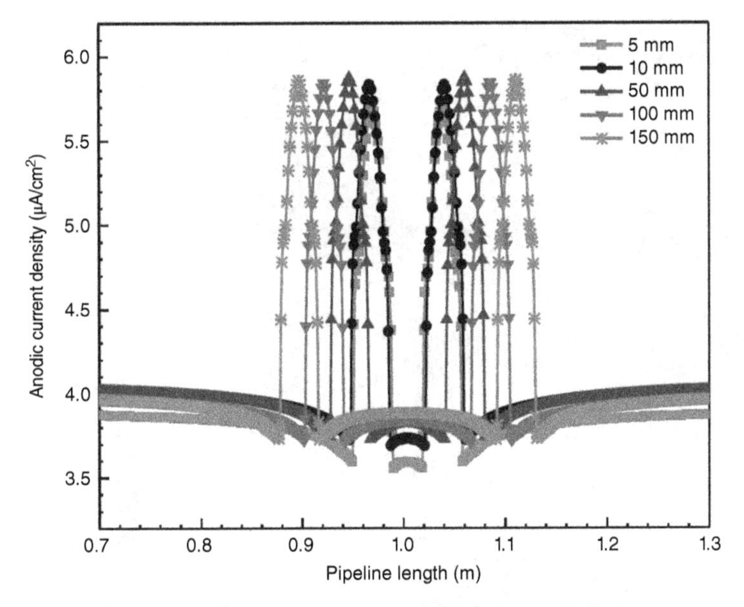

Figure 5.4 Linear distribution of anodic current density along the steel/solution interface with various longitudinal spacings from 5 to 150 mm and a fixed defect length of 60 mm under an axial tensile stress of 196 MPa in the soil solution. *Source:* From Sun and Cheng [2019a] / with permission from Elsevier.

adjacent corrosion defects. In other words, the longitudinal spacing between the defects is not important to affect the stress and corrosion distributions under the axial tensile stress.

5.2.1.3 Distributions of Stress and Anodic Current Density of a Pressurized Pipe Containing Two Corrosion Defects

The distribution of von Mises stress of the steel pipe containing two corrosion defects with various longitudinal spacings under an internal pressure of 15.3 MPa in the soil solution is shown in Figure 5.5. When the defect spacing is 150 mm, the stress distribution at both defects is of a bone shape, with a stress concentration at the defect center. As the corrosion defects become closer (i.e., a spacing of 50 mm), the high stress area expands to the adjacent zone between the defects, in addition to the defect center. As the defect spacing further decreases to 10 mm, the stress fields of the two defects start to merge. At the spacing of 5 mm, one stress field with a large bone shape is formed on the two corrosion defects, with the highest stress located at the defect centers. Obviously, reducing longitudinal spacing between the defects results in an increased stress concentration at the defects for pressurized pipes.

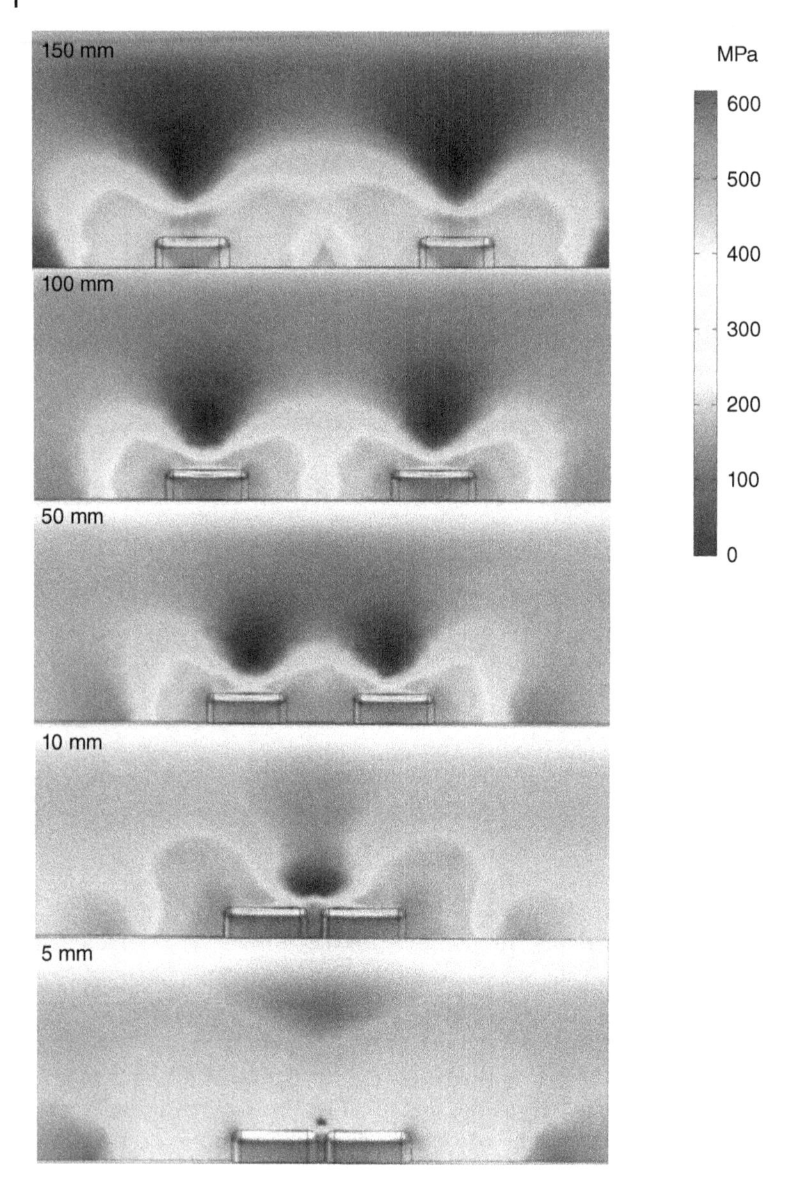

Figure 5.5 Distribution of von Mises stress of the steel pipe containing two corrosion defects with various longitudinal spacings under an internal pressure of 15.3 MPa in the soil solution. *Source:* From Sun and Cheng [2019a] / with permission from Elsevier.

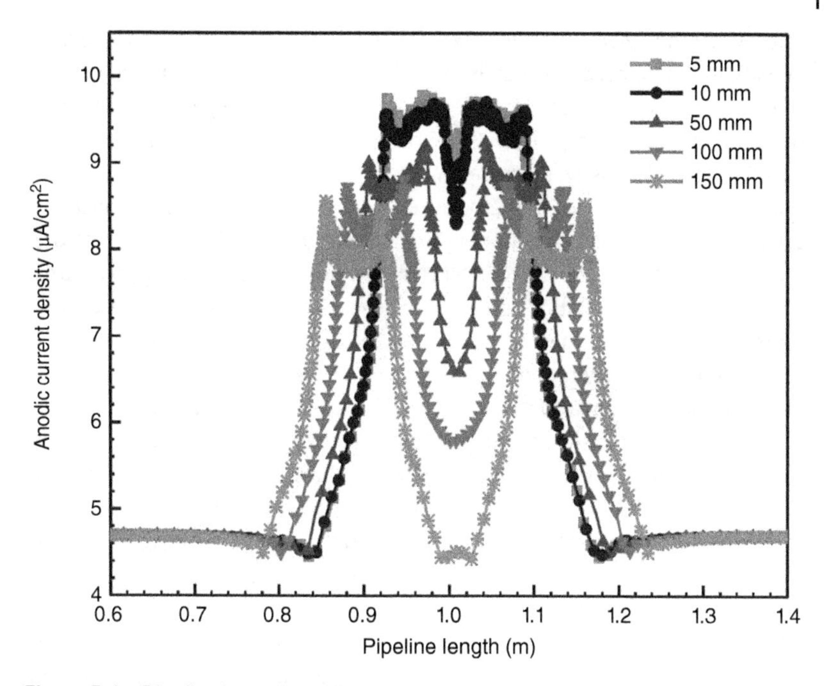

Figure 5.6 Distributions of anodic current density, i.e., corrosion rate, along the axial direction of the corrosion defects with various spacings under a 15.3 MPa internal pressure in the soil solution. *Source:* From Sun and Cheng [2019a] / with permission from Elsevier.

The distributions of anodic current density, i.e., corrosion rate, along the axial direction of the corrosion defects with various spacings under a 15.3 MPa internal pressure in the soil solution are shown in Figure 5.6. There is the greatest anodic current density at the defect center due to the stress concentration. When the defect spacing is large, such as 150 mm, the current density in the area between the defects is the smallest. As the spacing reduces, the current density increases. When the spacing is as small as 5 mm, the anodic current density at the defect center is the greatest. Moreover, the current density in the area between the defects also increases remarkably.

5.2.1.4 A Critical Longitudinal Spacing Criterion

The modeling results show that, depending on the spacing between the longitudinally aligned corrosion defects, their mutual interaction exists to affect not only the local stress distribution but also the electrochemical corrosion rate. When the defect spacing is 150 mm, as shown above, the anodic current density in the defect adjacency is identical to that of the uncorroded area on the pipe, indicating that the two defects can be assessed separately as there is no interaction between them. Once the spacing decreases to 100 mm, the anodic current density increases apparently. When the spacing between the defects is reduced to 5 mm, the anodic current densities at both the defect center and the adjacent area are approximately identical, showing a strong interaction between the defects to accelerate corrosion at the adjacent area. Therefore, a critical longitudinal spacing exists, below which there is an interaction between the corrosion defects to enhance corrosion rate by the so-called M–E interaction. Under the given condition, this value is determined to be between 100 and 150 mm. It is noted that the critical spacing value can be varied under other pipeline operating conditions.

To determine the critical spacing between adjacent corrosion defects, a criterion, i.e., the ratio of the anodic current density at the middle of adjacent corrosion defects to that far away from the defects ($i^{a}_{mid-defects}/i^{a}_{far-defects}$) is proposed. It is the first time using the anodic current density to evaluate whether an interaction exists between the adjacent corrosion defects considering the M–E interaction for accelerated corrosion. When the ratio is greater than 1, an interaction exists between the corrosion defects to enhance the anodic current density (i.e., corrosion rate) of the steel at the defect area. If the ratio is equal to 1, there is no interaction between the defects. The greater the ratio, the more significant the interaction between the defects for corrosion enhancement in aqueous solutions.

It is further established that the critical spacing between the longitudinal corrosion defects, as discussed above, depends on the defect length. Figure 5.7 shows the maximum longitudinal spacing between adjacent corrosion defects enabling an interaction between them as a function of the defect length. It is seen that the maximum interacting spacing between the corrosion defects increases as the defect length increases. For short defects such as those of 20, 40, and 60 mm in length, the maximum interacting spacing increases rapidly with the defect length. After the defect is longer than 80 mm, the increasing tendency slows down.

In summary, the spacing between adjacent corrosion defects is an important parameter affecting their mutual M–E interaction, and thus, the failure pressure of pipelines. While many factors such as stress and defect geometry affect the critical spacing, a criterion based on anodic current density resulting from corrosion of steel in corrosive enhancements is proposed. This is novel as the M–E interaction always exists at corrosion defects on pipelines. An ignorance of the M–E interaction will not be able to provide accurate assessment of multiple corrosion defects and the FFS of pipelines.

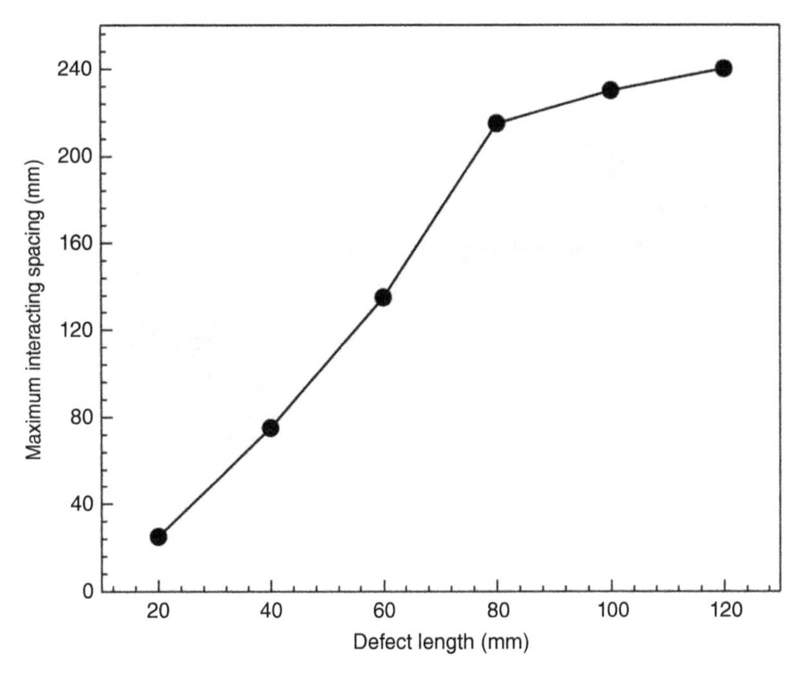

Figure 5.7 The maximum longitudinal spacing between adjacent corrosion defects enabling an interaction between them as a function of the defect length. *Source:* From Sun and Cheng [2019a] / with permission from Elsevier.

5.2.2 Circumferentially Aligned Corrosion Defects Under the M–E Interaction

In addition to the typical longitudinal orientation between adjacent corrosion defects, they can also present circumferentially on pipelines. This section introduces the progress in development of a numerical model for assessment of circumferentially aligned corrosion defects, by considering the mechanical and electrochemical factors and their interaction, on an X46 steel pipe exposed to a near-neutral pH soil solution.

5.2.2.1 The Model

An FE-based, multi-physics field coupling model is developed to investigate the M–E interaction between circumferentially aligned corrosion defects on an X46 steel pipeline, where the distributions of stress and anodic current density (i.e., corrosion rate) at the defects are determined under various internal pressures. A critical spacing criterion evaluating the existence of mutual interaction between the circumferential defects is defined. Half of

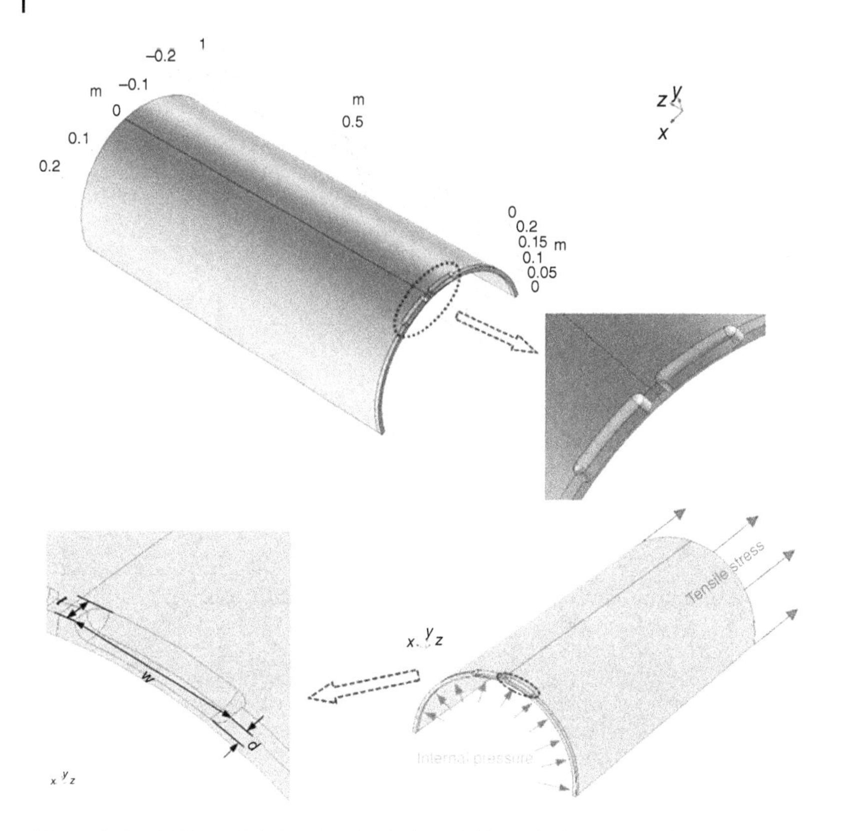

Figure 5.8 A 3D model showing half of an X46 steel pipe containing two circumferentially aligned corrosion defects, and the geometrical parameters of the defects. *Source:* From Sun and Cheng [2019b] / with permission from Elsevier.

the steel pipe containing two circumferentially aligned cubic corrosion defects is simulated by a 3D model, as shown in Figure 5.8, to reduce the implementation complexity and computational time considering the symmetry of the object. The geometrical parameters of the corrosion defects are also included.

5.2.2.2 Distributions of Stress and Anodic Current Density of the Pipe Under Axial Tensile Stresses

The distribution of von Mises stress at the corrosion defects with various circumferential spacings under a fixed axial tensile stress of 375 MPa is shown in Figure 5.9. The applied stress level is selected because the corrosion defects are under a plastic deformation under the stress [Sun and Cheng, 2021]. When the circumferential spacing between the defects is 1.8°, the

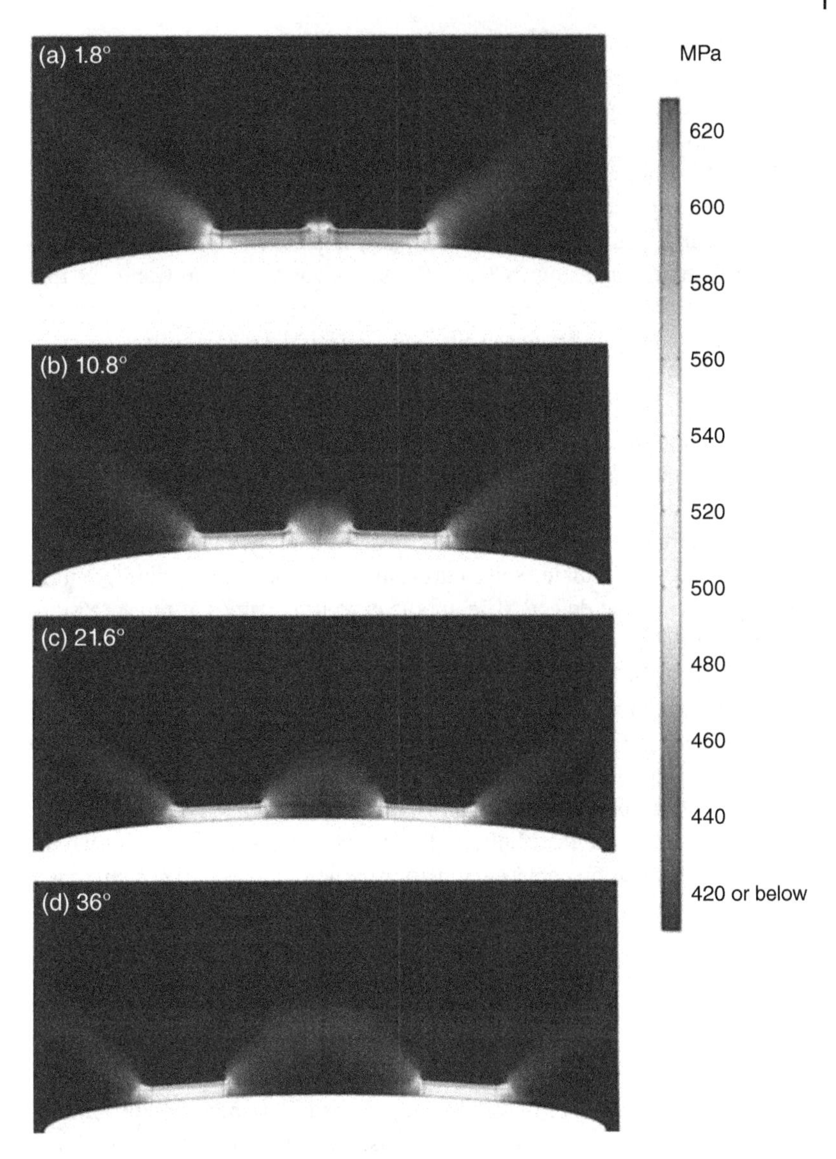

Figure 5.9 Distributions of von Mises stress at corrosion defects with various circumferential spacings under a fixed axial tensile stress of 375 MPa. (a) 1.8°, (b) 10.8°, (c) 21.6°, and (d) 36°. *Source:* From Sun and Cheng [2021] / with permission from Elsevier.

defects and their adjacent area are colored in red with a stress concentration over 620 MPa, indicating a high plasticity. As the spacing increases to 10.8°, the red zone is located at the defect only and the stress level decreases to about 580 MPa. When the defect spacing is 21.6°, the stress further decreases to 550 MPa, especially at the adjacent area between the defects, where the stress is below yielding strength of the steel. The two defects can be treated independently, while their mutual interaction is negligible. With the defect spacing further increasing to 36.0°, the stress concentration occurs at the defects only, but at a low stress level of about 540 MPa. Thus, an increasing circumferential spacing between corrosion defects decreases both stress concentration and the plasticity at both the defects and their adjacent area. When the spacing exceeds 21.6°, as determined under the given condition, the mutual interaction between the defects is negligible.

As stated, the anodic current density is an indicator of corrosion rate of steels when measured electrochemically. Figure 5.10 shows the distribution of anodic current density at the corrosion defects under a fixed tensile stress of 375 MPa in the simulated soil solution as a function of the circumferential spacing between the defects. The anodic current densities at the corrosion defects and the adjacent area continuously decrease as the circumferential spacing increases. Moreover, at small circumferential spacing such as 1.8°, there is a high anodic current density at the defect adjacency.

5.2.2.3 Distributions of Stress and Anodic Current Density of a Pressurized Pipe Containing Two Corrosion Defects

Figure 5.11 shows the distribution of von Mises stress at two circumferentially aligned corrosion defects with various spacings on the steel pipe under an internal pressure of 18 MPa. It is seen that, when the circumferential spacing of the two defects is 1.8°, the maximum stress concentration occurs at the defect sides, while the adjacent area between the defects also has a high stress concentration. Moreover, the stress level exceeds the yielding strength of the steel. As the circumferential spacing increases to 10.8°, the stress distribution inside the defects does not change, but the stress concentration at the adjacent area reduces to elastic range. This phenomenon is also found when the circumferential spacing increases to 21.6° and 36°. Therefore, when the defect spacing is sufficiently small (such as 1.8°), the adjacent area between the defects is under stress concentration, generating a plastic zone. With the increase of the circumferential spacing between defects, the interaction between them decreases and even disappears. The defects can thus be treated separately.

The distribution of anodic current density at the two circumferential corrosion defects on the pipe with an internal pressure of 18 MPa in the soil

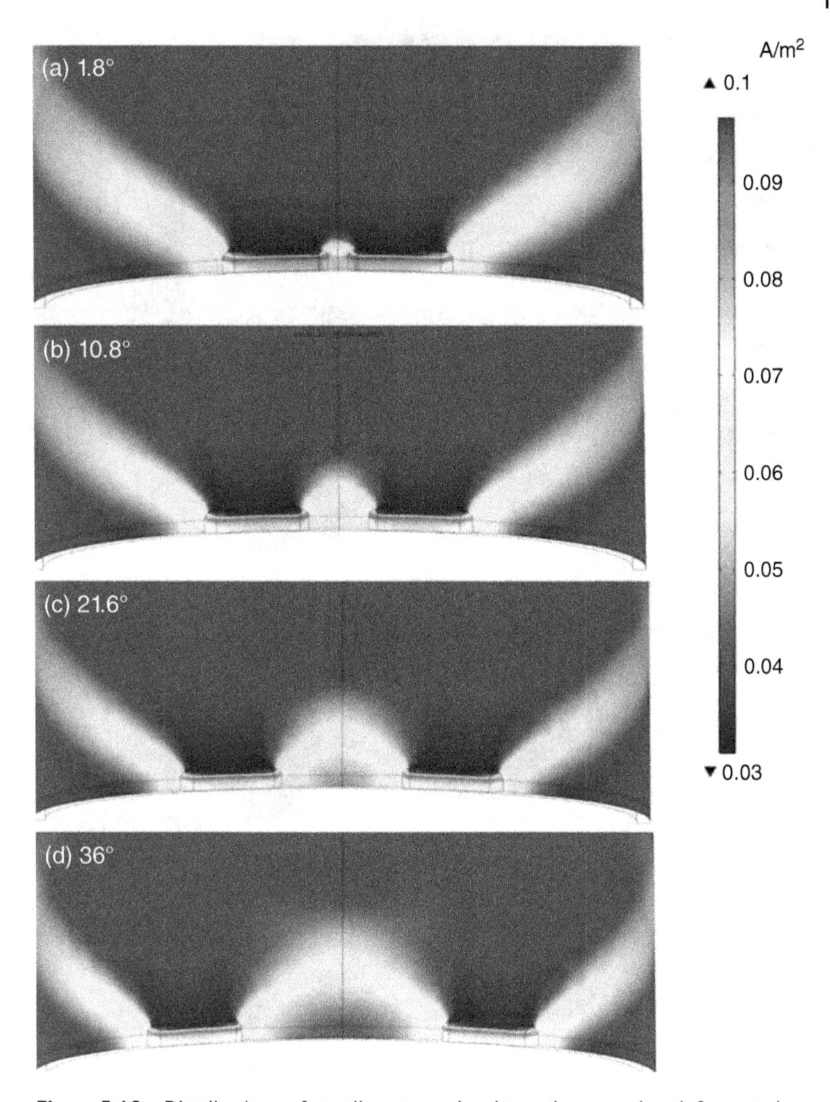

Figure 5.10 Distributions of anodic current density at the corrosion defects under a fixed axial tensile stress of 375 MPa in a simulated soil solution as a function of the circumferential spacing between defects. (a) 1.8°, (b) 10.8°, (c) 21.6°, and (d) 36°. *Source:* From Sun and Cheng [2021] / with permission from Elsevier.

solution as a function of the circumferential spacing is shown in Figure 5.12. When the circumferential spacing is 1.8°, the greatest anodic current density is located at inside sides of the defects of about 0.14 A/m^2. A bit lower anodic current density of about 0.11 A/m^2 is recorded at the adjacent area between

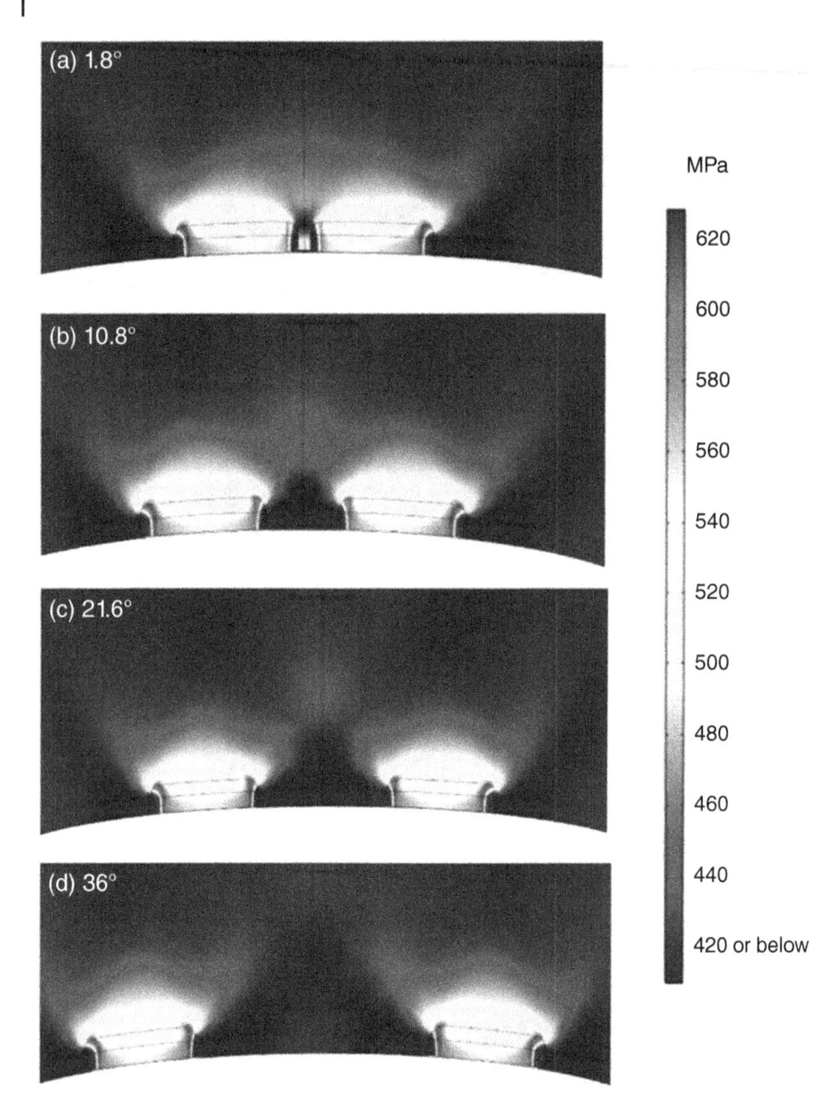

Figure 5.11 Distributions of von Mises stress at the circumferentially aligned corrosion defects with various spacings under an internal pressure of 18 MPa. (a) 1.8°, (b) 10.8°, (c) 21.6°, and (d) 36°. *Source:* From Sun and Cheng [2021] / with permission from Elsevier.

the defects. As the circumferential spacing increases, the maximum anodic current density at the defects does not change apparently. However, the anodic current density at the defect adjacency drops rapidly. For example, at the circumferential spacing of 10.8° and 21.6°, the anodic current densities

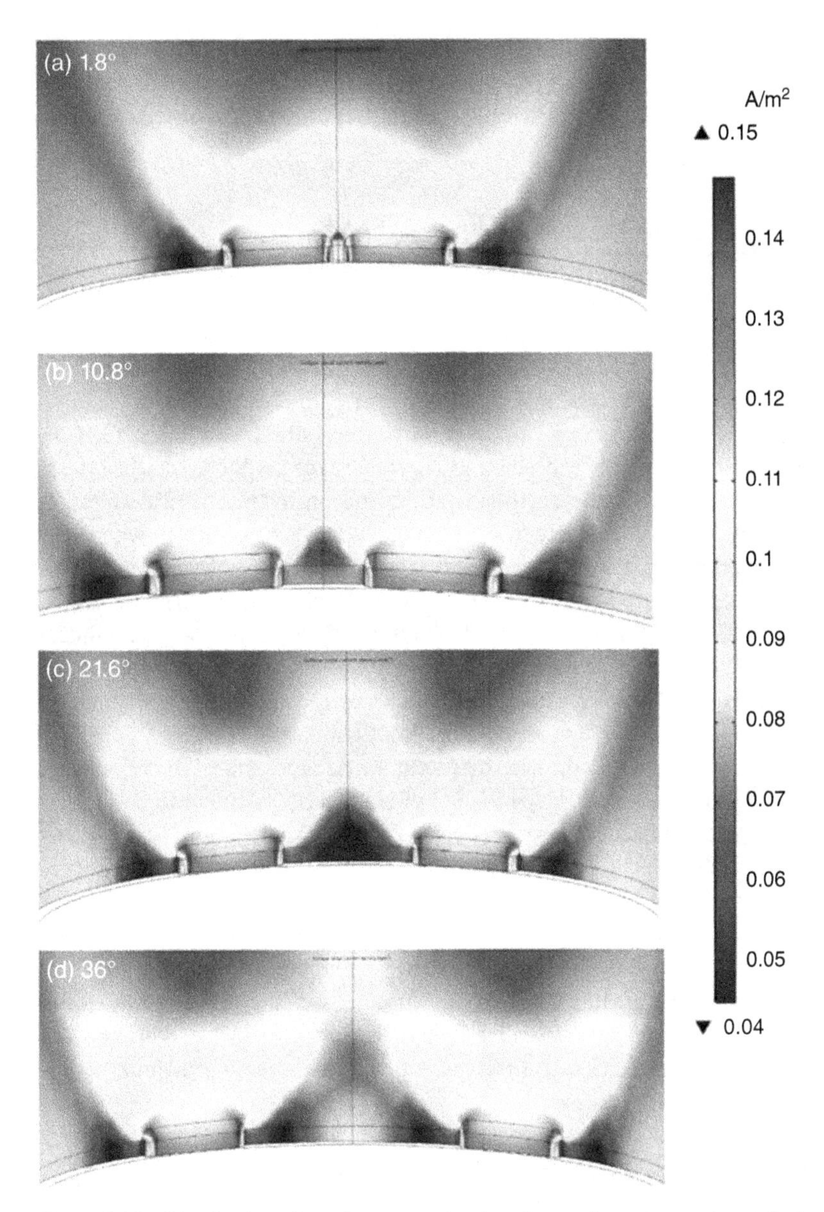

Figure 5.12 Distribution of anodic current density of two adjacent corrosion defects on the steel pipe under an internal pressure of 18 MPa in the soil solution as a function of the circumferential spacing. (a) 1.8°, (b) 10.8°, (c) 21.6°, and (d) 36°. *Source:* From Sun and Cheng [2021] / with permission from Elsevier.

at the defect adjacency are about 0.07 and 0.05 A/m^2, respectively. At the circumferential spacing of 36°, the two corrosion defects have their own current fields, which do not overlap each other, where the two defects are assessed separately.

5.2.2.4 A Critical Circumferential Spacing Criterion

When circumferentially aligned corrosion defects are present on pipelines, an M–E interaction exists between the defects when their spacing is within a certain range. The interaction results in a high local stress concentration and a great anodic current density (i.e., corrosion rate) at the defect adjacency, accelerating localized corrosion of the pipelines at both the defects and the adjacency. Thus, it is critical to define a proper criterion to evaluate if an interaction exists between adjacent corrosion defects during defect assessment. It is found that the electrochemical corrosion and mechanical stress factors contribute differently to the interaction. For example, when the circumferential spacing between the defects is up to 21.6°, the von Mises stress at the defect adjacency decreases to the same level as the stress of the uncorroded region. From the stress viewpoint, the defects are defined as noninteracting. However, the anodic current density at the adjacent area between the defects is still 25% greater than that of the uncorroded region at the same circumferential spacing [Sun and Cheng, 2019b], indicating that the interaction still exists between the corrosion defects in the electrochemical corrosion field.

A new criterion is proposed to determine the critical circumferential spacing between adjacent corrosion defects, within which an interaction exists between the defects. The criterion is defined as the ratio of the anodic current densities at the middle of the defect adjacency to that of uncorroded region, i.e., $i^{a}_{mid-defects}/i^{a}_{far-defects}$. When the ratio is greater than 1, an interaction exists, causing an increased anodic current density at the adjacency of the defects. The greater the ratio, the stronger the interaction between the defects. When the ratio is equal to 1, there is no interaction between the two corrosion defects, and they can thus be assessed as individual single defects.

It is noted that the proposed criterion can be influenced by defect geometrical parameters, i.e., defect width, length, and depth, in addition to the circumferential spacing. It is usually convenient to normalize the circumferential spacing as nondimensional by converting the angular angle to pipeline geometrical parameters including the pipe outer diameter and pipe wall thickness. For example, the corrosion defects with a circumferential width of 8° (i.e., $0.025\pi D$), 16° (i.e., $0.05\pi D$), and 32° (i.e., $0.1\pi D$) are used

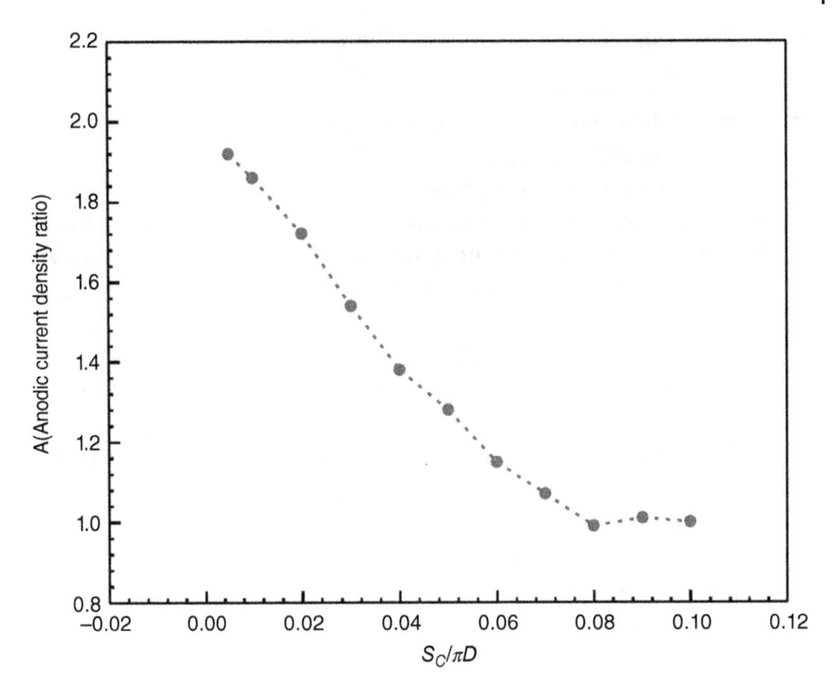

Figure 5.13 The ratio of the anodic current density at the defect adjacency to that of the uncorroded region as a function of the dimensionless circumferential spacing, S_C, for normally wide corrosion defects of $0.05\pi D$ in width.
Source: From Sun and Cheng [2019b] / with permission from Elsevier.

to represent narrow, normal, and wide defects, respectively, with the same length of $\sqrt{Dt}$ and depth of 0.7 t. Figure 5.13 shows the ratio of the anodic current density at the defect adjacency to that of the uncorroded region as a function of the dimensionless circumferential spacing, $S_C\,(\pi D)$, for normally wide corrosion defects of $0.05\pi D$. It is seen that the ratio decreases with increased S_C, indicating that the interaction between the corrosion defects decreases as the circumferential spacing increases. When the circumferential spacing reaches $0.08\pi D$, the ratio approaches 1, and the adjacent corrosion defects can be treated as two single isolated defects at this spacing. For narrow and wide corrosion defects with $0.025\pi D$ and $0.1\pi D$ in widths, respectively, the critical spacings are determined as $0.06\pi D$ and $0.08\pi D$, respectively. The results indicate that, after exceeding a certain width, the mutual interaction between the circumferentially aligned corrosion defects will not depend on the defect width.

5.2.3 Overlapped Corrosion Defects Under the M–E Interaction

Corrosion defects frequently overlap each other on pipelines. Investigation of the interaction between the overlapped corrosion defects also belongs to Level III defect assessment, relying on FE modeling and analysis of the corrosion defects under pipeline operating conditions. Although various methods and codes have been proposed for defect assessment and some apply for multiple corrosion defects on pipelines, none of them is applicable for overlapped defects. In a previous work, Sun and Cheng [2018] modeled and analyzed the stress distribution at two overlapped corrosion defects, i.e., one smaller corrosion defect located inside a larger corrosion defect, predicting the failure pressure of the pipeline. It was determined that the interaction of the overlapped corrosion defects decreased the failure pressure, as compared with the pipeline containing a single corrosion defect. Since the defects experience corrosion reaction, it is expected that the electrochemical corrosion interacts with the mechanical stress to generate the M–E interaction effect. The previous work did not integrate the M–E interaction in the FE model [Sun and Cheng, 2018]. As a result, the effect of corrosion reaction at the defects was ignored.

5.2.3.1 The Model

The M–E interaction at overlapped corrosion defects on an X46 steel pipe was modeled using the COMSOL Multiphysics® 4.2a software [Sun and Cheng, 2020], where the electrochemistry module provides an option to use experimentally measured potentiodynamic polarization parameters as electrochemical corrosion inputs [Yang and Cheng, 2016]. Due to symmetry, a quarter of the pipe containing two overlapped corrosion defects, as shown in Figure 5.14, was simulated to save computation time. The top (or big) defect is considered as the first layer defect, and the bottom (or small) defect is considered as the second layer defect. The two overlapping defects are aligned along the radial direction of the pipe. Their lengths are in the axial direction (z-axis) and the widths are in the circumferential direction (y-axis). The notch radius of the two defects is kept identical to their depths regardless of changes in length or width. To investigate the influence of defect length and depth on M–E interaction, it is assumed that the width of the defects keeps constant. This assumption is based on previous results that the defect width was usually not as important as the defect length and depth to affect the interaction of corrosion defects [Chandra and Mondal, 2016]. The mesh type in FE modeling is tetrahedral. The pipe body is under coarse meshing, and the corrosion defects are with mesh refinement. The mesh size

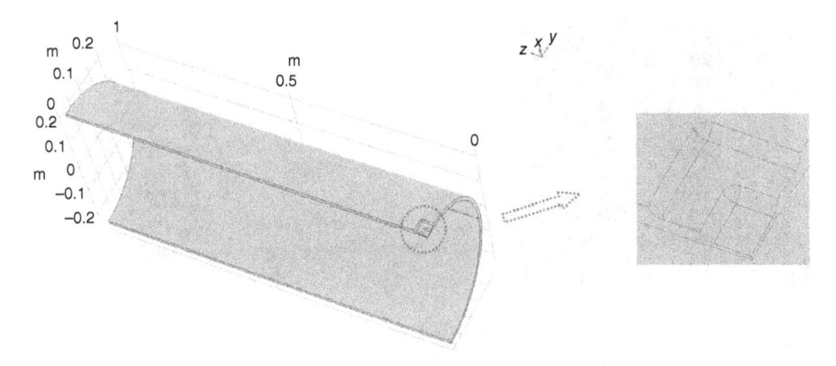

Figure 5.14 A 3D model showing a half of steel pipe containing two overlapped corrosion defects, where a quarter part of the defects is shown. *Source:* From Sun and Cheng [2020] / with permission from Elsevier.

ranges from 1 to 10 mm, as determined by a mesh sensitivity analysis conducted on mechanical stress and anodic current density.

The steel pipe is subject to an internal pressure, where both hoop stress and longitudinal stress are applied on the pipe. The corrosion defects are exposed to a near-neutral pH bicarbonate solution, which simulates the electrolyte trapped under disbonded coating. The FE modeling is conducted in three physical fields, i.e., mechanical stress, electrochemical corrosion, and the coupled multi-physics M–E fields. The distributions of elastoplastic stress and anodic current density are modeled, and the von Mises yielding criterion is used.

5.2.3.2 Modeling of Stress and Anodic Current Density at Overlapped Corrosion Defects Under Various Internal Pressures

The distribution of von Mises stress at the overlapped corrosion defects under various internal pressures is shown in Figure 5.15. Generally, an elevated internal pressure results in increased stress level at the defects. When the internal pressure is 11.65 MPa, the stress at the corrosion defects is almost identical to the stress of the pipe body, except for the corner and base of the bottom defect, where the local stress slightly exceeds the yield stress according to the stress-color bar. As the internal pressure increases to 13.65 MPa, a high stress concentration occurs at the bottom defect, and is more apparent at its corner. The top defect remains at the stress level below the yield stress. With further increase of the internal pressure to 15.65 MPa, the local stress at the corner of the bottom defect is close to the ultimate tensile strength of the steel, and the high stress area expands to the top defect. It is interesting to observe that the stress level at the top defect is

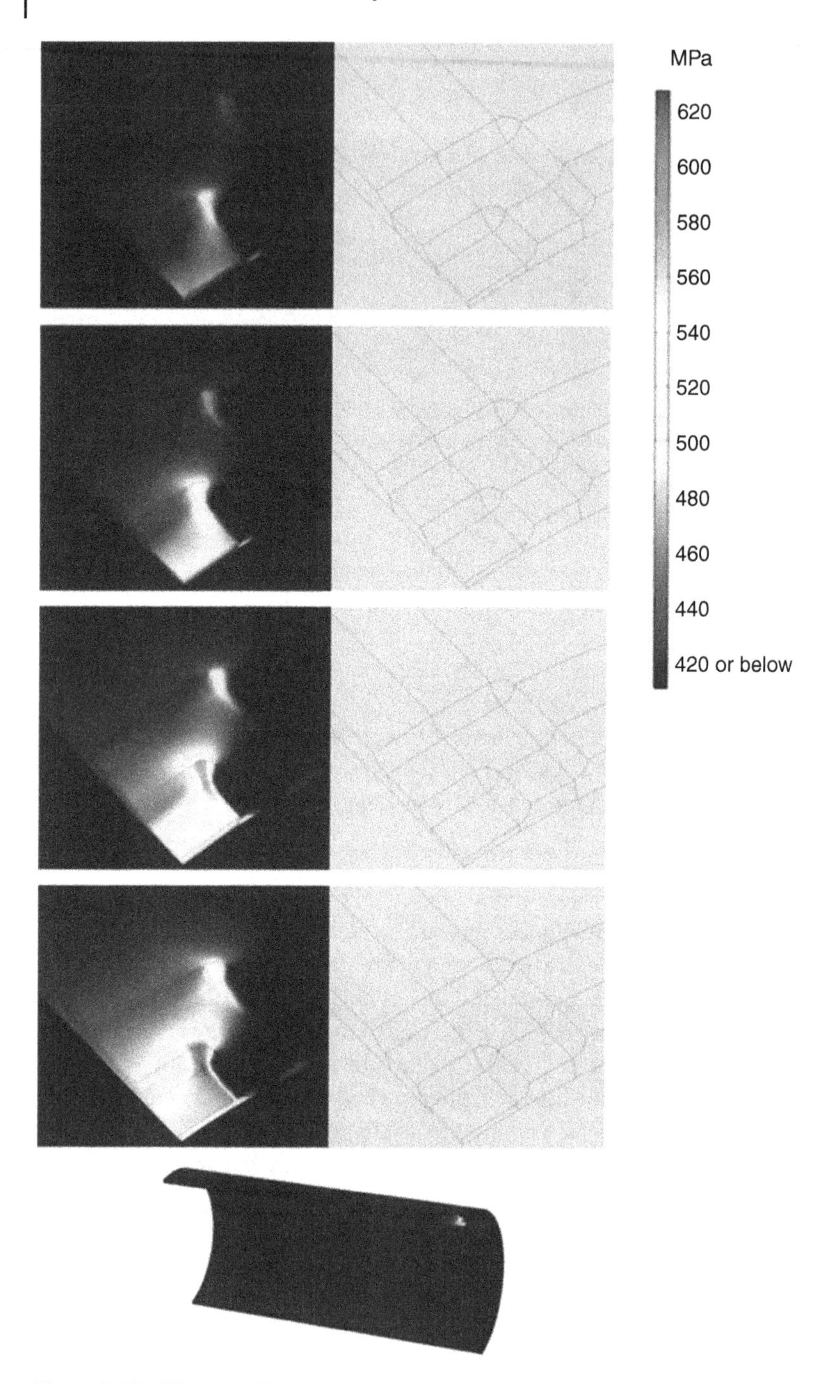

Figure 5.15 The von Mises stress distribution at the overlapped corrosion defects on the steel pipe under various internal pressures. *Source:* From Sun and Cheng [2020] / with permission from Elsevier.

still below the yield stress, except in its corner. When the internal pressure is up to 17.65 MPa, the local stress at the bottom defect is much greater than the yielding stress, with the maximum stress of about 620 MPa at the corner and side. The stress level at the top defect also exceeds the yield stress. Apparently, the bottom corrosion defect suffers from a much higher stress than the top defect.

Figure 5.16 shows the distribution of anodic current density at the overlapped corrosion defects in the near-neutral pH bicarbonate solution as a function of the internal pressure. When the internal pressure is 11.65 MPa, the anodic current density at the uncorroded region is about 0.04 A/m^2 (i.e., approximately 0.05 mm/y), while the corner of the bottom defect is up to 0.11 A/m^2 (i.e., 0.13 mm/y), showing an increased corrosion at the stress concentration area. When the internal pressure increases to 13.65, 15.65, and 17.65 MPa, the maximum anodic current densities at the corner of the bottom defect are up to 0.12, 0.13, and 0.15 A/m^2 (i.e., 0.14, 0.15, and 0.17 mm/y), respectively. However, the anodic current density at the uncorroded region on the pipe remains at 0.04 A/m^2. Obviously, a high anodic current density is always associated with high stress concentration, showing the stress-enhanced corrosion effect. Particularly, at the corner of the bottom defect, the corrosion rate is approximately three times that of the uncorroded region on the pipe.

5.2.3.3 Modeling of Stress and Anodic Current Density Distributions at Overlapped Corrosion Defects with Various Defect Depths

The distribution of von Mises stress at the overlapped corrosion defects with various depths of d_1 and d_2 under a fixed internal pressure of 17.65 MPa is shown in Figure 5.17, where d_1 and d_2 refer to the depths of top and bottom corrosion defects, respectively. It is seen that, while the depth of the bottom defect, d_2, is fixed at 4 mm, the stress concentration develops mainly at the corner of the bottom defect as d_1 increases. When $d_1 = 1$ mm, the stress distribution is approximately uniform, except for the longitudinal edge of the bottom defect. The overall stress level is below the yield stress. As d_1 increases to 2 and 3 mm, there is an increased stress concentration at the corner of the bottom defect, while the other region is still in a low stress level. When d_1 further increases to 4 mm, the maximum stress of about 630 MPa is obtained at the corner of the bottom defect. Similarly, while the depth of the top defect, d_1, is fixed at 4 mm, the increased depth of the bottom defect would increase the stress concentration of the bottom defect, while the average stress level at the top defect is low. As d_2 decreases from 4 to 3 mm, while the stress level at the corrosion defects reduces, the maximum stress is still

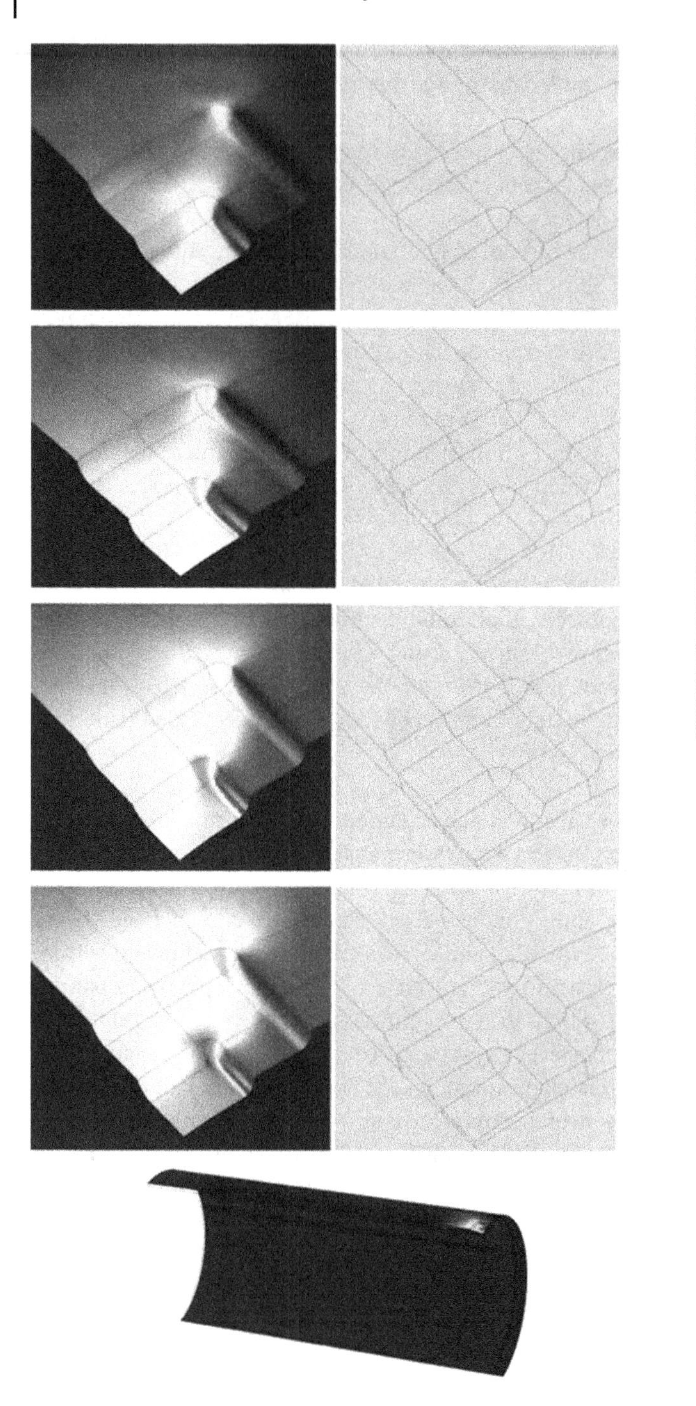

Figure 5.16 Distribution of anodic current density at the overlapped corrosion defects on the pipe in the test solution as a function of the internal pressure. *Source:* From Sun and Cheng [2020] / with permission from Elsevier.

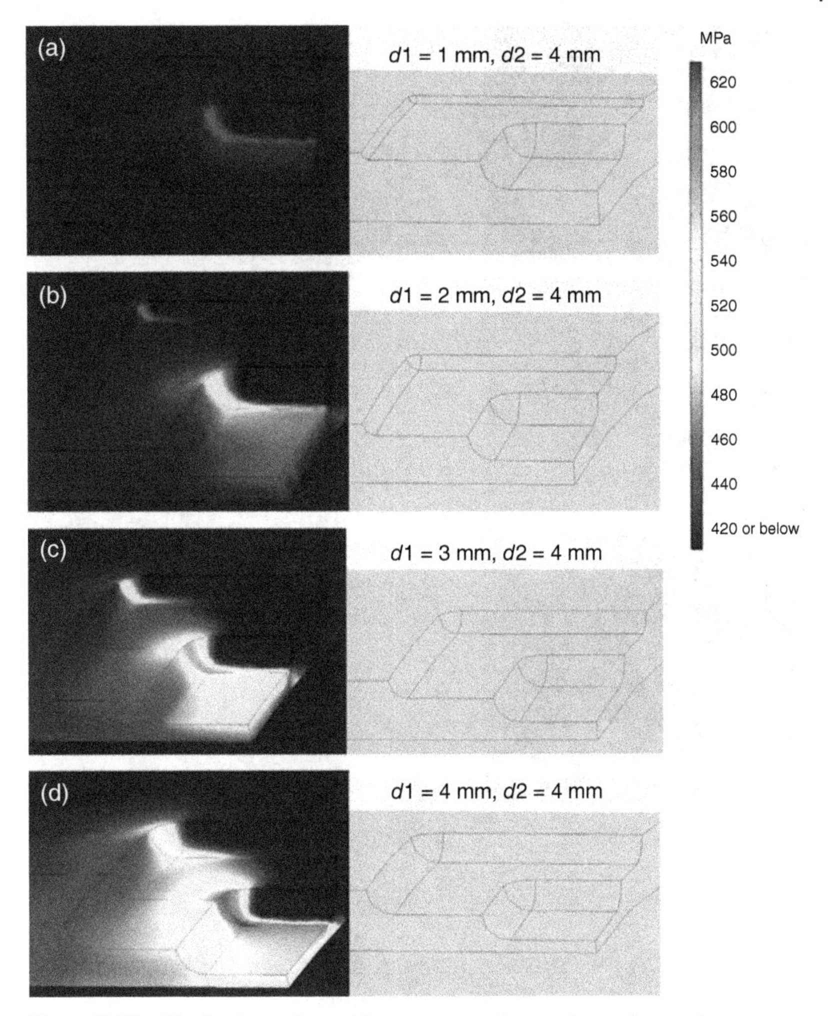

Figure 5.17 Distributions of von Mises stress at the overlapped corrosion defects with various depths of d_1 and d_2 under a fixed internal pressure of 17.65 MPa. (a–d) d_1 = 1, 2, 3, 4 mm, d_2 = 4 mm, (e–g) d_1 = 4 mm, d_2 = 3, 2, 1 mm. *Source:* From Sun and Cheng [2020] / with permission from Elsevier.

close to the ultimate tensile stress of 630 MPa at the longitudinal edge of the bottom defect. When d_2 is 2 mm, the high stress concentrates at the edge of the bottom defect only, while the other area is under a decreased stress level. When d_2 is decreased to 1 mm, a high stress concentration still exists, but is located at a very narrow zone at the corner of the bottom defect while the

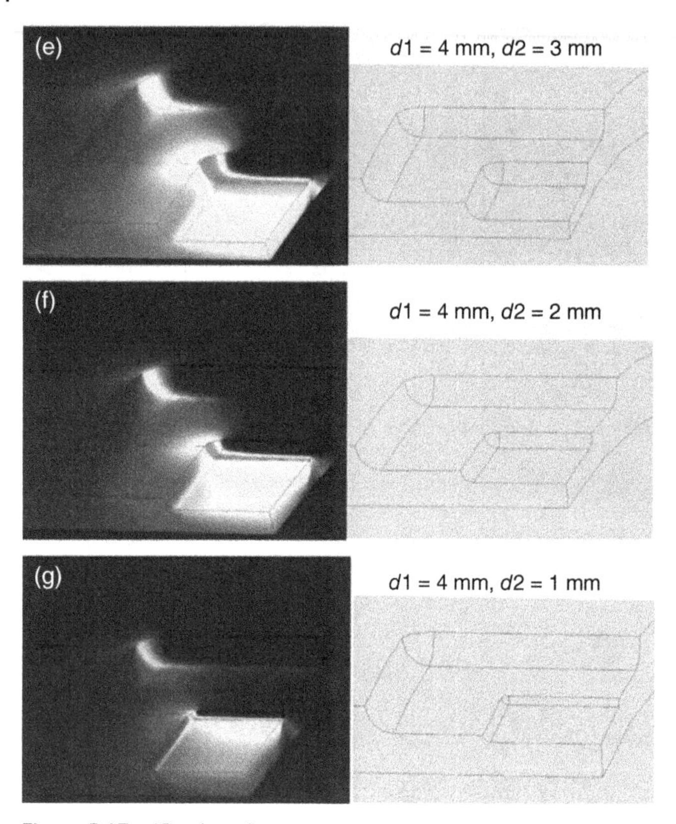

Figure 5.17 (Continued)

whole stress level decreases apparently. It is thus seen that, with the depth increase for either defect, the stress concentration develops mainly at the corner of the bottom defect, while the stress level of the top defect is low. Even when the top defect is in the elastic stress range, a high stress concentration can exist at the bottom defect. The stress concentration and the maximum von Mises stress at overlapped corrosion defects highly depend on the depth of the top defect. If the top defect is deep, no matter how small the bottom defect depth is, a high stress concentration can develop at the bottom defect.

Figure 5.18 shows the distributions of anodic current density at the overlapped corrosion defects with various depths of d_1 and d_2 under a fixed internal pressure of 17.65 MPa in the test solution. The dependence of the anodic current density on the top defect depth, d_1, is same as the distribution of von

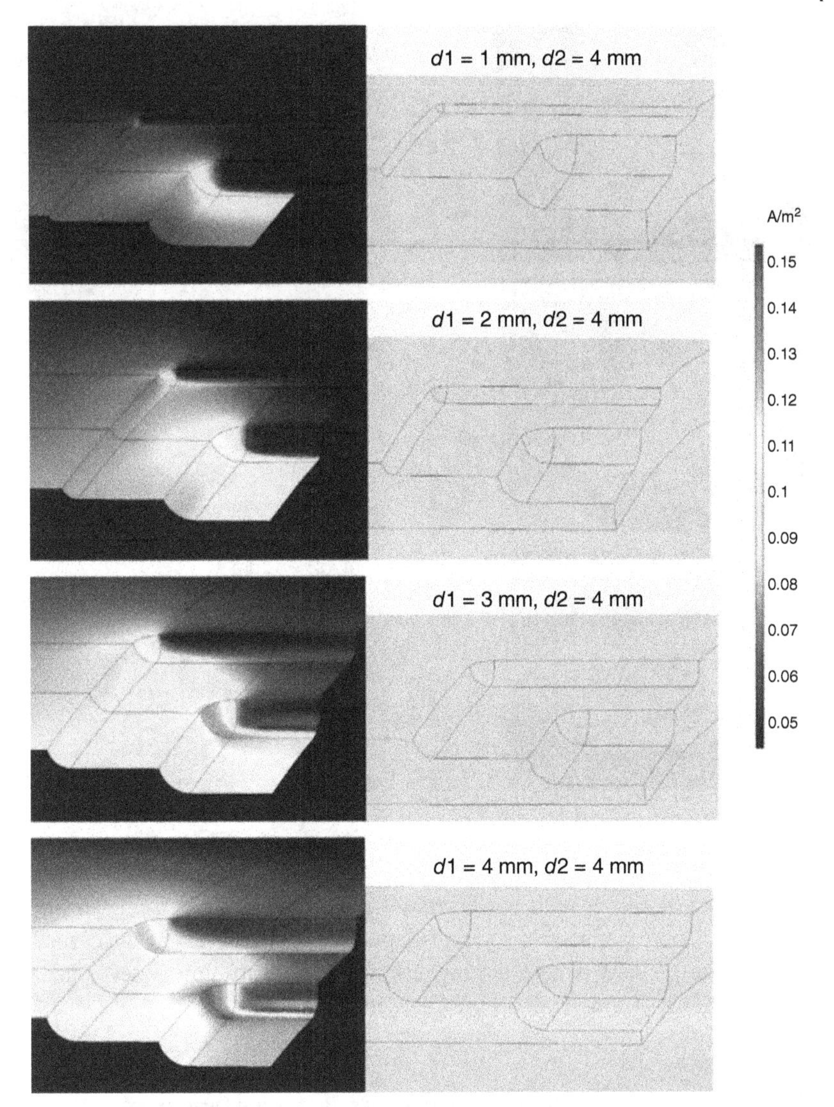

Figure 5.18 Distributions of anodic current density at the overlapped corrosion defects with various depths of d_1 and d_2 under a fixed internal pressure of 17.65 MPa in the test solution. *Source:* From Sun and Cheng [2020] / with permission from Elsevier.

Mises stress shown in Figure 5.17. When the bottom defect depth (d_2) is fixed at 4 mm, with the increase in d_1, the average anodic current density increases. The effect is especially apparent at the bottom defect, where the greatest anodic current density is observed. When the top defect depth

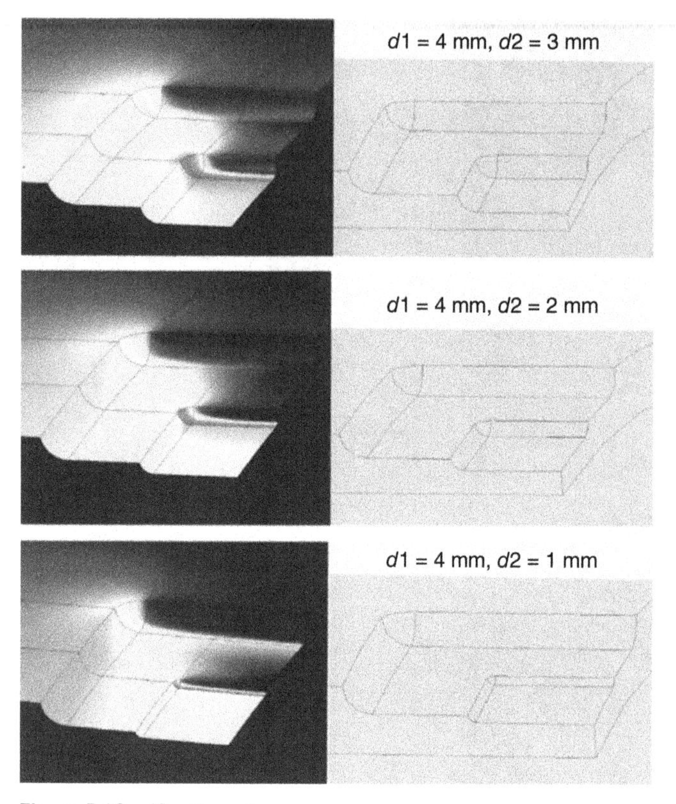

Figure 5.18 (Continued)

(d_1) is fixed at 4 mm, with the decrease in d_2, the anodic current density decreases. There is always the greatest anodic current density at the corner of the bottom defect, where the highest stress concentration is developed, showing the M–E interaction for increased corrosion.

5.2.3.4 Implications on Integrity of Pipelines Containing Overlapped Corrosion Defects

The presence of overlapped corrosion defects on pipelines results in a local stress concentration and enhanced M–E interaction at the defects. This scenario is unique compared with the situations where multiple corrosion defects are oriented either circumferentially or longitudinally [Sun and Cheng, 2019a, b]. The important findings include that the maximum stress always generates at the corner of the bottom defect, where the resulting M–E interaction induces an accelerated corrosion locally. By comparison with the work conducted on the overlapped corrosion defects in a mere mechanical field [Sun and Cheng, 2018], an inclusion of the M–C interaction in the model does not affect the mechanical performance of the steel pipe, while representing the realistic integrity-related issues. In addition to the defect depth, as discussed in the chapter, the defect length also affects the local stress concentration and M–E interaction [Sun and Cheng, 2020]. Generally, as the length of either top or bottom corrosion defect increases, the overall stress level increases, but the effect is more apparent at the bottom defect. Similarly, the increased M–E interaction due to the increasing length of the corrosion defects results in accelerated corrosion at the bottom defect rather than the top one. The effect of the defect geometry on both stress and corrosion is more apparent at the bottom defect than at the top defect, demonstrating that the bottom corrosion defect presents the weak site to cause pipeline perforation due to the synergism of mechanical stress and electrochemical corrosion.

5.3 Interactions of Multiple Corrosion Defects with Irregular Orientations

As stated above, methods and codes have been developed for assessment of multiple corrosion defects on pipelines, where the relative orientations of the defects are ideally modeled as longitudinal, circumferential, and radially aligned (i.e., overlapping, or inner and outer wall aligned). Corrosion defects may not be ideally aligned but are frequently located in random orientations. To date, there has been rare work investigating the interaction between corrosion defects with random orientations and the effect on performance and burst pressure of the pipelines. The following sections introduce the latest method developed for the purpose.

5.3.1 The Model Development

The FE modeling is based on the following conclusions drawn from previous work to ensure its adaptability:

- The effect of geometric differences of corrosion defects on their interaction is negligible [Zhang and Zhou, 2022].
- The interaction effect is independent of the steel grade but depends on the geometry and spacings of corrosion defects [Sun and Cheng, 2018].

- The mechanical properties and electrochemical corrosion parameters of pipeline steels are obtained using uniaxial tensile testing and corrosion measurements on the steel specimen in a near-neutral pH bicarbonate solution [Xu and Cheng, 2013].

A full-scale FE-based multi-physical field coupling model for an X100 steel pipe containing two identical external corrosion defects which are in varied orientations is shown in Figure 5.19. The pipe outer diameter D was 914 mm, and the pipe wall thickness t was 16 mm. The length of the pipe segment was set as 5.5 m, which was 6 times the outer diameter, avoiding the influence of boundary conditions on the modeling results at the defects. The corrosion defects were rectangular, with longitudinal, circumferential, and radial dimensions defined as length L (x-axis), width W (y-axis), and depth d (z-axis), respectively. The relative position of the defects depended on the circumferential spacing (S_C) and longitudinal spacing (S_L) between the defects. According to Figure 5.19h, the S_C between corrosion defects is defined by the arc length corresponding to the circular angle θ between the two defects, and $S_C = \pi D \times \theta/360°$. The following orientations of the corrosion defects are modeled. (i) Two corrosion defects are tangent to each other on the pipe surface as their longitudinal or circumferential spacing is 0, as shown in Figure 5.19a,b. (ii) The two defects overlap in either longitudinal or circumferential direction so that the S_C or S_L is negative. The value of −0.5 refers to the overlapping of a half dimension of the defects, as shown in Figure 5.19c,d. (iii) A complete overlapping of the defects so that S_C or S_L is −1 (Figure 5.19e,f). (iv) The defects are separated with positive values of S_C and S_L, as seen in Figure 5.19g.

The modeled pipe material is X100 steel, where an isotropic hardening function applies for the constitutive relation of the steel. The yield and ultimate tensile strengths of X100 steels are 806 and 891 MPa, respectively [Xu and Cheng, 2013]. The corrosion of X100 steel in near-neutral pH bicarbonate solutions is in active dissolution state, with the anodic and cathodic reactions, as well as electrochemical corrosion parameters defined previously [Xu and Cheng, 2012b, 2013]. A multi-physics field coupling effect is established at the corrosion defects by considering electrical and mechanical stress, and electrochemical corrosion fields.

According to Sun and Cheng [2019a], the ratio of $i^{a}_{mid-defects}$ to $i^{a}_{far-defects}$ defines the M–E interaction between adjacent corrosion defects. When the ratio ($i^{a}_{mid-defects}/i^{a}_{far-defects}$) > 1, the M–E interaction exists between the defects. The interaction degree is positively related to the ratio. There is no interaction between the corrosion defects when the ratio = 1.

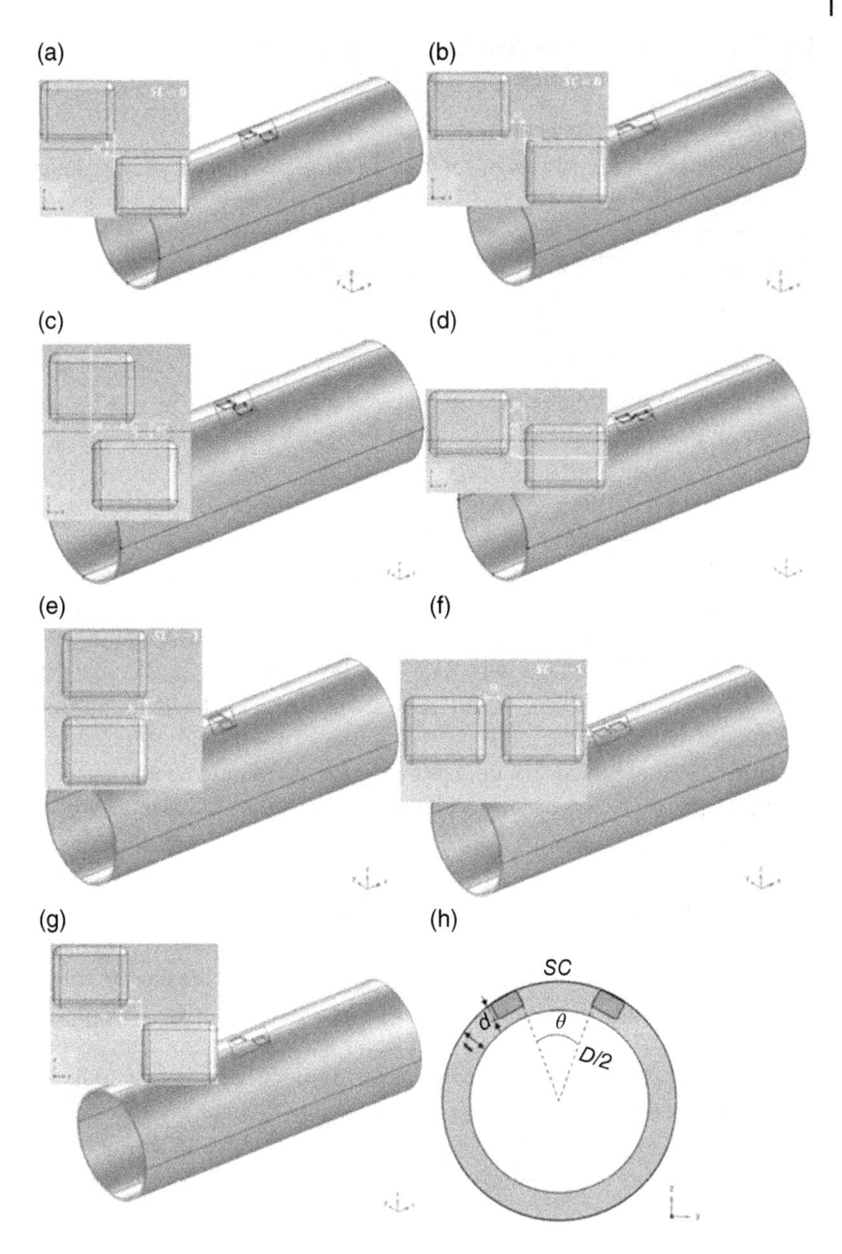

Figure 5.19 3D FE model for corrosion defects with different orientations on a steel pipe (a) $S_L = 0$, $S_C > 0$, (b) $S_C = 0$, $S_L > 0$, (c) $S_L = -0.5$, $S_C > 0$, (d) $S_C = -0.5$, $S_L > 0$, (e) $S_L = -1$, $S_C > 0$, (f) $S_C = -1$, $S_L > 0$, (g) $S_C > 0$, $S_L > 0$, and (h) the cross-sectional view of the circumferential spacing between the corrosion defects. *Source:* From Qin et al. [2023] / with permission from Elsevier.

5.3.2 Effects of Relative Positions and Spacing of the Corrosion Defects on M–E Interaction

5.3.2.1 Relative Longitudinal Positions and Spacing

At a fixed circumferential spacing of $0.2\sqrt{Dt}$, corrosion defects with various longitudinal spacings, as defined in Figure 5.19a,c,e,g, are modeled to determine the maximum von Mises stress and anodic current density, as well as the ratio of $i^{a}_{mid-defects}$ to $i^{a}_{far-defects}$, and the results are shown in Figure 5.20, where the dashed line in Figure 5.20b represents the ratio of 1 (i.e., no interaction exists between the corrosion defects). It is seen that both the maximum von Mises stress and the maximum anodic current density are recorded where the longitudinal spacing of the defects is 0. The von Mises stress and anodic current density then decrease as the longitudinal spacing increases positively until $0.8\sqrt{Dt}$ (i.e., 96 mm). After that, both values keep almost unchanged. It is seen that the interaction between the defects does not exist at this longitudinal spacing. When the defects overlap in the longitudinal direction (i.e., the spacing value is negative), the M–E interaction between the defects decreases, causing reduction in both von Mises stress and anodic current density. When the defects completely overlap in the longitudinal direction (i.e., $S_L = -1$), the von Mises stress and anodic current density decrease by 3.8% and 29.45%, respectively, compared with the values at the longitudinal spacing of 0.

Figure 5.21 shows the distributions of von Mises stress and anodic current density at the corrosion defects with various longitudinal spacings while under a fixed circumferential spacing of $0.2\sqrt{Dt}$. As the longitudinal overlaps increase (i.e., S_L decreases from 0 to −1), the von Mises stress gradually decreases. This may be attributed to a shielding effect of the overlapped defects [Zhang, 2021]. Moreover, the overlapped defects in the longitudinal direction cause a decrease in anodic current density at both the defect center and the adjacent area. The shielding effect also occurs in the electrochemical corrosion field. The greatest M–E interaction resulting in the highest von Mises stress and anodic current density levels occurs when the longitudinal spacing between the two defects is 0 (i.e., $S_L = 0$). The interaction between the defects gradually decreases as their longitudinal spacing increases. After exceeding $0.8\sqrt{Dt}$ (i.e., 96 mm), the corrosion defects can be assessed individually.

5.3.2.2 Relative Circumferential Positions and Spacing

Varied circumferential positions of corrosion defects are defined by the circumferential spacing, while the longitudinal spacing between them is fixed

(a)

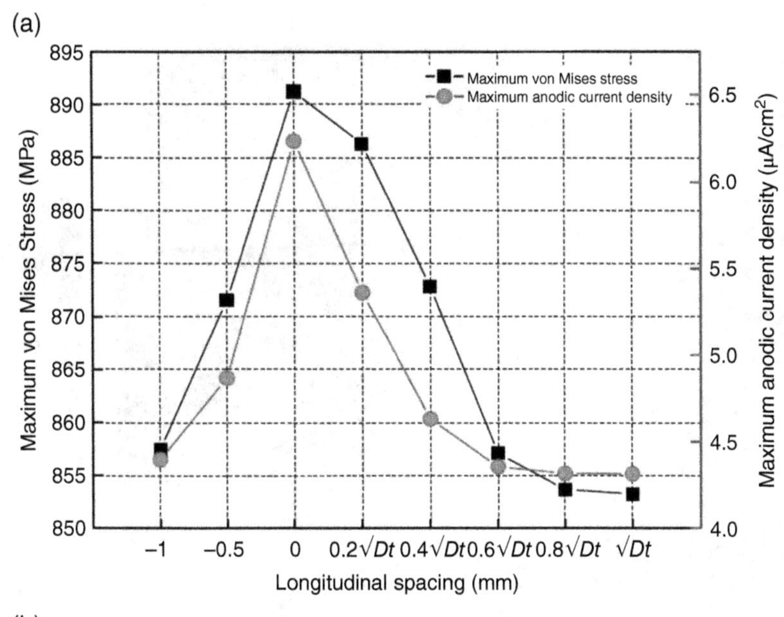

(b)

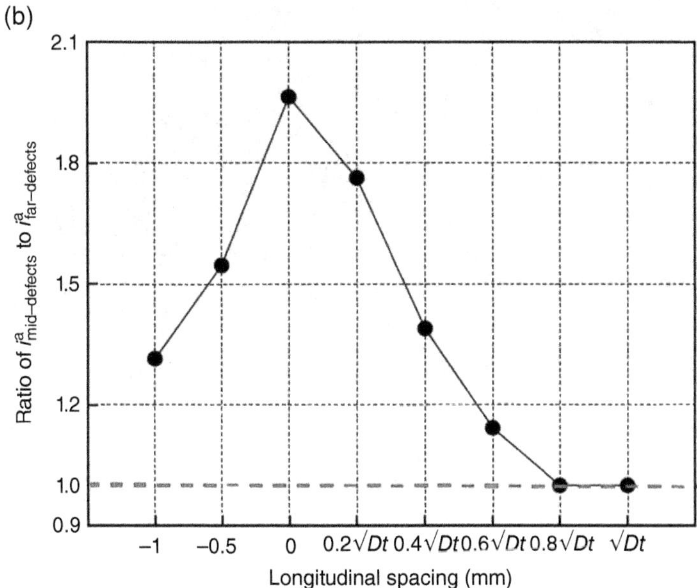

Figure 5.20 Effect of the longitudinal spacing between corrosion defects, which are at a fixed circumferential spacing of $0.2\sqrt{Dt}$, on (a) maximum von Mises stress and maximum anodic current density and (b) the ratio of $i^{a}_{mid-defects}$ to $i^{a}_{far-defects}$, where the dashed line refers to 1. *Source:* From Qin et al. [2023] / with permission from Elsevier.

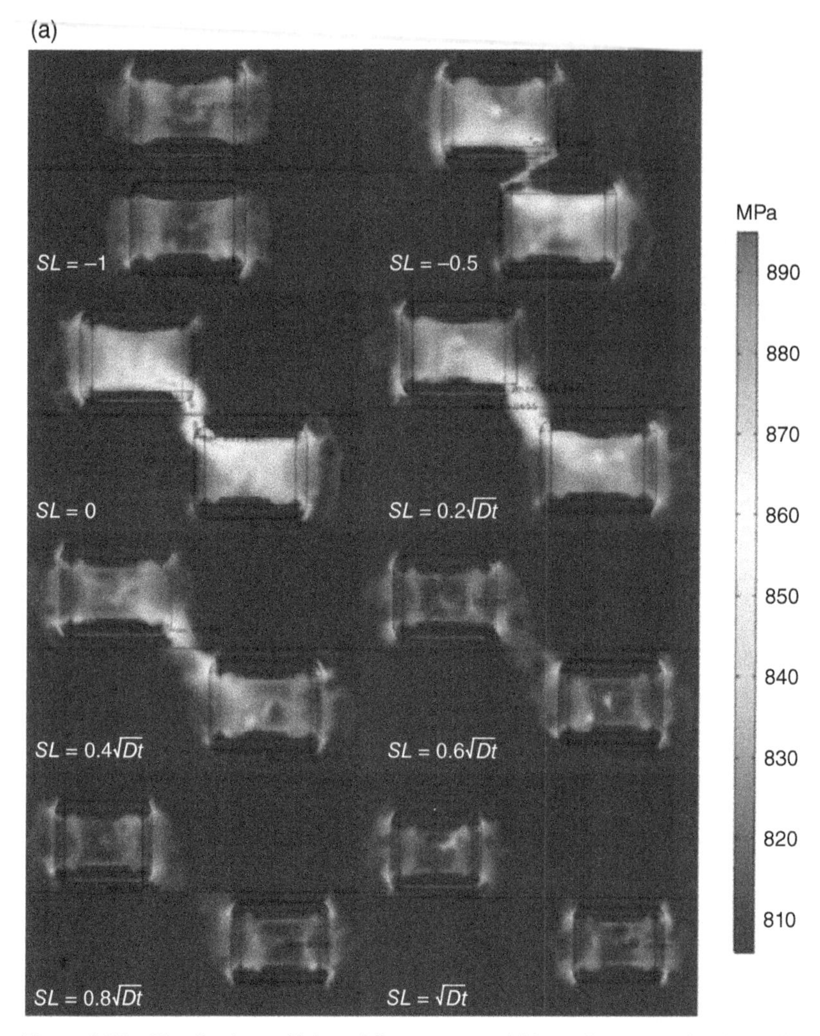

Figure 5.21 Distributions of (a) von Mises stress and (b) anodic current density at the corrosion defects with various longitudinal spacings. *Source:* From Qin et al. [2023] / with permission from Elsevier.

at $0.2\sqrt{Dt}$, as defined in Figure 5.19b,d,f,g. Figure 5.22 shows the effect of circumferential spacing on the maximum von Mises stress and maximum anodic current density, as well as the ratio ($i^{a}_{mid-defects}/i^{a}_{far-defects}$). In general, the maximum von Mises stress and maximum anodic current density decrease with the increased circumferential spacing. A critical value of

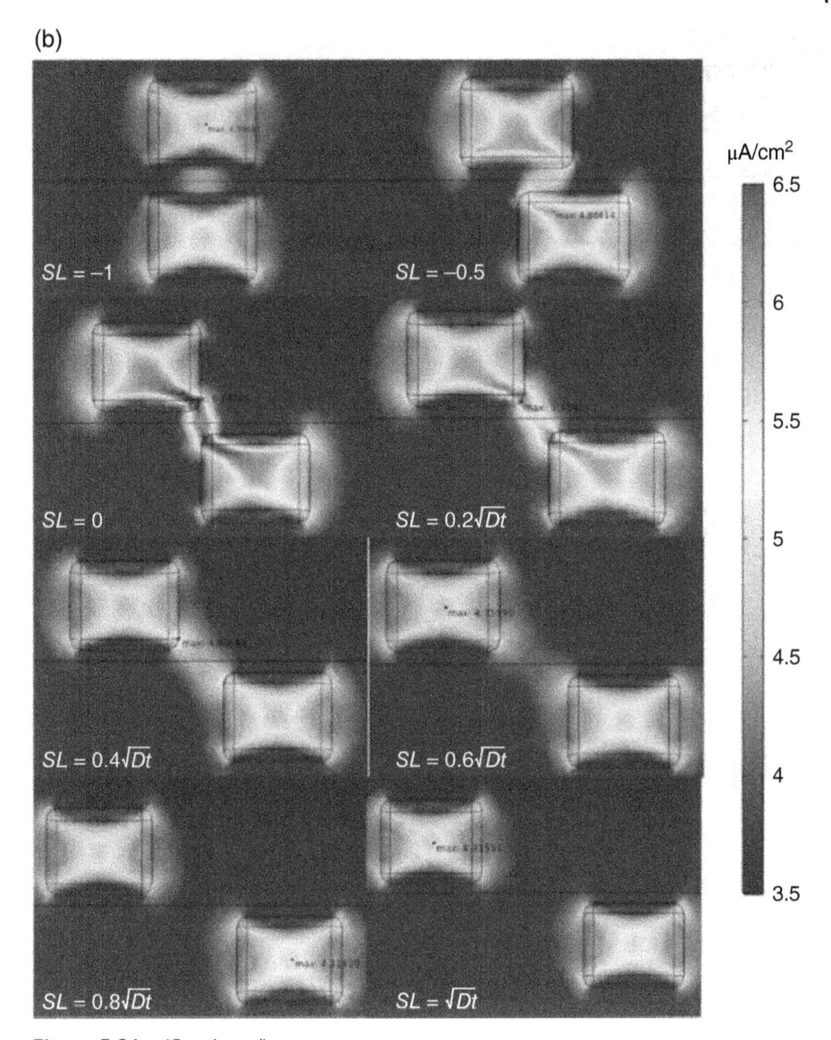

Figure 5.21 (Continued)

$0.6\sqrt{Dt}$ (i.e., 72 mm) exists, after which the maximum von Mises stress and anodic current density approximately keep unchanged. Also, the ratio $i^{a}_{mid-defects}/i^{a}_{far-defects} = 1$, i.e., the interaction between the defects is negligible at this circumferential spacing. The maximum von Mises stress reaches the ultimate tensile strength of the steel when the defects overlap in the circumferential direction (i.e., $-1 \leq S_C \leq 0$). The maximum anodic current density and the M–E interaction decrease when the defects completely overlap (i.e., $S_C = -1$).

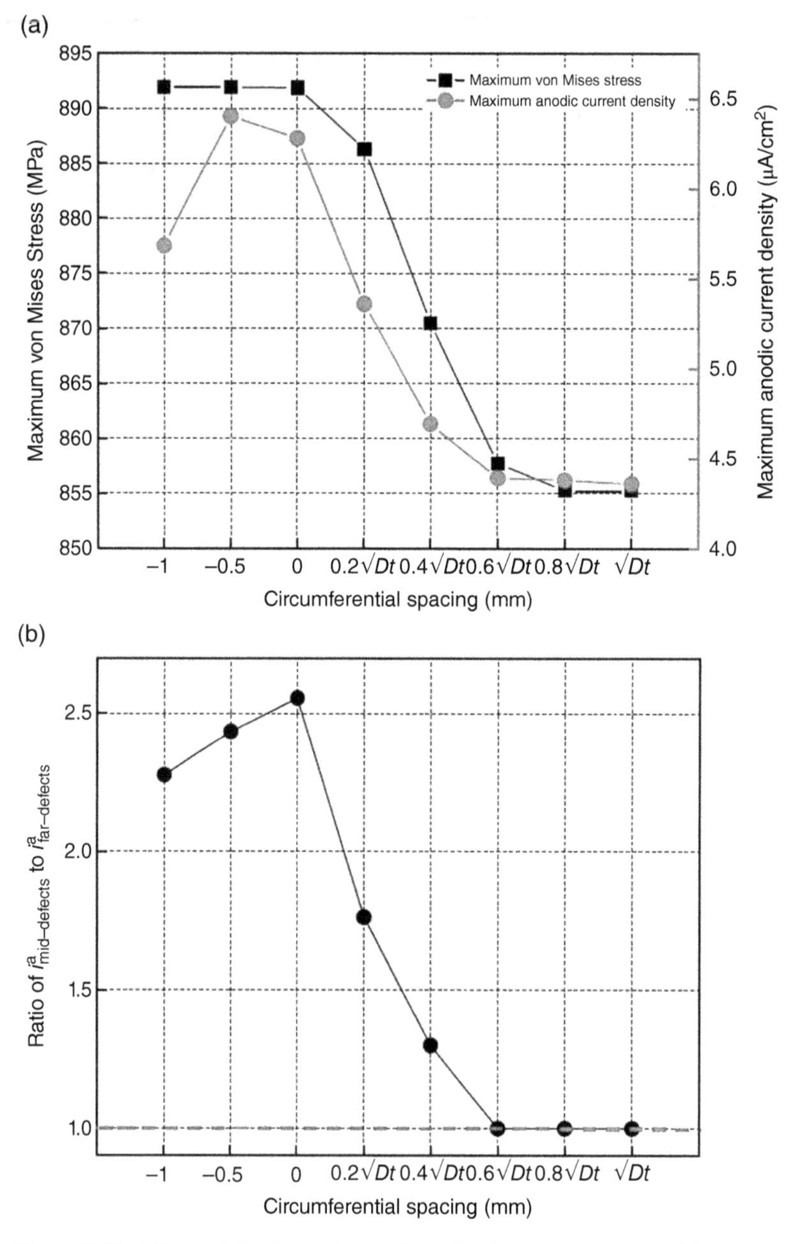

Figure 5.22 Effect of the circumferential spacing between corrosion defects, which are at a fixed longitudinal spacing of $0.2\sqrt{Dt}$, on (a) maximum von Mises stress and maximum anodic current density and (b) the ratio of $(i^a_{mid-defects}/i^a_{far-defects})$, where the dashed line refers to the ratio of 1. *Source:* From Qin et al. [2023] / with permission from Elsevier.

Figure 5.23 shows the distributions of von Mises stress and anodic current density at two corrosion defects with various circumferential spacings and a fixed longitudinal spacing of $0.2\sqrt{Dt}$ under an internal pressure of 17 MPa. When the overlapping level of the defects increases, the maximum von Mises stress occurs at the adjacent area, rather than the defects. The

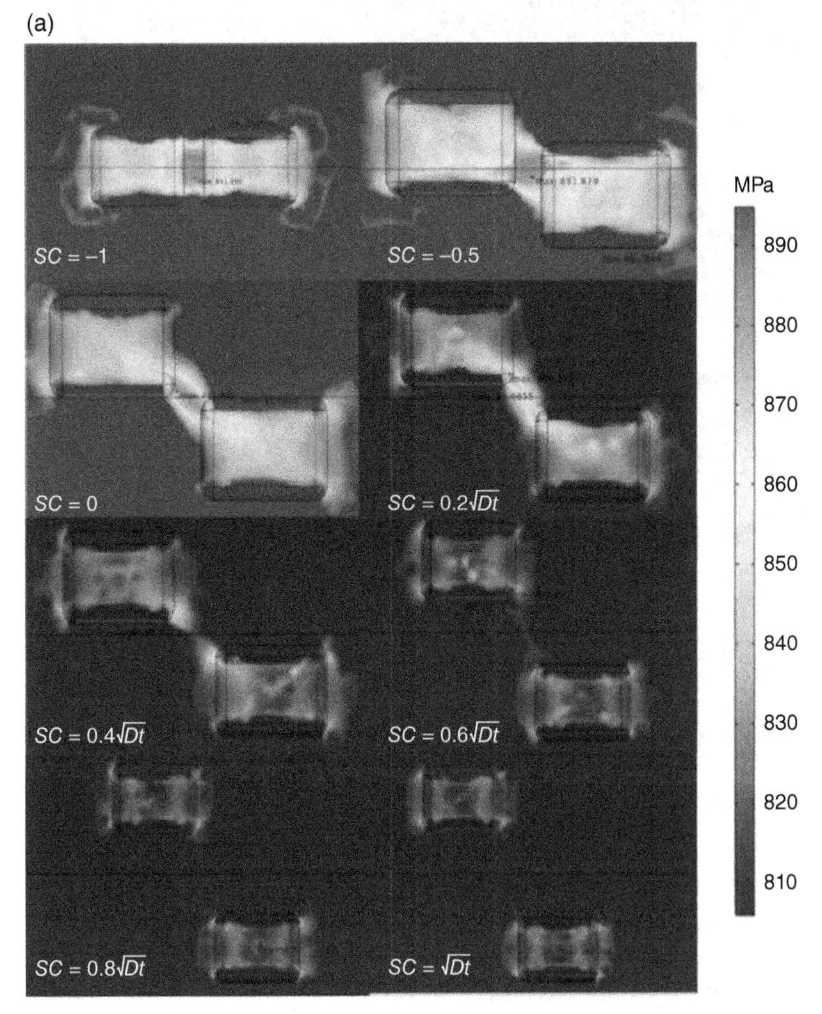

Figure 5.23 Distributions of (a) von Mises stress and (b) anodic current density at two corrosion defects with various circumferential spacings and a fixed longitudinal spacing of $0.2\sqrt{Dt}$ under an internal pressure of 17 MPa. *Source:* From Qin et al. [2023] / with permission from Elsevier.

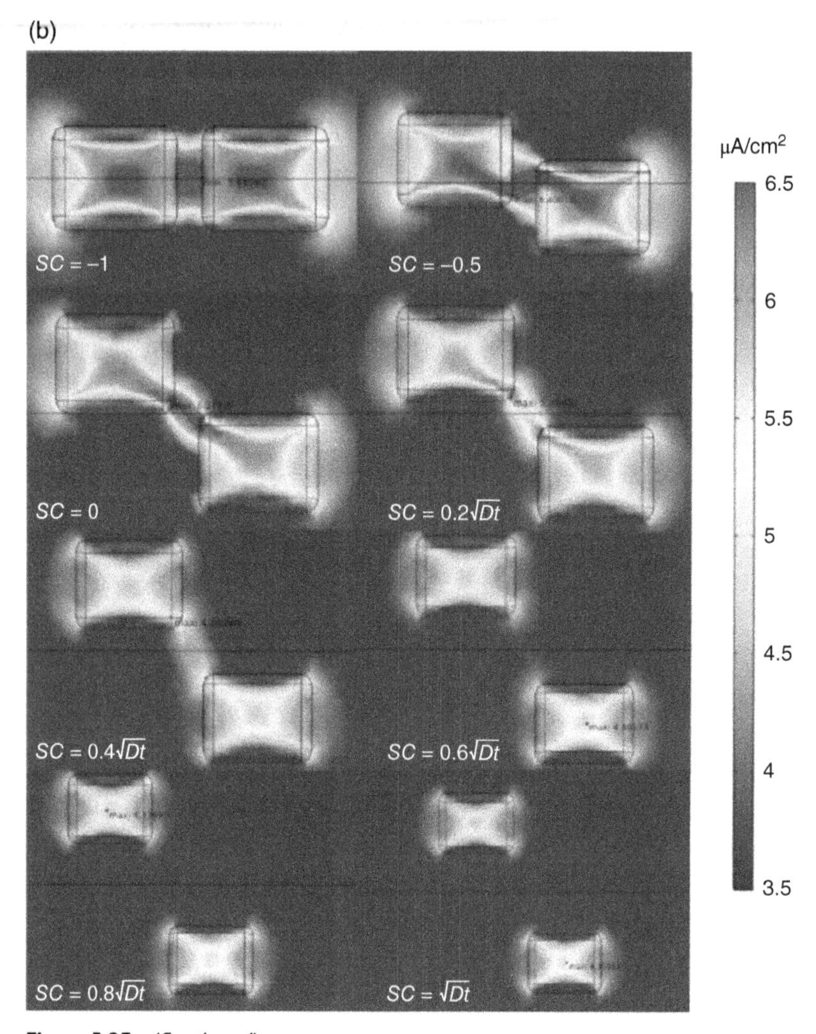

Figure 5.23 (Continued)

maximum von Mises stress exceeds yield strength of the steel when the circumferential spacing is smaller than $0.6\sqrt{Dt}$. As the spacing increases, the von Mises stress levels at both the defects and the adjacent area decrease due to reduced M–E interaction. When the circumferential spacing reaches $0.8\sqrt{Dt}$, the von Mises stress level at the adjacent area between the defects is in elastic range, and the maximum von Mises stress is shifted to the defect center. The von Mises stress at the defect is independent of the circumferential spacing in the absence of M–E interaction, where the corrosion defects can be assessed separately. Furthermore, the M–E interaction enhances the anodic current density at the defects when they overlap. However, when the defects completely overlap in the circumferential direction (i.e., $S_C = -1$), the maximum anodic current density decreases due to shielding effect on the M–E interaction. The increase in circumferential spacing between the defects reduces the anodic current density. When the circumferential spacing exceeds $0.6\sqrt{Dt}$, the electrochemical corrosion fields of the two defects are separated, and the local anodic current density is not related to the circumferential spacing. The anodic current density in the defect adjacent area is much lower, indicating a lower corrosion rate.

5.3.3 Implication on Pipeline Integrity in the Presence of Multiple, Irregularly Oriented Corrosion Defects

A parameter sensitivity analysis method is used to define the contributions of specific parameters such as defect spacing, defect depth, and length, and internal pressure to the M–E interaction. The used method is the delta moment independent measure. Figure 5.24 shows the parameter sensitivity on the degree of M–E interaction between corrosion defects with

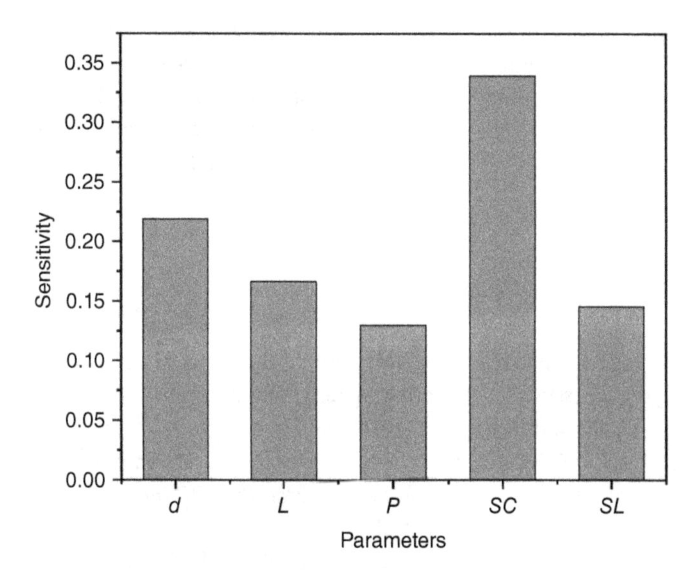

Figure 5.24 Parameter sensitivity on the M–E interaction between corrosion defects with various orientations on X100 steel pipelines. *Source:* From Qin et al. [2023] / with permission from Elsevier.

various orientations [Qin et al., 2023]. It is seen that the M–E interaction between the defects is most sensitive to the circumferential spacing (S_C), followed by defect depth (d), length (L), longitudinal spacing (S_L), and internal pressure (P). Thus, there is the strongest influence of the relative circumferential orientation of corrosion defects on the M–E interaction.

The M–E interaction of adjacent corrosion defects affecting the pipeline performance condition is associated with the critical spacing between the defects. For example, Sun and Cheng [2019a] determined that the maximum spacing between longitudinally aligned corrosion defects (120 mm in length) was 240 mm under the given condition. This work established that when the circumferential spacing was 24 mm, the M–E interaction between two 120 mm-long defects does not exist after their longitudinal spacing is up to 96 mm. This shows that the maximum spacing under which the M–E interaction exists between adjacent defects depends on both circumferential and longitudinal orientations. Thus, the M–E interaction between corrosion defects with irregular orientations should be considered in pipeline defect assessment program. However, existing interaction rules do not include this, and should be updated for improved pipeline integrity.

The work was conducted on X100 steel pipe, and the results can be extended to other steel grades if conditions permit. It is also possible to study pipeline conditions under different external environments, such as offshore and marine environments. While the deterministic approach is used to investigate the M–E interactions between corrosion defects with various orientations, a probabilistic method can be developed to capture parametric uncertainties, such as the variations in steel's properties and corrosion defect parameters [Adumene et al., 2021] by combining with the ILI data.

References

Adumene, S., Khan, F., Adedigba, S., Zendehboudi, S., Shiri, H. (2021) Offshore pipeline integrity assessment considering material and parametric uncertainty, *J. Pipeline Sci. Eng.*, 1, 265–276.

Benjamin, A.C., Cunha, D.J.S. (2007) New method for the prediction of the failure pressure of interacting corrosion defects, *Proc. the 17th Int. Offshore Polar Eng. Conf.*, ISOPE, Lisbon, Portugal.

Benjamin, A.C., Freire, J.L.F., Vieira, R.D., Diniz, J.L., De Andrade, E.Q. (2005) Burst tests on pipeline containing interacting corrosion defects, *Proc. Int. Conf. Offshore Mechanics Arctic Eng.*, paper no. OMAE2005.67059, Halkidiki, Greece.

Benjamin, A.C., Freire, J.L.F., Vieira, R.D., Diniz, J.L., De Andrade, E.Q. (2006) Burst tests on pipeline containing interacting corrosion defects, *Proc. Int. Conf. Offshore Mech. Arctic Eng.*, Hamburg, Germany.

Benjamin, A.C., Freire, J.L.F., Vieira, R.D., Cunha, D.J.S. (2016a) Interaction of corrosion defects in pipelines – part 1: fundamentals, *Int. J. Press. Vessel Pip.*, 144, 56–62.

Benjamin, A.C., Freire, J.L.F., Vieira, R.D., Cunha, D.J.S. (2016b) Interaction of corrosion defects in pipelines – part 2: MTI JIP database of corroded pipe tests, *Int. J. Press. Vessel Pip.*, 145, 41–59.

Chandra B., Mondal B.C. (2016) Burst pressure assessment for pipelines with multiple corrosion defects, *Can. Soc. Civil Eng. Resilient Infrastructure*, London, OB, Canada.

DNV (2004) *Corroded Pipelines–Recommended Practice*, RP-F10, Norway.

Liu, Z.Y., Li, X.G., Cheng, Y.F. (2011) Effect of strain rate on cathodic reaction during stress corrosion cracking of X70 pipeline steel in a near-neutral pH solution, *J. Mater. Eng. Perform.*, 20, 1242–1246.

Qin, G.J., Huang, Y.F., Wang, Y.H., Cheng, Y.F. (2023) Pipeline condition assessment and finite element modeling of mechano-electrochemical interaction between corrosion defects with varied orientations on pipelines, *Tunnel. Underground Space Technol.*, 136, 105101.

Sun, J.L., Cheng, Y.F. (2018) Assessment by finite element modeling of the interaction of multiple corrosion defects and the effect on failure pressure of corroded pipelines, *Eng. Struct.*, 165, 278–286.

Sun, J.L., Cheng, Y.F. (2019a) Modelling of mechano-electrochemical interaction of multiple longitudinally aligned corrosion defects on oil/gas pipelines, *Eng. Struct.*, 190, 9–19.

Sun, J.L., Cheng, Y.F. (2019b) Investigation by numerical modeling of the mechano–electrochemical interaction of circumferentially aligned corrosion defects on pipelines, *Thin Wall. Struct.*, 144, 106314.

Sun, J.L., Cheng, Y.F. (2020) Modelling of mechano-electrochemical interaction at overlapped corrosion defects and the implication on pipeline failure prediction, *Eng. Struct.*, 213, 110466.

Sun, J.L., Cheng, Y.F. (2021) Modeling of mechano-electrochemical interaction between circumferentially aligned corrosion defects on pipeline under axial tensile stresses, *J. Petro. Sci. Eng.*, 198, 108160.

Wang, Y., Wharton, J.A., Shenoi, R.A. (2016) Mechano-electrochemical modelling of corroded steel structures, *Eng. Struct.*, 128, 1–14.

Xu, L.Y., Cheng, Y.F. (2012a) Reliability and failure pressure prediction of various grades of pipeline steel in the presence of corrosion defects and pre-strain,
Int. J. Press. Vessel Pip., 89, 75–84.

Xu, L.Y., Cheng, Y.F. (2012b) An experimental investigation of corrosion of X100 pipeline steel under uniaxial elastic stress in a near-neutral pH solution, *Corros. Sci.*, 59, 103–109.

Xu, L.Y., Cheng, Y.F. (2013) Development of a finite element model for simulation and prediction of mechano-electrochemical effect of pipeline corrosion, *Corros. Sci.*, 73, 150–160.

Yang, Y., Cheng, Y.F. (2016) Stress enhanced corrosion at the tip of near-neutral pH stress corrosion cracks on pipelines, *Corrosion*, 72, 1035–1043.

Zhang, S. (2021) *Burst capacity evaluation of corroded pipelines under internal pressure and internal pressure combined with longitudinal compression*. PhD thesis, Western University, London, ON, Canada.

Zhang, S., Zhou, W. (2022) Interaction assessment of diagonally spaced identical corrosion defects and fully aligned unequal-sized corrosion defects under combined loads, *J. Press. Vessel Technol.*, 144, 041803.

6

Assessment of Dents on Pipelines

6.1 Introduction

Dent, a permanent inward plastic deformation on pipe walls, is a common type of mechanical damage introduced during pipeline construction and excavation activities. A dent does not cause metal loss on the pipe body. Statistics showed that dents have caused frequent pipeline failures [Liu and Francis, 2004; Hyde et al., 2007; Abdelmoety et al., 2022]. It was reported that 50% of pipelines in service contained over 10 dents after excavation and inspection [Dawson et al., 2006]. The ILI data collected on 7022 km of pipelines in Europe found 9851 dents, with an average of 1.4 dents per kilometer [Dawson et al., 2006]. In China, a total of 193 dents, with an average of 1.6 dents per kilometer, were detected on a 120 km pipeline of the Western Pipeline Company. In particular, the depths of 112 dents exceeded 6% of the pipe outer diameter, and the maximum dent depth was up to 15% [Allouti et al., 2012]. Moreover, 60% of pipeline failure cases were related to dents or dents combined with other types of defects [Tian and Zhang, 2017a]. Development of accurate models and reliable techniques for dent assessment has been paid extensive attention for pipeline FFS determination [Zhao et al., 2022a,b].

The main standards and methods used today for dent assessment include ASME B31.8, ASME B31G, API 579, API 1160, CSA Z662, and the United Kingdom Onshore Pipeline Association (UKOPA) [Bernard et al., 2013; Gao and Krishnamurthy, 2015; Dawson et al., 2018; Tee and Wordu, 2020]. Principally, there are two types of criteria for dent assessment, i.e., depth-based criterion and strain-based criterion, where a critical depth and a critical maximum strain, respectively, at the center of a dent are used as indicators to determine the damage associated with the dent [Noronha et al., 2010; Wu et al., 2015]. The standards based on

Defect Assessment for Integrity Management of Pipelines, First Edition. Y. Frank Cheng.

the critical dent depth were developed to evaluate dent-induced pipeline damages based on dimension of the dent depth [Adeeb and Horsley, 2006; Allouti et al., 2012; Arumugam et al., 2018]. For the strain-based criterion, the maximum equivalent strain at the dent center was calculated based on developed formulas and the dent profile information obtained from ILI tools [Lukasiewicz et al., 2006; Gao and Krishnamurthy, 2015]. When a dent is combined with other types of defects, such as gouge, corrosion, and cracks, which are also commonly encountered on pipelines, the pipelines tend to suffer from a reduced burst strength and pressure-bearing capability [Błachut and Iflefel, 2008; Freire et al., 2019; Zhao et al., 2021]. Methods and standards are thus required to assess threats induced by the combined defects to the pipelines [Alexander and Brownlee, 2007; Gao et al., 2008; Lukasiewicz et al., 2006].

Experimental testing and numerical modeling, along with comparisons with field-collected data, have been used to validate and improve the assessment methods and criteria for dents on pipelines. Typical procedures for both testing and modeling methodologies include denting conducted by a proper indenter on a selected pipe segment, removal of the indenter, and application of a load (such as internal pressure) on the dented pipe [Cunha et al., 2014; Rezaee et al., 2018; Zhang et al., 2020a]. When a dent is combined with other types of defects such as gouge, corrosion, or a crack, a defect combination should be introduced on the pipe surface and then the denting process is conducted. Similar processes are followed in numerical simulation to obtain a model that approximates the experimental condition. It is generally accepted [Choi et al., 2003; Ghaednia et al., 2015a, b] that experimental testing is the best way to generate true results. However, the experiments are usually costly and time-consuming, with requirements of specific facilities. Numerical modeling such as FE analysis provides a promising alternative for dent assessment [Zhao and Cheng, 2022a, b], while the accuracy and reliability of the models must be verified by testing results or field data.

6.2 Standards and Methods for Dent Assessment

6.2.1 Existing Dent Assessment Standards

A plain dent, the simplest form of mechanical damage on pipelines, is the dent with a smooth transition of the curvature on pipe body [Allouti et al., 2012; Bratton et al., 2012]. Generally, the plain dent induces a significant plastic deformation but does not reduce failure pressure of the pipelines [Kec and Cerny, 2017]. Typical procedures in assessment of a plain dent include: (1) analysis of ILI data to define geometrical parameters of the dent, such as its depth, length, and width; (2) calculations of 3D strains at the dent or other dent features such as the critical depth; and (3) prediction of fatigue life of the pipelines by applying a failure criterion as described in various standards [Noronha et al., 2010; Gao and Krishnamurthy, 2015; Okoloekwe et al., 2020].

The depth-based criterion for assessment of plain dents is considered inaccurate, although it is convenient to obtain the depth parameter and conduct pipeline failure evaluation based on the information [Tian et al., 2020]. In contrast, the strain-based criterion is believed more accurate but relies on specific calculation methods. Table 6.1 lists commonly used standards and methods for assessment of plain dents on pipelines.

During long-term service of dented pipelines in the field, other types of defects can be generated in the dent area [Błachut and Iflefel, 2008; He and Zhou, 2021; Zhao et al., 2021]. Figure 6.1 shows schematically various types of dent-defect combinations on pipelines [Zhao et al., 2022a]. Compared with plain dents, a dent combined with other defects like a gouge, corrosion, or a crack incur a more serious threat to integrity of the pipelines but is much more complicated to assess. It was confirmed that the presence of other defects at the dent area could decrease the pressure-bearing capacity of the pipelines [MacDonald and Cosham, 2005; MacDonald et al., 2007; Shaik, 2015], while a plain dent usually does not affect the failure pressure. A dent subject to fatigue or interacting with corrosion or SCC also induces an increased threat to pipeline integrity. The main standards for assessment of a dent combined with another defect include ASME B31.8, ASME B31G, and API 579. The 3D geometrical parameters of the gouge, corrosion, and crack, along with the dent dimension, are required to perform the assessment [Qin and Cheng, 2021]. The standards applied on dent-defect combinations are also included in Table 6.1.

6.2.2 Principles of the Dent Assessment Standards

For almost all the dent assessment standards used today, the failure criteria applied for plain dents and the dent-defect combinations are similar or even identical. These standards are usually not mandatory to follow for pipeline operators. For example, although the ASME B31.8 provides a relatively complete set of methods for dent assessment, it also states that engineers can select other proper methods, if applicable, when assessing a dented pipeline [American Society of Mechanical Engineering, 2020]. The same notes are also included in other standards. Obviously, the dent assessment methods included in the standards are not regarded as absolutely accurate by the standard developers.

Table 6.1 Standards and methods used for assessment of plain dents and a dent combined with other types of defects on pipelines.

Standard	Plain dent	Dent with gouge	Dent with corrosion	Dent with crack
ASME B31.4	Dent depth < 6% OD	Repair	Corrosion defect is accessed by ASME B31.G	Repair
ASME B31.8	Circumferential bending strain $\varepsilon_1 = (1/2)t\left(\frac{1}{R_0} - \frac{1}{R_1}\right)$ (6.1) Longitudinal bending strain $\varepsilon_2 = -(1/2)t/R_2$ (6.2) Longitudinal extensional strain $\varepsilon_3 = (1/2)(d/L)^2$ (6.3) Equivalent strain on the inside pipe surface $\varepsilon_i = \left[{\varepsilon_1}^2 - \varepsilon_1(\varepsilon_2 + \varepsilon_3) + (\varepsilon_2 + \varepsilon_3)^2\right]^{1/2}$ (6.4) Equivalent strain on the outside pipe surface $\varepsilon_o = \left[{\varepsilon_1}^2 + \varepsilon_1(-\varepsilon_2 + \varepsilon_3) + (-\varepsilon_2 + \varepsilon_3)^2\right]^{1/2}$ (6.5) Maximum equivalent strain < 6% Or dent depth < 6% OD	Repair	Repair	Repair

CSA Z662	Dent depth < 6% OD and length/depth < 20 or the maximum equivalent strain <6%	Repair	Accessed by ASME B31.G for corrosion between 10% and 40% of wall thickness	Repair
49CFR 192	Dent depth < 6% OD	Repair	6% OD and metal loss per corrosion criterion	Repair
49CFR 195	Dent depth < 6% OD	Repair	Repair	Repair
API 1160	Dent depth < 6% OD	Repair	Immediate repair, unless engineering evaluation shows not an immediate risk	Repair
UKOPA	Dent depth < 7% OD or strain <6%	Repair	Corrosion depth < 20% wall thickness	Repair

where *OD* is outer pipe diameter; R_o, R_1, and R_2 are initial outer radius of the pipe, outer radius of curvature of the pipe in transverse plane across the sent, and outer radius of curvature of the pipe in longitudinal plane through the dent, respectively; t is pipe wall thickness; L is dent length in longitudinal direction; d is dent depth; ε_1 and ε_2 are bending strains in the circumferential and longitudinal directions, respectively; ε_3 is extensional strain in the longitudinal direction; and ε_i and ε_o are inside and outside equivalent strains, respectively.

Source: From Bernard et al. [2013], Gao and Krishnamurthy [2015], Dawson et al. [2018], and Tee and Wordu [2020].

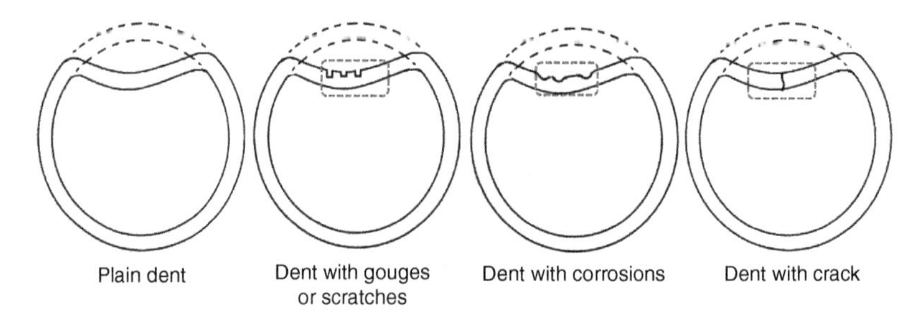

Figure 6.1 Schematic diagram illustrating different types of dent-defect combinations that can be present on pipelines. *Source:* From Zhao et al. [2022a].

As stated above, the failure criteria defined in various standards for dent assessment mainly include the depth-based criterion and the strain-based criterion [Adeeb and Horsley, 2006; Wu et al., 2016]. The depth is a direct geometric parameter to define a plain dent, and the depth-based criterion has become a convenient and popular method to evaluate dented pipelines for failure prediction [Tian et al., 2020]. However, field experiences showed that the dent depth barely affected the pressure-bearing capacity of a dented pipeline when the depth exceeded 20% of the pipe outer diameter [Shuai et al., 2018; Tian et al., 2020]. Even when the dent depth is less than 6% of the pipe outer diameter, the pipeline can still suffer from fatigue failure originating from the dent [Arumugam et al., 2018]. Thus, it is generally believed that the depth-based criterion is not sufficiently accurate for dent assessment on pipelines.

Compared to the depth-based criterion, the strain-based criterion is considered more accurate for assessment of plain dents on pipelines [Wu et al., 2015; Okoloekwe et al., 2020]. The strain-based criterion requires input parameters such as the dent length (L) in the axial direction, the initial pipe surface radius (R_0), the external surface radius of curvature in the transverse plane (R_1) through the dent, and the external surface radius of curvature in the longitudinal plane (R_2) through the dent, and the dent depth (d). The 3D strains and the equivalent strain are calculated using formulas, i.e., Eqs. (6.1)–(6.5), in ASME B31.8 [American Society of Mechanical Engineering, 2020], as shown in Table 6.1. The surface radius of the curvature (R_d) can be calculated by:

$$\frac{1}{R_\text{d}} = \frac{\dfrac{d^2y}{dx^2}}{\left[1 + \left(\dfrac{dy}{dx}\right)^2\right]^{\frac{3}{2}}} \tag{6.6}$$

where R_d can be either R_1 or R_2, and y is the function of contour of the dented area with the variable x. The profile of the dented area should be measured to determine the function of the geometrical shape and then the radius of curvature by Eq. (6.6). Various ILI tools are effective in measuring and estimating the profile of the dented area [Coramik and Ege, 2017; Xie and Tian, 2018].

The strains obtained from calculations in ASME B31.8 standard include bending strains in the circumferential (ε_1) and longitudinal (ε_2) directions, respectively, a membrane strain in the longitudinal direction (ε_3), and equivalent strains on the inside (ε_i) and outside (ε_o) pipe surfaces. It is noted that Eqs. (6.1)–(6.5) in Table 6.1 were developed based on a thin plate model [Gao et al., 2008], where the total strain is divided into circumferential and longitudinal bending strains and membrane strains, respectively. A further assumption is that the circumferential membrane strain is ignored due to negligible change of transverse girth for the dented pipelines. However, this assumption is questionable. From FE analysis and full-scale tests on dented pipelines, it was found that the maximum circumferential membrane strain was 20% [Rafi, 2011], which was significant compared to other strains. For deep dents, the circumferential membrane strain should not be ignored. Investigations indicated that the accuracy of the formula for membrane strain in ASME B31.8 was poor due to an inadequate analogy to radial strains in a circular plate [Lukasiewicz et al., 2006]. Modifications are required to overcome the problem for improved accuracy in strain determination.

According to the large-deformation thin shell theory, the strains of dented pipelines were derived by Lukasiewicz et al. [2006], and the equivalent strain, ε_{eq}, was calculated by:

$$\varepsilon_{eq} = \frac{2}{\sqrt{3}}\left[{\varepsilon_x}^2 + \varepsilon_x\varepsilon_y + {\varepsilon_y}^2\right]^{\frac{1}{2}} \tag{6.7}$$

where ε_x and ε_y are strains in the axial and circumferential directions, respectively, which can be calculated by:

$$\varepsilon_x = \varepsilon_x^{m} \pm \varepsilon_x^{b}, \quad \varepsilon_y = \varepsilon_y^{m} \pm \varepsilon_y^{b} \tag{6.8}$$

where the superscripts "m" and "b" denote membrane strain and bending strain, respectively. The sign is positive for the strain of inner pipe surface and negative for the outer pipe surface. Thus, the equivalent strain should contain the membrane strain in the circumferential direction. Eq. (6.7) is derived based on the plastic strain theory with the presupposition that the volume of the material remains constant during deformation (i.e., $\varepsilon_x + \varepsilon_y + \varepsilon_z = 0$), where ε_z is the strain in the radial direction of the pipe.

6.2.3 Limitations of the Existing Standards and Improved Strain Determination for Dent Assessment

Most of the present standards for dent assessment on pipelines focus on plain dents, rather than dent-defect combinations. Moreover, the existing standards and methods tend to be conservative and contain certain limitations in the assessment process. A further improvement of the dent assessment techniques is required for integrity management of pipelines.

Nowadays, modifications on plain dent assessment techniques have mainly focused on implementation of the strain-based criterion to replace the depth-based criterion that has been commonly used in industry. Various depth-based methods are mostly empirical and do not have sufficient theoretical support. Thus, they are used as the Level 1 assessment only. As a comparison, the strain analysis at a dent can provide more information for evaluation of FFS of the pipelines, such as fatigue life prediction [Dawson et al., 2018]. Improvements have been achieved in three aspects, i.e., new methods for strain calculation at a dent, detailed definition of the dent profile, and accurate pipeline failure criteria.

While the ASME B31.8 standard introduces the strain-based criterion for dent assessment, many problems and limitations still exist. First, the criterion assumes that the maximum bending and membrane strains are located at the center of a dent [Noronha et al., 2010]. However, both experimental testing and FE analysis indicate that the maximum strain is not always at the dent center, especially for deep dents [Rafi, 2011]. As the dent depth increases, the location of the maximum strain moves from the center of a dent to its side [Shuai et al., 2020]. It means that the pipeline failure assessment based on strain calculations at the dent apex can give misleading information. This is particularly true for deep dents. Second, the formulas in ASME B31.8 are sometimes confusing. For example, the formulas used for calculation of bending and membrane strains are obtained from empirical methods. The FE analyses show that the formulas cause predictable erroneous strains [Shahzamanian et al., 2021]. Moreover, there is no explanation for the origin of the empirical equivalent strain equations [Gao et al., 2008]. Third, there is an important assumption in the standard that the membrane strains in circumferential and radial directions and the shear strain are negligible [Noronha et al., 2010]. Nevertheless, denting is a plastic deformation on pipelines, where the plane strain assumption is not proper. Finally, for unconstrained dents, there will be a re-rounding process after denting, during which the dent depth and strain will decrease. As a result, the maximum deformation during the denting cannot be accurately estimated by ASME B31.8 [Zhang et al., 2020b]. In summary, the strain-based method used in ASME B31.8 improves the dent assessment on pipelines, but still cannot provide accurate results for failure prediction of the dented pipelines.

By analyzing the problems existing in strain calculations in ASME B31.8, Lukasiewicz et al. [2006] proposed a more accurate formula based on the theory of large deformation of a cylindrical shell. Using the dent profile data derived from ILI tools, the bending strains in longitudinal and circumferential directions, i.e., $\varepsilon_x^{\mathrm{b}}$ and $\varepsilon_y^{\mathrm{b}}$, can be calculated by:

$$\varepsilon_x^{\mathrm{b}} = \frac{t}{2}\frac{\partial^2 w}{\partial x^2}, \quad \varepsilon_{\mathrm{y}}^{\mathrm{b}} = \frac{t}{2}\frac{\partial^2 w}{\partial y^2} \tag{6.9}$$

where w is pipe wall deflection in the radial direction. The profile functions in the longitudinal and circumferential directions are defined as u and v. The membrane strains $\varepsilon_x{}^{\mathrm{m}}$ and $\varepsilon_y{}^{\mathrm{m}}$ and the shear strain γ_{xy} are then derived by:

$$\begin{aligned}\varepsilon_x^{\mathrm{m}} &= \frac{\partial u}{\partial x} + \frac{1}{2}\left(\frac{\partial w}{\partial x}\right)^2 + \varepsilon_x^{\mathrm{o}}, \quad \varepsilon_y^{\mathrm{m}} = \frac{\partial u}{\partial y} + \frac{w}{R} + \frac{1}{2}\left(\frac{\partial w}{\partial y}\right)^2 \\ &+ \varepsilon_y^{\mathrm{o}}\gamma_{yx} = \frac{\partial u}{\partial y} + \frac{\partial v}{\partial x} + \left(\frac{\partial w}{\partial x}\right)\left(\frac{\partial w}{\partial y}\right)\end{aligned} \tag{6.10}$$

Compared with the method in ASME B31.8, the Lukasiewicz method improves the accuracy in strain determination at the dent. The ASME B31.8 method underestimates the equivalent strain by a factor of about 2 [Gao et al., 2008]. Moreover, the membrane and shear strains cannot always be ignored since they may have values similar to the longitudinal strains.

Other methods calculating the strain at a dent based on derivation of the elasticity mechanics were proposed by Dubyk and Seliverstova [2019], but with complicated forms. The strain evaluation depends on measurements of the dent profile to calculate the parameters like the curvature of the dent area. To approximate the dent profile, the most widely studied method is the B-spline curve method, i.e., a piece-wise polynomial interpolation function based on data from ILI tools or FE analysis [Okoloekwe et al., 2018]. A fourth-order B-spline curve was used by Noronha et al. [2010] to approximate the dent profile in the longitudinal and circumferential directions. The method was proved to simulate the dent profile at a reasonable accuracy.

6.3 Assessment of Dent-Defect Combinations on Pipelines

Although it is quite common that a dent coexists (or is in adjacency) with another defect of different types on pipelines, there has been no reliable method available for accurate assessment of the dent-defect combination.

This section reviews and summarizes the established work on assessment of a dent combined with a gouge, corrosion, or a crack.

6.3.1 Dent with a Gouge

A gouge (or a scratch) is a mechanical damage causing steel removal from pipe surface. A high length-to-width ratio and a sharp notch profile are main geometric features of a gouge [Pluvinage et al., 2011]. A cold-worked hardened layer usually forms at the gouge to decrease local ductility [MacDonald and Cosham, 2005]. A gouge can be generated along with a dent, greatly reducing the pressure-bearing ability of a pipeline [Błachut and Iflefel, 2007, 2008]. The combination of a dent and a gouge is not allowed to present on pipelines according to the standards listed in Table 6.1. The combined defects should be repaired once detected. In addition, gouges on pipelines are frequently associated with cracks [Ma et al., 2013].

For a pipeline containing a plain dent accompanied with a gouge, the burst pressure of the pipe is lower than the pipeline containing a plain dent only or an equivalent gouge in the absence of the dent. For example, MacDonald et al. [2007] found that the burst and fatigue strengths of a pipe containing a dent combined with a gouge were remarkably lower than the strengths of the pipe containing an equivalent plain dent. The failure mode during burst testing on a pipe containing the combined dent-gouge defect was rupture associated with a brittle crack propagation [Błachut and Iflefel, 2007]. A large tearing through the gouge was found on dented pipe, where the failed area propagated symmetrically on the dent shoulder. With further microscopic examination and magnetic particle inspection, a hardening effect and microcracks were identified in the dent-gouge area [Zarea et al., 2014]. The burst pressure was mainly dependent on the geometrical parameters of the dent and the gouge, such as the dent depth and shape, and the gouge size [Naghipour et al., 2018; Tian and Zhang, 2017a; Zhao et al., 2021]. As the depth and length of the gouge increased, the burst pressure of the pipe decreased. A plastic strain was concentrated at the dent area, and the circumferential strain was larger than the axial strain at the dent center. In most studies, the gouge was usually located at the center of a dent, where the length of the gouge was aligned with either the longitudinal or the circumferential direction of the pipeline. In general, the location and orientation of a gouge on pipelines are not certain. The relative position of the gouge at the dent area plays an important role in affecting the burst pressure of pipelines [Lancaster and Palmer, 1996]. It was found that, when the gouge was located at the dent center and the gouge length was short relative to the dent size, the dent depth did not apparently influence the burst pressure of the pipe. However, when the gouge was on the flank of the dent area, or the gouge was sufficiently long, the burst pressure would reduce remarkably as the dent depth increased.

Nowadays, a standard including a definitive criterion for assessment of a pipeline containing a dent and gouge combination has not been available. Some standardized methods that have been used for assessment of pipelines containing a gouge are reviewed herein. For pipelines that contain a longitudinally oriented gouge under static loading, the NG-18 equation has been widely used to determine the failure stress (σ_F) of the pipelines [Bernard et al., 2013]:

$$\sigma_F = 1.15\sigma_y \left[\frac{1 - \frac{d_g}{t}}{1 - \frac{d_g}{(Mt)}} \right] \tag{6.11}$$

$$M = \sqrt{1 + 0.26\left(L_g/\sqrt{R_p t}\right)^2} \tag{6.12}$$

where d_g is maximum depth of the gouge, and L_g is length of the gouge.

In ASME B31G standard, the method for determination of the failure stress of gouged pipelines is similar [American Society of Mechanical Engineering, 2017]:

$$\sigma_F = \frac{\sigma_Y + \sigma_U}{2} \left[\frac{1 - \frac{d_g}{t}}{1 - \frac{d_g}{(Mt)}} \right] \tag{6.13}$$

$$M = \sqrt{1 + 0.8\left(L_g/\sqrt{Rt}\right)^2} \tag{6.14}$$

The modified ASME B31G equation is also included in ASME B31G with an improved accuracy of the parameter M:

$$\sigma_F = \frac{\sigma_y + \sigma_U}{2} \left[\frac{1 - \frac{d_g}{t}}{1 - \frac{d_g}{(Mt)}} \right] \tag{6.15}$$

$$M = \sqrt{1 + 0.314\left(L_g/\sqrt{Rt}\right)^2 - 0.0084\left(L_g/\sqrt{Rt}\right)^4} \tag{6.16}$$

Other standards like DNV RP-101 have similar equations. These methods attempt to estimate the failure pressure of pipelines containing a gouge with the assumption that the plastic deformation at the gouge is constrained [Allouti et al., 2014].

6.3.2 Corrosion in Dent

Corrosion is one of the primary mechanisms causing pipeline failures [Cosham and Hopkins, 2004; Leis, 2021; Qin and Cheng, 2021; Zhang et al., 2021a, b]. As the main form of metal loss defect, corrosion in either uniform or localized form is commonly present on pipelines. Moreover, pipeline corrosion can happen internally or externally. A decrease of pipe wall thickness due to corrosion will greatly influence the mechanical performance of the pipelines, such as burst pressure, fatigue life, and fracture resistance [Shuai et al., 2008]. Qin and Cheng [2021] reviewed corrosion defect assessment techniques for pipelines. For pipelines containing a dent, corrosion tends to preferentially occur in the dent area due to a high local stress and, sometimes, an aggressive environment generated at the dent [Babbar and Clapham, 2009; Hafez, 2021].

The ASME B31.8 standard suggests that, for a dent-corrosion combination, the safety assessment should be performed separately. The dent, which is regarded as a plain dent, is assessed by the strain-based criterion, and the corrosion defect is assessed by ASME B31G [Shuai et al., 2017]. In standards like CSA Z662 and UKOPA, the depth-based criterion is used for dent assessment. Of various standards developed today, the ASME B31G is most used, where the burst pressure, P_b, is determined by:

$$P_b = \sigma_{flow} \left[\frac{1 - \frac{2}{3}\left(\frac{d}{t}\right)}{1 - \frac{\frac{2}{3}\left(\frac{d}{t}\right)}{M}} \right] \quad M = \sqrt{1 + 0.8z} \tag{6.17}$$

where $z = L^2/Dt$, L is longitudinal length of the corrosion defect, and d is corrosion depth. The equation is applicable for $z \leq 20$. When $z > 20$, the burst pressure is determined by:

$$P_b = \sigma_{flow}\left(1 - \frac{d_c}{t}\right) \tag{6.18}$$

The flow stress, σ_{flow}, is set as $1.1 \times$ SMYS.

A modified version of the ASME B31G defines that, for $z \leq 50$,

$$\sigma_F = \sigma_{flow} \left[\frac{1 - 0.85\frac{d}{t}}{1 - \frac{0.85d}{(Mt)}} \right] \quad M = \sqrt{1 + 0.6275z - 0.003375z^2} \tag{6.19}$$

For $z > 50$, $M = 0.032\,z + 3.3$. It is noted that the assessment methods listed above are categorized as Level 1 methods, where the profile of the corroded area is treated as parabolic ($z \leq 20$) or rectangular ($z > 20$). In the modified method, the corrosion profile is regarded as a mixed type of geometric shape [Qin and Cheng, 2021]. The circumferential geometry of the corrosion area is usually ignored. In addition to ASME B31G, the standards such as DNV-RP-F101, API 579, and BS 7910 assess corrosion defects using similar formulas but with different M factors and flow stresses.

As stated, the Level II method for assessment of corroded pipelines is provided in ASME B31G using an effective area method [American Society of Mechanical Engineering, 2017; Adib-Ramezani et al., 2006). The metal loss area in the longitudinal plane A and the original area A_0 should be measured to calculate the burst pressure by:

$$P_b = \sigma_{flow} \left[\frac{1 - \dfrac{A_1}{A_0}}{1 - \dfrac{\left(\dfrac{A_1}{A_0}\right)}{M}} \right] \tag{6.20}$$

The accuracy of the corrosion area is critical to the Level II assessment.

The Level III assessment relies on FE modeling and numerical calculations and provides more accurate results compared with Levels I and II assessments.

6.3.3 Dent with Cracks

Cracks represent the most dangerous feature threatening pipeline integrity [Okodi et al., 2020]. The cracks usually initiate at irregularities, such as dents, on pipe surface, serving as stress risers. The cracks propagate due to steel yielding and plastic deformation induced by a high stress concentration. For pipelines containing cracks, the burst pressure is affected by the crack geometry, such as crack depth, width, and length, as well as the crack propagation direction. Of cracks with various orientations on the pipelines, the longitudinal cracks, which are perpendicular to the primary stress (i.e., hoop stress) resulting from internal pressure, are generally more dangerous [Okodi, 2021]. Pipelines can be failed by either fatigue cracking or SCC [Liu et al., 2011; Tang and Cheng, 2011; Cheng, 2013; Okodi et al., 2021], both of which can initiate at a dent. Commonly used standards for crack assessment include API 579 and BS 7910 [Bedairi et al., 2012], where a failure assessment diagram (FAD) is used to determine if the cracks are acceptable to pipelines. Figure 6.2 shows a typical FAD recommended by API 579, where K_r is toughness ratio and L_r^P is load ratio [American Petroleum Institute/American Society of Mechanical Engineering, 2016]. If a point is on or inside

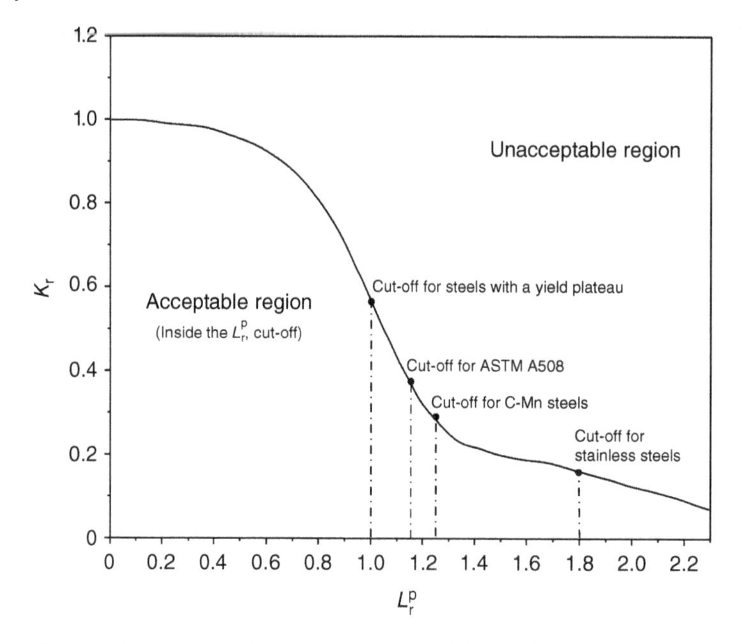

Figure 6.2 A typical FAD recommended by API 579 for crack assessment. *Source:* From American Petroleum Institute/American Society of Mechanical Engineering [2016].

the FAD boundary, the crack size under the given conditions is acceptable. Sometimes, the critical crack length can be predicted by FAD. There are three levels of crack assessment. In API 579, Levels I and II assessments are applicable for cracks that become arrested under given loading conditions and service environments. The critical length and depth of the cracks are considered in Level I to estimate severity of the damage. In Level II, the stress intensity factor is determined for assessment. The crack dimension obtained from ILI data is used in calculations of local stress and the corresponding stress intensity factor.

If a crack is expected to grow, the Level III assessment should be conducted, where true stress–strain results are used in FAD. It is assumed that the growth of a pre-existing crack is controlled by stress intensity factor at the crack tip. Four types of crack growth mechanisms are assessed for prediction of the remaining service life, including fatigue cracking, corrosion fatigue cracking, SCC, and HIC. The BS 7910 standard also used FAD as the major method for crack assessment by a similar procedure to API 579. A basic difference between API 579 and BS 7910 is the method used to determine reference stress and the stress intensity factor [Bedairi et al., 2012].

Cracks can initiate at a dent, where a high strain is generated during formation of the dent or due to preferential corrosion, mechanical damage, or cyclic loading [Ghaednia et al., 2013]. Field experiences on dented pipelines showed that cracks could initiate even when the dent was shallow [Arumugam et al., 2018]. Although a plain dent contributes little to failure of pipelines, the cracks initiating in the dent area remarkably degrade the pipeline integrity [Alexander, 1999]. It was found [Luo et al., 2020] that penetrating cracks located in the dent center possessed typical fatigue characteristics. In addition, the external coating can be damaged during denting. As a result, a corrosive environment will generate, which, combined with the stress concentration at the dent, causes SCC [Cheng, 2013]. The M–E interaction explains pipeline SCC initiating at a dent due to a synergism of local stress or strain concentration and electrochemical corrosion reactions [Xu and Cheng, 2012a, b]. When a dent is combined with a gouge, the micro-cracks generated at the base of the gouge can lead to propagating cracks.

Combination of a dent and a crack is not allowed in most standards. Once one or more cracks are detected at a dent, remediation, or replacement will be required. Due to the limited resolution and accuracy of ILI tools, most cracks cannot be easily found until the pipelines leak, or other failure modes happen. Standards such as 49 Code of Federal Regulations (CFR) 192 (gas pipelines) and 195 (liquid pipelines) do not include sufficient methods for assessment of the dent-crack combination. For gas pipelines, an immediate repair is recommended. For liquid pipelines, cracks combined with a dent on the upper two-third of the pipeline need to be repaired immediately and otherwise, a service for 60 days is acceptable [Gao and Krishnamurthy, 2015]. The ASME B31.8 states that a dented pipeline with cracks less than 12.5% of pipe wall thickness can be removed by grinding [Ghaednia et al., 2015a]. It was also found [Ghaednia et al., 2014] that the burst pressure of pipelines was rarely affected when the crack depth was small; while when the crack exceeded 40% of the pipe wall thickness, the burst pressure decreased by 55%.

6.4 Fatigue Failure of Pipelines Containing Dents

Serving as a local stress riser, the dent strongly affects fatigue failure of pipelines. While some methods are proposed to assess fatigue of dented pipelines, as listed in the standards in Table 6.1, they are basically inadequate. Cyclic loading due to pressure fluctuations of pipelines remarkably affects stress and strain at the dent, and thus, fatigue failure of the pipelines [Cunha et al., 2009]. It has been established that the constrained dent influences fatigue life. Generally, a constrained dent, where the indenter keeps contacting with the pipe surface, will enable a longer fatigue life than a dent-free pipeline of the identical specification since the indenter supports the pipeline against internal pressure and its fluctuations [Gao and Krishnamurthy, 2015]. For deep-constrained dents (i.e., the dent depth exceeds 4% of the pipe outer diameter), the crack initiation location deviates from the dent center. However, for shallow constrained dents, the crack is usually located at the dent center.

The API RP 1183 provides assessment methods for fatigue life of dented pipelines [American Petroleum Institute, 2020], where three levels of assessment provide rankings of the dent geometry severity (Level I), of the dent geometry and load severity (Level II), and of the dent fatigue life assessment (Level III). In Level I assessment, the geometrical parameters of the dent are used to estimate fatigue life of the pipeline by:

$$N = a(\mathrm{SP})^b \tag{6.21}$$

where N is estimated fatigue life of the dented pipe in cycles, SP is the dent shape parameter, and a and b are geometrical coefficient and exponent, respectively.

Different from Level I assessment, the Level II assessment considers historical data of operating pressure and constrained conditions of the dent. The number of stress cycles is estimated by Eq. (6.21) under different pressures. The assessment methods of Levels I and II are basically empirical and conservative. The Level III assessment based on FE modeling provides a more accurate method for fatigue life estimation, where the crack growth is simulated by nonlinear numerical analysis.

The method for fatigue assessment of pipelines containing dents, as included in API 1156, is based on limited conditions, i.e., geometrical characteristics and stress concentration level of the dents [Cunha et al., 2014]. The formulas used to determine the fatigue life based on stress-number of cycle (S-N) curves include:

$$N = 4.424 \times 10^{23}(K_F \Delta P)^{-4} \tag{6.22}$$

$$N = \exp\left(43.944 - 2.971 \ln\left(K_F \frac{\Delta P}{2}\right)\right) \tag{6.23}$$

where K_F is fatigue stress concentration factor (SCF), and ΔP is the pressure variation. In this method, the conversion factor used for calculating the range of nominal hoop stress from the internal pressure considers the values of indenter shape, d/t, i.e., the ratio of the dent depth to pipe wall thickness, i.e., 34 and 68 [Alexander, 1999]. It is thus unclear how the conversion factor can be computed for other d/t values, especially when the values are outside the range of 34 and 68. The method was thought conservative by a factor of two with respect to stress, and 20 with respect to number of cycles [Alexander, 1999].

In both ASME FFS-1 and API 579-1, the fatigue assessment on dented pipelines is performed by Level II method [Shirband et al., 2020]. Based on a semi-empirical S-N model, the remaining fatigue life of a pipeline containing unconstrained dents is calculated by:

$$K_{\mathrm{d}} = 1 + 2\sqrt{d^{1.5} \cdot \frac{t}{D}} \tag{6.24}$$

$$N_{\mathrm{c}} = 5622 \cdot \left(\frac{\sigma_{\mathrm{U}}}{2\sigma_{\mathrm{eq}} K_{\mathrm{d}}}\right)^{5.26} \tag{6.25}$$

where σ_{eq} is the equivalent stress, and K_{d} is SCF associated with the dent. It should be noted that a safety factor of 10 is used to ensure a conservative estimation.

6.5 Failure Criteria of Pipelines Containing Dents

The critical strain-based criterion in ASME B31.8 standard for failure prediction of pipelines containing dents is an empirical recommendation. It is assumed that the dented pipelines are safe under cyclic loading when the maximum equivalent strain is smaller than 6% [Rafi, 2011]. However, the strain limit of 6% is arbitrarily selected. To improve the assessment accuracy, new failure criteria should be developed.

6.5.1 Oyane's Plastic Failure Criterion and Ductile Fracture Damage Index (DFDI) Criterion

Ductile damage resulting from accumulation of plastic deformation is one of the mechanisms applicable for pipelines to initiate cracks and cause fracture [Alashti et al., 2015]. Oyane et al. [1980] considered the cumulative damage generated during plastic deformation and proposed a plastic failure criterion for dented pipelines:

$$I = \frac{1}{C_2} \int_{0}^{\varepsilon_{\mathrm{eq}}} \left| \frac{\sigma_{\mathrm{m}}}{\sigma_{\mathrm{eq}}} + C_1 \right| d\varepsilon_{\mathrm{eq}} \tag{6.26}$$

where I is an integral value used as the damage indicator, C_1 is a constant obtained through burst test on a non-indented pipe, C_2 is elongation rate of pipe steel measured in uniaxial tensile testing, σ_m is mean stress, and σ_{eq} and ε_{eq} are equivalent stress and strain, respectively. When the indicator I is equal to and exceeds 1, ductile fracture will occur on the pipe [Wu et al., 2019].

Similarly, the DFDI criterion assumes that the growth of microcracks on ductile solids causes the materials to fracture [Li and Dang, 2017]. The DFDI criterion is expressed as:

$$\mathrm{DFDI} = \int_0^{\varepsilon_{eq}} \frac{d\varepsilon_{eq}}{1.65\varepsilon_{crit} \exp\left(-\frac{3\sigma_m}{2\sigma_{eq}}\right)} \tag{6.27}$$

where ε_{crit} is critical strain of ductile materials to initiate cracks, as measured by uniaxial tensile testing [Gao et al., 2013; Arumugam et al., 2016]. Plastic fracture takes place when the DFDI value is equal to or exceeds 1.

Both Oyane's criterion and the DFDI criterion were verified experimentally and numerically to effectively predict cracking of materials including steels [Arumugam et al., 2016, 2018; Wu et al., 2016]. The generation of dents on a pipeline is a plastic deformation process, where both criteria are suitable for failure assessment of pipelines [Li and Dang, 2017].

6.5.2 Strain Limit Damage (SLD) Criterion

The SLD criterion uses elastic–plastic FE analysis to estimate the accumulated plastic damage on pressure vessels, including pipelines [Gao and Krishnamurthy, 2015]:

$$D_{et} = D_{eform} + \sum_{k=1}^{M} D_{e,k} \leq 1 \tag{6.28}$$

where D_{et} is an indicator of the limit state for a structure such as a pipeline to carry no further load, D_{eform} is the damage resulted from deforming, and $D_{e,k}$ is the damage during the kth load increment, which is associated with the total plastic strain and stress. When D_{et} exceeds 1, the load capacity of the structure reaches its limit. The SLD criterion is based on the minimum reduction in area and elongation to failure, while the properties of the material are not required. The stress and strain data used in the criterion is obtained from FE analysis.

6.5.3 Net Section Failure Criterion and Plastic Collapse Strain Criterion

The net section failure criterion is a stress-based criterion, where the minimum von Mises stress in the dent area should not exceed the flow stress, $\sigma_{flow} = (\sigma_y + \sigma_u)/2$ [Liu et al., 2017]. The plastic collapse strain criterion is commonly used in pressurized structures like pipelines. The plastic collapse state is defined as the point of intersection on the load–displacement curve between the peak and a line drawn from the origin with a slope twice of the elastic slope [Zhao et al., 2020]. It is effective to use the plastic collapse criterion to assess dented pipelines where a large plastic deformation occurs [Baek et al., 2012].

6.5.4 Remaining Fatigue Life Criterion

Methodologies for fatigue life assessment of dented pipelines are summarized and commented in [Cunha et al., 2014], where European Pipeline Research Group (EPRG) 1995 and EPRG 2000 are thought as the best empirical methods in terms of the quality-of-fit to published full-scale test data [Gao and Krishnamurthy, 2015]. According to the EPRG methods, fatigue life of a pipeline containing unconstrained dents is determined by the number of cycles of circumferential stress, N, and the SCF at the dent, K_d:

$$K_d = 2.871\sqrt{d\frac{t}{D}} \tag{6.29}$$

$$N = 1000\frac{\sigma_u - 50}{2\sigma_{FS}K} \tag{6.30}$$

where σ_{FS} is fatigue strength.

Stress-life fatigue design equations were used by Petronas to estimate the fatigue life of dented pipelines [Cunha et al., 2014], where the SCF was limited to the linear elastic stage. The fatigue limit of pipeline steels is assumed to be attained at 10^6 cycles.

$$K_d = A_P + B_P\left(\frac{d}{D}\right)\left(\frac{D}{t}\right)^{1.14} \tag{6.31}$$

$$N = \left[\frac{\sigma_a}{C_P\left(1 - \left(\frac{\sigma_m}{Su}\right)^2\right)}\right]^{\frac{1}{b}} \tag{6.32}$$

where A_P, B_P, and C_P are coefficients, which depend on dent geometry, pipe dimension, and steel properties, respectively, and σ_a and σ_m are alternating stress and mean stress, respectively. It is noted that both EPRG methods and the Petrobras model are stress-based.

A strain-based model was proposed for fatigue life assessment of dented pipelines as follows [Gao and Krishnamurthy, 2015]:

$$\Delta\varepsilon = 3.5\frac{\sigma_U}{E}N^{-0.12} + \varepsilon_0 N^{-0.6} \tag{6.33}$$

where $\Delta\varepsilon$ is the cyclic strain range, E is Young's modulus, and ε_0 is true strain to failure. The cyclic strain range at the dent can be obtained by FE analysis.

6.6 Finite Element Modeling for Dent Assessment on Pipelines

The FE modeling and analysis enables integration of multi-physics field coupling effect for assessment of defects on pipelines such as dents and a dent combined with other types of defects. The FE modeling process generally includes the following procedure.

6.6.1 Simulation of the Denting Process

Denting is an elastoplastic deformation process. A typical numerical modeling and analysis process follows the steps, as described in Figure 6.3, for assessment of a dent on a pipeline [Baek et al., 2012; Li and Dang, 2017; Han et al., 2018; Kainat et al., 2019; Pinheiro et al., 2019; Wu et al., 2019; Zhu and Wang, 2019; Shuai et al., 2020].

- Denting: An indenter is placed on the pipe surface without application of a force. A downward displacement load is then applied on the indenter by

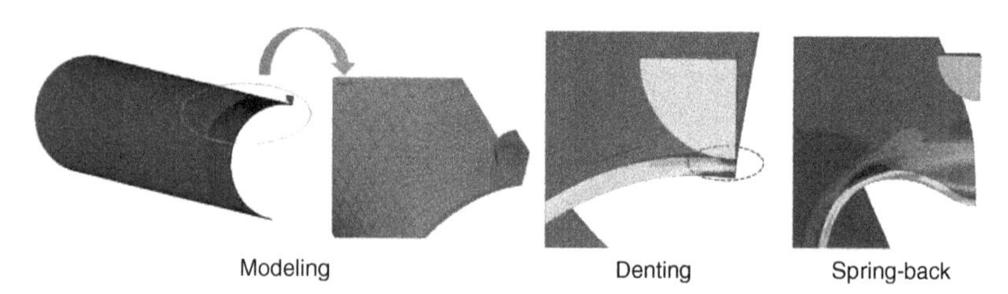

Figure 6.3 Schematic diagram showing the steps for modeling and analysis of a denting process on pipe. *Source:* From Shuai et al. [2020] / with permission from Elsevier.

small increments until a set value of depth penetrating in the pipe body is reached.

- Re-rounding: For an unconstrained dent, when the indenter is removed upward, the dent area will have a re-rounding process. The unconstrained dent has a final depth after a complete re-rounding. For a constrained dent, the indenter will be maintained at the dent and does not move.
- Loading: A static pressure or a cyclic pressure is applied on the interior of the pipe to determine the burst pressure or fatigue life of the dented pipe. Other types of loads such as bending moment can also be applied.

6.6.1.1 Materials Model

When a permanent plastic deformation occurs on the pipe body during denting, modeling of the denting process requires a good alignment with the true stress–strain relationship of the pipe steel. The commonly used elastoplastic materials models include plastic hardening model, R-O stress–strain rule (i.e., Eqs. 3.2 and 3.3), and power-law model (i.e., Eq. 3.4) [Qin and Cheng, 2021]. The true stress–strain data measured by tensile testing are used as inputs to develop the materials model [Arumugam et al., 2016; Luo et al., 2020]. The plastic hardening model includes isotropic hardening (i.e., Eq. 3.1) [Han et al., 2018], kinematic hardening [Tiku et al., 2012; Li and Dang, 2017], and a combination of isotropic and kinematic hardening [Pinheiro et al., 2019].

6.6.1.2 Model Development

Numerical models are developed by 3D solid elements [Iflefel et al., 2005; Kainat et al., 2019], where a quarterly symmetric model is often used to reduce the density of grids and save computational time. The advantages of using solid elements include a good approximation of dented pipelines and convenient calculations of the mechanical parameters in axial, circumferential, and radial directions. Shell element is also used for FE analysis [Lockey et al., 2014; Pinheiro et al., 2019; Pournara et al., 2019], performing well in nonlinear analysis where large inelastic deformations occur on steel cylinders. To eliminate the effect of structural edges on modcling, the pipe segment selected for modeling should be long enough and the ratio of pipe length to pipe outer diameter is more than 3 [Shuai et al., 2018; Zhang et al., 2020b]. Contact elements and target elements should be assigned to the indenter surface and the dent area, respectively. The grid density for the pipe body that is far away from the dent area can be sparse, while the density should be increased in the dent area to satisfy the requirement of accuracy.

In experimental testing, caps should be placed at both ends of the pipe segment to enable the pipe with an internal pressure-bearing capacity. However, the FE modeling can simplify the structure without caps at both ends. Instead, the equivalent tensile stress due to internal pressure is applied on the ends of the pipe. The equivalent stress is calculated by the shell theory [Iflefel et al., 2005]:

$$\sigma_{\mathrm{eq}} = \frac{PD_i^2}{D^2 - D_i^2} \tag{6.34}$$

6.6.1.3 Modeling Verification

Generally, a comparison between the modeling results and testing data is the direct method to verify the accuracy of a model. Strain gauges cannot be attached on pipe body during denting. Other parameters that can be used to verify the FE modeling include the relationship between reaction force and displacement of the indenter during denting and removal of the indenter [Han et al., 2018; Zhao et al., 2020].

6.6.2 Modeling for Dent Assessment of Pipelines

Generally, burst tests provide direct data to prove the accuracy of developed models. Due to strict requirements of testing facility, space, and safety regulations, as well as high costs, modeling is regarded as a promising alternative for pipeline failure assessment. Table 6.2 lists typical burst modeling by FE analysis on dented pipelines.

Figure 6.4 shows the distributions of hoop and axial strains at a dent during indentation, which is applied with a spherical indenter of 100 mm in diameter, and spring-back by FE modeling on an X52 steel pipe [Zhao and Cheng, 2022a]. It is seen that both hoop and axial strains decrease in response to the spring-back. Generally, the hoop strain on the outer surface of the dent is tensile (positive) and the inner surface is under compressive strain. For axial strain, the inner surface is in tension and the outer surface is in compression. The circumferential and axial strains as a function of circumferential angle and axial distance, respectively, after spring-back are shown in Figure 6.5, where the origin point (i.e., 0° in circumferential angle and 0 mm in axial distance) refers to the dent apex. It is noted that the circumferential angle cannot be directly determined from FE modeling and is to be calculated using a relationship between the arc length (obtained from software) and the radius of curvature. As expected, the largest circumferential and axial strains are recorded at the dent apex. However, for the equivalent von Mises strain, the maximum value is not at the dent apex but

Table 6.2 Typical burst and fatigue modeling by FE analysis on dented pipelines.

Method	Pipeline steel	Materials model	Dent	Failure criterion	Dent depth
Allouti et al. (2012)	A37 steel	Isotropic strain hardening	Spherical	Burst pressure	$0.1 \sim 0.28$ OD
Liu et al. (2017)	API X60	Stress–strain relationship	Spherical	Burst failure	$0.02 \sim 0.2$ OD
Shuai et al. (2018)	API 5L X52	R-O model	Spherical	Burst failure	0.09 OD
Cunha et al. (2014)	Low carbon steel	Combined isotropic and kinematic hardening	Spherical and cylindrical	Failures ranging from around 6,000 to more than 10^6 cycles	$0.02 \sim 0.12$ OD
Lockey et al. (2014)	\	Linear-elastic material curve	\	S-N curve in standard	Less than 0.02 OD
Tiku et al. (2012)	X52	Nonlinear kinematic hardening	Ellipsoidal shape	Fatigue failure	$0.01 \sim 0.10$ OD
Kainat et al. (2019)	API 5L X70	Kinematic hardening	Spherical	S-N curve in a standard	Multiple depth
Pournara et al. (2019)	X52	flow plasticity model	Wedge-type	$\Delta\varepsilon \sim N$ fatigue curve	Multiple depth

Note: OD refers to pipe outer diameter.
Source: From Zhao et al. [2022a].

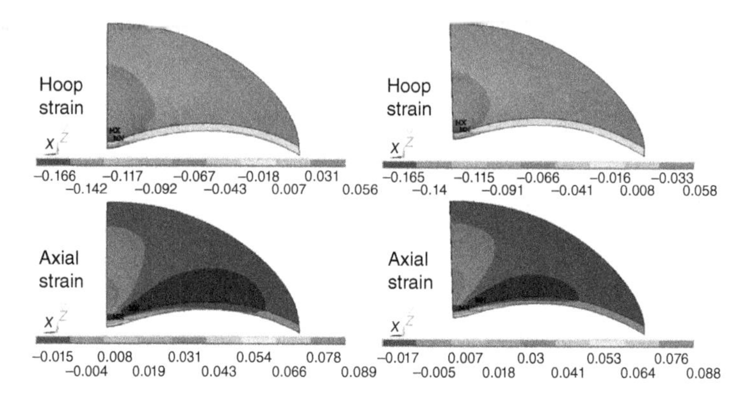

Figure 6.4 Distributions of hoop and axial strains at the dent during indentation (left) and spring-back (right) by FE modeling. *Source:* From Zhao et al. [2022a] / with permission from Elsevier.

deviates a bit to the dent side. This phenomenon is related to the dent depth. With an increased indentation displacement, the location with the maximum equivalent strain deviates from the dent center. This makes the existing standard such as ASME B31.8 cannot give accurate strain results, especially for deep dents.

Recently, Zhao and Cheng [2022a] have proposed a new method based on the DFDI criterion combined with improved strain determination at a dent by FE modeling for dent assessment and pipeline failure prediction. In the method, the DFDI formula considers the influence of spring-back upon removal of the indenter from the created dent and the difference between equivalent strain at the dent apex and maximum equivalent strain at the entire dent area. Generally, the spring-back can reduce the equivalent strain at dent apex and the DFDI value. Ductile damage is cumulative in nature. For example, the microcracks generated before spring-back do not disappear as the strain decreases. Thus, it is required to determine strain and the DFDI before spring-back, which are then used to correct both the strain and DFDI after the indenter is removed.

Figure 6.6a shows the relationship between the dent depth after spring-back and the initial displacement, h_o, of the indenter applied on the pipe, where a linear relationship is observed. The spring-back of the pipe is not affected by the radius of curvature of the indenter. The equivalent strain at the dent apex after spring-back is also approximately linearly related to the strain before spring-back, as seen in Figure 6.6b. Linear equations are fitted from the results in Figure 6.6:

$$h_0 = 0.79h - 8.21 \tag{6.35}$$

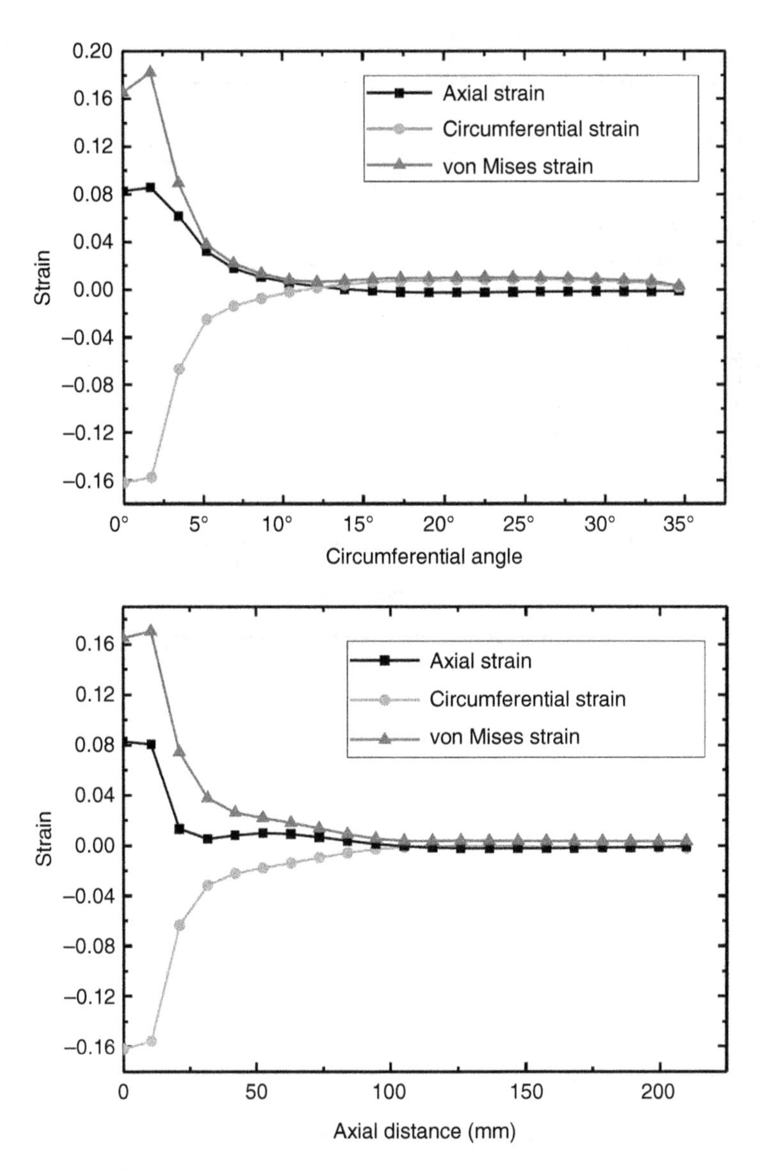

Figure 6.5 Circumferential and axial strains as a function of circumferential angle and axial distance, respectively, after spring-back. The origin point of both figures refers to the dent apex. *Source:* From Zhao et al. [2022a] / with permission from Elsevier.

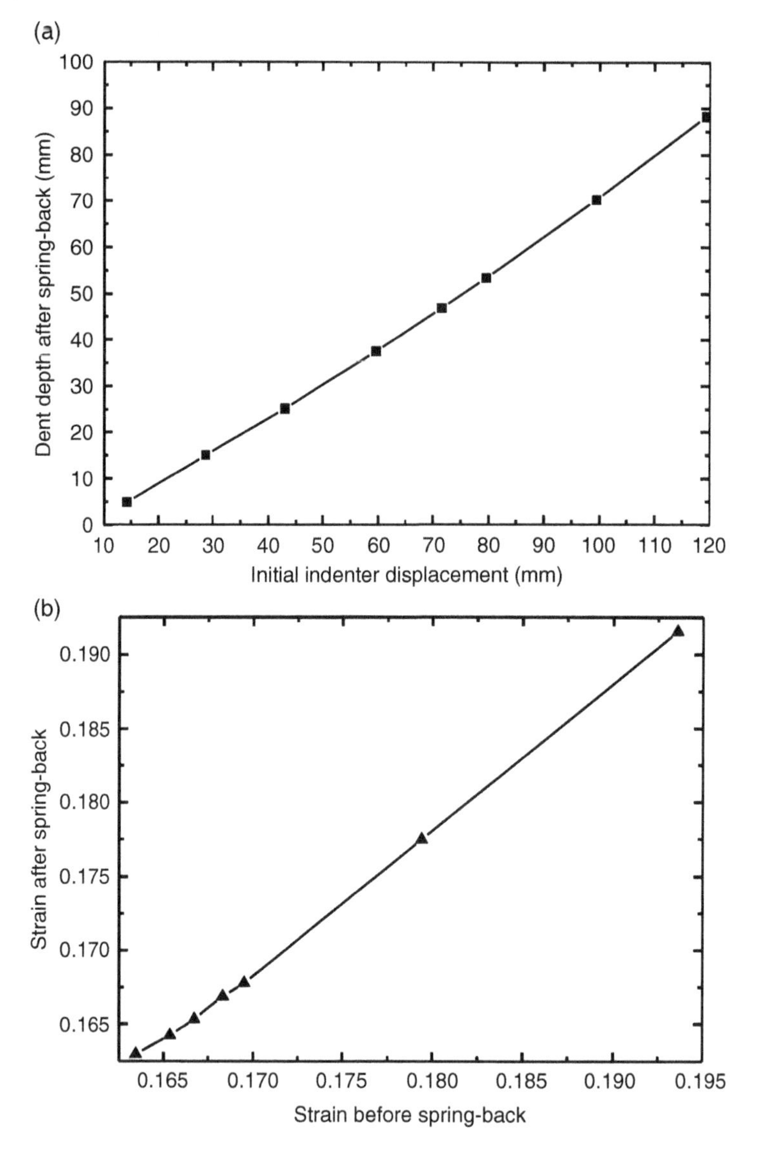

Figure 6.6 (a) Relationship between the dent depth after spring-back and the initial displacement of the indenter and (b) equivalent strain at the dent apex before and after spring-back. *Source:* From Zhao et al. [2022a] / with permission from Elsevier.

$$\varepsilon_{\text{as}} = 0.92\varepsilon_{\text{ini}} + 0.01 \tag{6.36}$$

where h is the final depth of the dent measured after removal of the indenter, ε_{as} is the strain at the dent apex after spring-back, and ε_{ini} is the strain at the dent apex before spring-back.

The critical strain $\varepsilon_{\text{crit}}$ to initiate cracks is 51.2% for X52 pipeline steel [Arumugam et al., 2016]. To further simplify the formula for DFDI calculation from Eq. (6.27), a bi-axial loading is assumed, which implies $\sigma_1 = \sigma_2 \neq 0$ and $\sigma_3 = 0$ [Arumugam et al., 2012], where σ_1, σ_2, and σ_3 are triaxial stresses, respectively, on the pipe. The simplified DFDI value before spring-back, D_{e0}, can be calculated by:

$$D_{\text{e0}} = \frac{\varepsilon_{\text{eq}} - \text{const.}}{\left(\frac{\varepsilon_0}{1.65}\right) \cdot f_1} \tag{6.37}$$

where f_1 is a factor representing the difference of strains after and before spring-back. Although the equivalent strain at the dent apex after removal of the indenter, i.e., ε_{eq}, can be obtained by ILI tools, the factor f_1, which is important to obtain an accurate DFDI, should be obtained by FE modeling for specific cases.

It is established [Zhao and Cheng, 2022a] that, when the indenter displacement is smaller than 8% of the pipe outer diameter, the maximum DFDI is located at the center of the dent. Otherwise, the location with the maximum DFDI deviates from the dent center. Moreover, when the spring-back process is considered, the strain distribution becomes more complicated at the dent, as seen in Figure 6.6. Therefore, determination of the DFDI value at the dent apex only is not sufficient for fracture damage evaluation of dented pipelines. The FE modeling can provide equivalent strain at the dent apex and the maximum equivalent strain at the dent area. The relationship between the two strains, as shown in Figure 6.7, is derived as:

$$\varepsilon_{\max} = 2.45\varepsilon_{\text{apex}} - 0.22 \tag{6.38}$$

where $\varepsilon_{\max}$ is the maximum equivalent strain at the dent area, and $\varepsilon_{\text{apex}}$ is equivalent strain at the dent apex. Eq. (6.37) is further modified as:

$$D_{\text{em}} = \begin{cases} \dfrac{\varepsilon_{\text{eq}} - A}{\left(\frac{\varepsilon_o}{1.65}\right) \cdot f_1}, \text{when } h_o \leq 0.08\,D \\[2ex] \dfrac{\varepsilon_{eq} - A - B \cdot f_1}{\left(\frac{\varepsilon_o}{1.65}\right) \cdot f_1 \cdot f_2}, \text{when } h_o > 0.08\,D \end{cases} \tag{6.39}$$

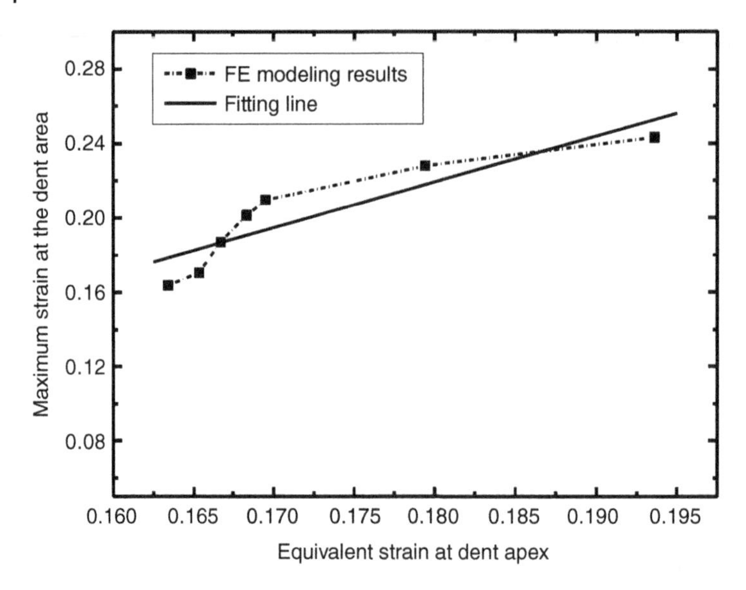

Figure 6.7 Relationship between the equivalent strain at the dent apex and the maximum strain at the dent area obtained by FE modeling. *Source:* From Zhao et al. [2022a] / with permission from Elsevier.

where D_{em} is the maximum DFDI at a dent, A and B are constants, and f_2 is a factor which is determined in Eq. (6.38) as the slope (i.e., 2.45 under the given condition).

Eq. (6.39) is the final form of the criterion for dent assessment on pipelines with an improved accuracy for failure prediction. When the initial displacement of the indenter, h_o, is smaller than 8%D, the D_{em} can be determined from the equivalent strain at the dent apex. However, when h_o exceeds 8%D, the maximum equivalent strain at the dent area should be determined by the developed FE model. A new method is thus developed for dent assessment, as compared with the industry practice used today, based on the exact geometrical profile of the dent present on pipelines.

6.6.3 Modeling Assessment for Dent-Corrosion Combinations on Pipelines

The corrosion defect can be numerically modeled as a single pit, a long groove, or a large metal loss area with various shapes in FE analysis [Shuai et al., 2008]. Dimensional parameters required to model a corrosion feature include the corrosion depth, width, and length. Rectangular, spherical, semi-elliptical, and long-bunt notch shapes [Choi et al., 2003; Adib-Ramezani et al., 2006; Ma et al., 2013] have been chosen to represent corrosion defects. Solid element is assigned in the model, and grids at the vicinity of the corrosion defect are refined. The pipe wall thickness is reduced at the corrosion defect, causing a reduction in burst pressure of the pipelines. Therefore, the burst pressure is usually used as the main criterion for assessment of the residual strength of a corroded pipeline. Although there have been numerous works for assessment of pipelines containing either a dent or corrosion defect(s), there is limited work to model and assess the dent-corrosion combinations on pipelines.

6.6.3.1 Corrosion in Dent

Corrosion often initiates at a dent on pipelines [Allouti et al., 2014; Zhao et al., 2021]. When the pipe steel is exposed to a corrosive environment, preferential corrosion tends to preferentially occur at the dent due to the so-called M–E interaction under a locally high stress and strain concentration [Xu and Cheng, 2013; Wang et al., 2021]. Corrosion is an electrochemical reaction, where the anodic reaction is normally iron oxidation, and the cathodic reaction depends on the environmental conditions. For example, buried pipelines can suffer from corrosion in near-neutral pH electrolytes under disbonded coatings, where the cathodic reaction is mainly the hydrogen evolution [Cheng and Norsworthy, 2017]. Different from dent, a corrosion defect can continuously grow to cause increased metal loss with time.

Defect assessment methods available today usually consider the dent and the corrosion defect separately [Gao and Krishnamurthy, 2015]. Although some failure forms such as fatigue and internal pressure-induced burst were investigated in a dented pipeline [Heggen et al., 2014; Gossard et al., 2016], the denting process was applied on the pipeline where a corrosion defect was already present. This is not representative of the reality. Zhao and Cheng [2022b] developed a FE model to assess corrosion in dent on an X52 steel pipe by considering both mechanical and electrochemical corrosion factors and their interaction at an unconstrained dent. Figure 6.8 shows the model for a pipe segment, an indenter, and the electrolyte the pipe steel is exposed to, where x, y, and z axes refer to circumferential, radial, and longitudinal directions, respectively. The spherical indenter of 100 mm in diameter was placed on the top of the pipe and moved along the radial direction toward the pipe body. The electrolyte (i.e., a near-neutral pH bicarbonate solution) was modeled as a block (colored blue), covering the dent area. It is believed that the dimension of the electrolyte in the developed model does not affect corrosion. However, corrosion occurs at the interface between the electrolyte and the pipe steel. As a result, the area where corrosion occurs should be large enough to cover the entire dent, controlling the scale of the grids. The indenter volume was a part of the solution during

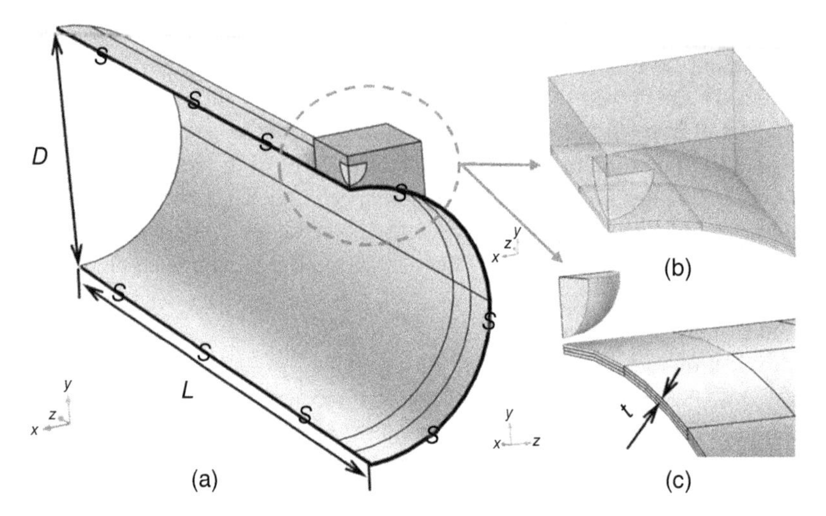

Figure 6.8 A model for (a) a pipe segment, (b) an indenter, and (c) the electrolyte the pipe steel is exposed to. *Source:* From Zhao and Cheng [2022b] / with permission from Elsevier.

electrochemical modeling. Direct solvers were used for both solid mechanics and electrochemical corrosion analysis.

The modeling results of von Mises stress, anodic current density, and net current density at a dent with various initial depths after spring-back are shown in Figure 6.9, where the net current density is the sum of anodic and cathodic current densities and is present by the density of arrows in the solution. Generally, as the initial dent depth h_0 increases, the maximum von Mises stress deviates from the dent center. The area with a high stress expands toward the dent sides. The maximum net current density and anodic current density are observed at the dent center at various initial dent depths. Moreover, the anodic reaction area, i.e., the corrosion area, increases as the initial dent depth increases.

The stress level at the dent is directly associated with the anodic current density (i.e., corrosion rate), demonstrating the essential role of M–E interaction for corrosion. The local stress concentration at the dent always results in an increased anodic current density (i.e., an increased corrosion rate). Moreover, the stress distribution at the dent area is consistent with the distribution of anodic current density. For assessment of corrosion in dent, the important factor affecting both stress and corrosion is the dent depth, while the internal pressure has a limited effect on the corrosion rate [Zhao and Cheng, 2022b]. In both the circumferential and longitudinal directions at the dent, as the dent depth increases, the anodic current density increases

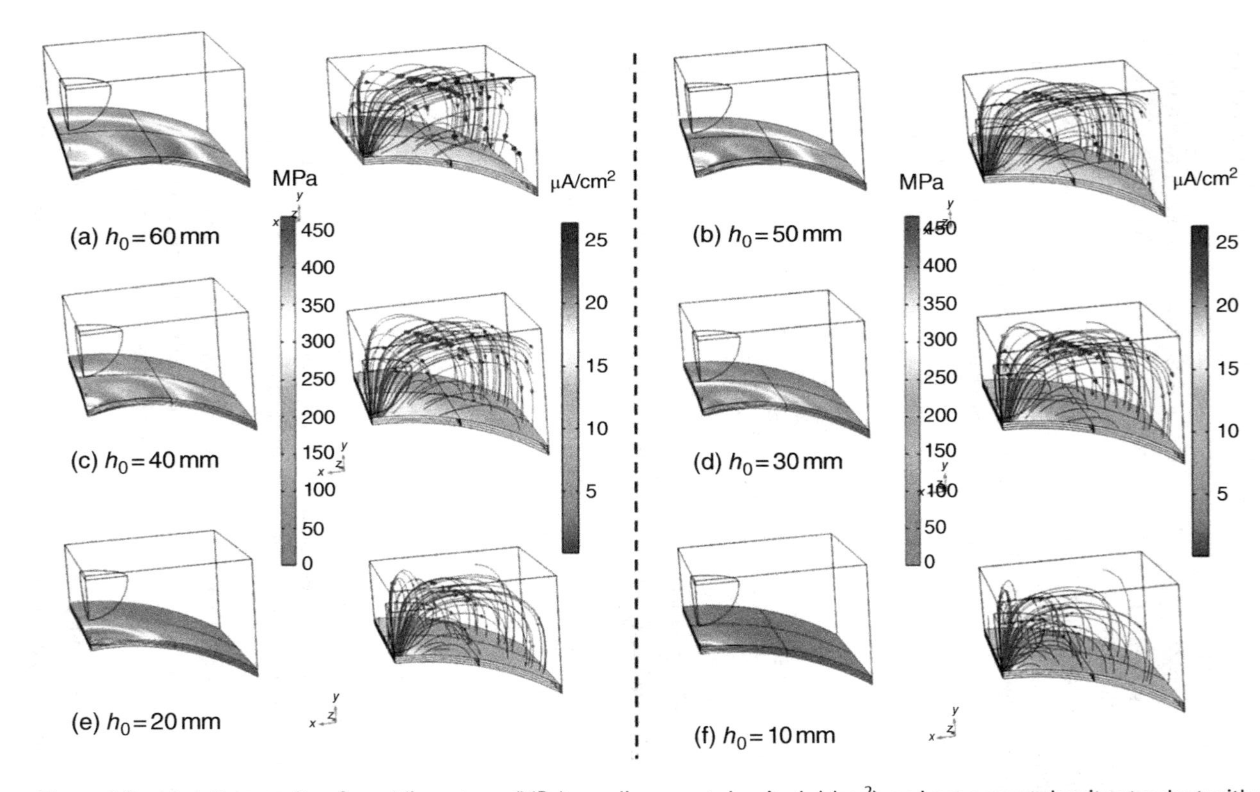

Figure 6.9 Modeling results of von Mises stress (MPa), anodic current density (μA/cm^2), and net current density at a dent with various initial dent depths (a–f) after spring-back, where the net current density is present by the density of arrows in the solution. *Source:* From Zhao and Cheng [2022b] / with permission from Elsevier.

remarkably. Moreover, a deeper dent results in a larger area to be affected by the dent for both stress and corrosion. The maximum anodic current density at the dent center is about 2.5–3 times of the anodic current density at the dent sides, showing an accelerated corrosion. The corrosion scenario will cause corrosion growth preferentially at the dent, especially at the dent center where the pipeline will leak eventually.

Furthermore, Zhao et al. [2022b] developed a new method for assessment of the burst pressure of corroded pipelines containing a dent, while considering the mutual interaction between corrosion defect and the dent, as well as the evolution of corrosion defect with time. The method enables determination of both the pressure-bearing capability of the pipeline and its remaining service life due to the dent-enhanced corrosion growth prior to leakage. For an X80 steel pipe containing corrosion in dent, the pressure capacity can be determined by a criterion considering the length of corrosion defect, dent depth, and the service time. The time-dependent failure pressure, $P(\text{Time})$, of the corroded pipe at a certain time is calculated by:

$$P(\text{Time}) = \frac{2t}{D-t} \cdot \sigma_{\text{flow}} \cdot \left[\frac{1 - \dfrac{d(\text{Time})}{t}}{1 - \dfrac{d(\text{Time})}{t\,M(Time)}} \right] \tag{6.40}$$

$$M(\text{Time}) = \sqrt{1 + 0.31 \frac{L^2(Time)}{Dt}} \tag{6.41}$$

where d and L are depth and longitudinal length of the corrosion defect. Figure 6.10 shows the relationship between failure pressure and service time for an X80 steel pipe containing corrosion in dent with various initial corrosion depths and dent depths. As expected, the failure pressure decreases with time. For deep corrosion defects such as 5.4 mm in depth, the failure pressure is lower at specific times than the pipe containing a corrosion defect of 2 mm in depth. Moreover, the co-existence of a deeper dent (i.e., over 30 mm in depth) makes the failure pressure decrease much more rapidly with time. However, an opposite tendency is observed for the pipe containing shallow corrosion defects due to different stress-enhanced corrosion rates [Zhao et al., 2022b].

In practice, while defect assessment is conducted on the defects such as dents and corrosion, the dent-corrosion interaction is usually ignored, especially the continuous growth of corrosion defects with time. To accurately determine the pipeline's pressure capability and remaining service life, the new assessment method recommends the following steps be taken. First, measure geometric parameters of both corrosion defect and dent,

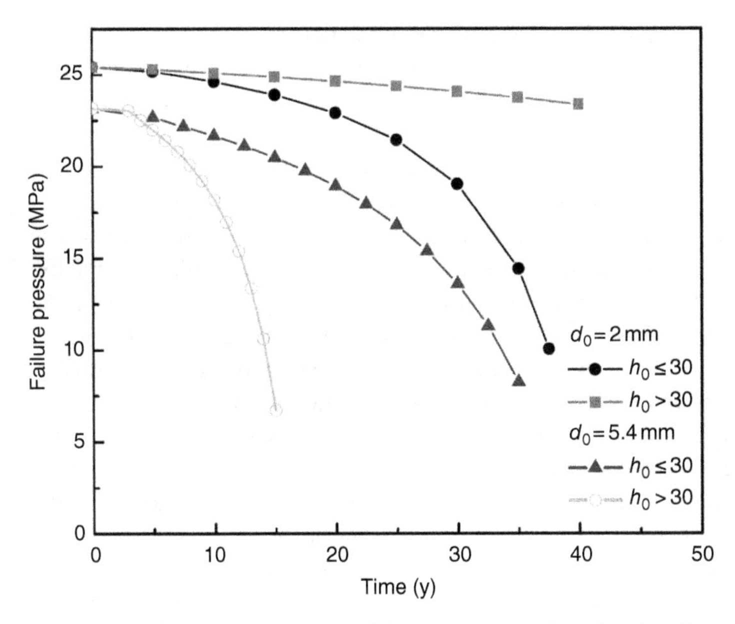

Figure 6.10 Relationship between failure pressure and service time for a corroded X80 steel pipe with various initial corrosion depths and dent depths. The initial corrosion defect length is $L_0 = 39.6$ mm. *Source:* From Zhao et al. [2022b] / with permission from Elsevier.

and determine their initial depths based on some empirical equations. Second, estimate the initial pressure capacity, $P(0)$, of the pipeline using the formula developed by Det Norske Veritas [2010] before the corrosion defect starts to grow. If $P(0) < \text{SF} \times P_b$ (where SF is safety factor and P_b is burst pressure), the pipeline may fail during service. The pipe segment containing the defects should be repaired or replaced. Finally, if $P(0) > SF \times P_b$, the pipeline can bear the operating pressure and continue to operate. However, the corrosion defect grows over time, which can be accelerated by the stress effect induced by the dent. If the pressure capacity decreases and becomes smaller than P_b after a certain time, the pipeline should be properly maintained.

Following the assessment steps, the failure pressure of pipelines containing corrosion in dent with their mutual interaction can be determined. Moreover, the remaining service life is estimated by predicting the corrosion growth rate. It is realized that the remaining life of a pipeline is not only affected by corrosion and dent but also by other threats such as fatigue and cracking. Thus, a proper pipeline integrity management program should be composed of modeling and assessment of various threats to the pipelines.

6.6.3.2 A Dent in Adjacency with a Corrosion Defect

When a dent interacts with an adjacent defect such as a corrosion feature and a crack, the pipelines suffer from increased threat of integrity degradation [Wang et al., 2012; Ghaednia et al., 2015b; Cai et al., 2018]. To date, a gap exists in assessment of the interaction of coincident corrosion defect and a dent. Specifically, there has been no criterion available to identify the existence of an interaction between a dent and an adjacent corrosion defect. The challenge was noted by API that a generalized dent-corrosion interaction criterion has not been developed yet [American Petroleum Institute, 2020]. The FE-based method proposed by Sun et al. [2021] is the first of its kind enabling evaluation of the interaction between a dent and an adjacent corrosion defect and the impact on pipeline integrity.

A 3D model showing the indentation on an X46 steel pipe to create a dent and the adjacent rectangularly shaped corrosion defect is shown in Figure 6.11, where the effects of the dent depth and length, as well as the corrosion depth, are considered in modeling. Figure 6.12 shows the von Mises stress contour of the pipe containing a dent adjacent to a corrosion defect with a spacing of 100 mm under an internal pressure of 6.8 MPa, where the dent depth, corrosion length and corrosion depth are 20.0 mm, 100 mm, 50% *t*, respectively. It is seen that the maximum von Mises stress occurs at the longitudinal edge of the corrosion defect, where the local stress is up to 630 MPa (i.e., the ultimate tensile strength of X46 steel). The failure pressure of the pipe containing both the dent and corrosion defect is estimated to be 6.80 MPa. As a comparison, when the pipe segment contains a corrosion defect only, the failure pressure is 7.75 MPa. Thus, the presence of a dent in adjacency to the corrosion defect decreases failure pressure of the pipe due to their interaction.

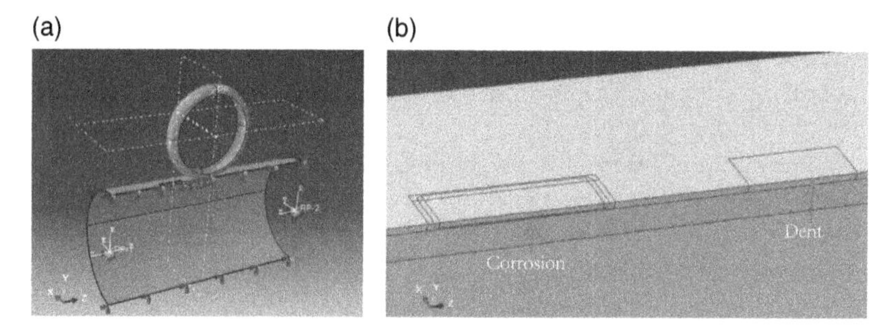

Figure 6.11 Schematic diagrams showing (a) a 3D model for a steel pipe and an indenter and (b) a dent adjacent to a corrosion defect. *Source:* From Sun et al. [2021] / American Society of Civil Engineers.

(a)

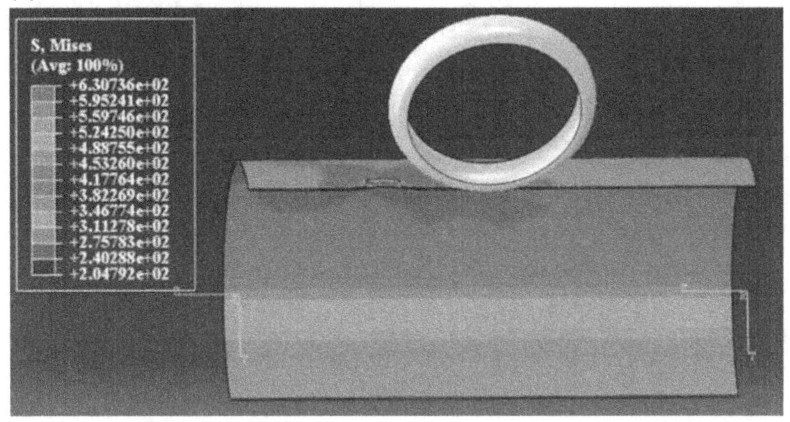

(b)

Figure 6.12 Von Mises stress contour of the pipe containing a dent adjacent to a corrosion defect with a spacing of 100 mm under an internal pressure of 6.8 MPa (a) 3D view and (b) 2D cross-sectional view. *Source:* From Sun et al. [2021] / American Society of Civil Engineers.

Furthermore, the interaction between the dent and the corrosion defect depends on their spacing. As shown in Table 6.3, with the increase of the spacing between them, the interaction effect decreases. When the spacing exceeds a critical value, the interaction is negligible, and the failure pressure of the pipe does not change with a further increase of the spacing. The dent and the corrosion defect can be assessed independently. To quantitatively define the interaction, the failure pressure of a pipeline containing a single corrosion feature is used as a reference. When the ratio of the failure pressure in the presence of both a dent and corrosion defect to the reference failure pressure is equal to 1, there is no mutual interaction between the two

Table 6.3 Effect of the spacing between a dent and a corrosion defect on failure pressure of the steel pipe, where the dent depth, corrosion length, and corrosion depth are fixed.

Dent-corrosion feature spacing (mm)	Failure pressure of the pipe (MPa)
100	7.75 × 87.7%
110	7.75 × 92%
150	7.75 × 100%
200	7.75 × 100%
250	7.75 × 100%

Source: From Sun et al. [2021] / American Society of Civil Engineers.

defects. If the ratio is smaller than 1, an interaction exists to further decrease the failure pressure of the pipeline.

6.6.4 Assessment for Dent Combined with Other Defects on Pipelines

6.6.4.1 Dent Combined with a Gouge

In experimental testing, a gouge is machined on a pipe segment, and an indenter is applied on the pipe to make a combination of a dent with the gouge. Numerical modeling by FE is conducted with the similar procedure to previously described. To model the gouge on the pipe wall, the solid element is selected. The shape of the gouge can be semi-elliptical [Allouti et al., 2014], cylindrical [Zhao et al., 2021], semi-cylindrical [Tian and Zhang, 2017b], or rectangular [Cosham et al., 2007]. The stress concentration at the corner of the gouge is usually reduced. Additionally, grids in the defect area should be refined to ensure accuracy. Orientally, the gouge can be located at the center of the dent, across the dent area, or at the edge of the dent. After denting, loads like internal pressure or a bending moment are applied on the pipe [Tian and Zhang, 2017b; Bao and Zhou, 2021].

6.6.4.2 Dent Combined with Cracks

Modeling of cracks on pipelines by FE is challenging, although some progress has been made in the recent years. Bedairi et al. [2012] investigated cracks existing in corrosion area by modeling the crack tip as a blunt notch with a specific radius of one-thousandth of the size of the plastic zone. V-shape notches were used by Ghaednia et al. [2015a, b] to model cracks in pipelines, where the grids in the crack area were properly adjusted. Another method to model pipeline cracks is the so-called extended FE method (XFEM) [Okodi, 2021; Zhang et al., 2021b]. The XFEM is a built-in module in ABAQUS software platform that can simulate structural discontinuities such as cracks and holes. A dented pipe containing a crack was modeled by shell and solid elements with the XFEM technique, where the shell element was used to model the pipe body, and the solid element modeled the crack [Okodi, 2021]. The shell-solid coupling constraint was utilized to attach the shell elements with the solid elements. A dent was produced by applying an indenter with a certain load.

References

Abdelmoety, A.K., Kainat, M., Yoosef-Ghodsi, N., Li, Y., Adeeb, S. (2022) Strain-based reliability analysis of dented pipelines using a response surface method, *J. Pipeline Sci. Eng.*, 2, 29–38.

Adeeb, S.M., Horsley, D.J. (2006) A numerical procedure to establish a safe working pressure during excavation of a pipeline in a rock ditch, *Int. J. Press. Vessel. Pip.*, 83, 488–497.

Adib-Ramezani, H., Jeong, J., Pluvinage, G. (2006) Structural integrity evaluation of X52 gas pipes subjected to external corrosion defects using the SINTAP procedure, *Int. J. Press. Vessel Pip.*, 83, 420–432.

Alashti, R.A., Jafari, S., Hosseinipour, S.J. (2015) Experimental and numerical investigation of ductile damage effect on load bearing capacity of a dented API XB pipe subjected to internal pressure, *Eng. Fail. Anal.*, 47, 208–228.

Alexander, C.R. (1999) Review of experimental and analytical investigations of dented pipelines, *Press. Vessel Pip. Conf.*, ASME, Boston, MA, USA.

Alexander, C., Brownlee, K. (2007) Methodology for assessing the effects of plain dents, wrinkle bends, and mechanical damage on pipeline integrity, *Corrosion' 2007*, paper no. 07139, NACE, Houston, TX, USA.

Allouti, M., Schmitt, C., Pluvinage, G., Gilgert, J., Hariri, S. (2012) Study of the influence of dent depth on the critical pressure of pipeline, *Eng. Fail. Anal.*, 21, 40–51.

Allouti, M., Schmitt, C., Pluvinage, G. (2014) Assessment of a gouge and dent defect in a pipeline by a combined criterion, *Eng. Fail. Anal.*, 36, 1–13.

American Petroleum Institute (2020) *Assessment and Management of Pipeline Dents*, API RP 1183, Washington, DC, USA.

American Petroleum Institute/American Society of Mechanical Engineering (2016) *Fitness for Service*, API/ASME 579-1/FFS-1, Washington, DC, USA.

American Society of Mechanical Engineering (2017) *Manual for Determining the Remaining Strength of Corroded Pipelines*, ASME B31G, New York, USA.

American Society of Mechanical Engineering (2020) *Gas Transmission and Distribution Piping Systems*, ASME B31.8, New York, USA.

Arumugam, U., Gao, M., Krishnamurthy, R., Wang, R., Kania, R. (2012) Root cause analysis of dent with crack: a case study, *Int. Pipeline Conf.*, Calgary, AB, Canada.

Arumugam, U., Gao, M., Krishnamurthy, R., Wang, R., Kania, R. (2016) Study of a plastic strain limit damage criterion for pipeline mechanical damage using FEA and full-scale denting tests, *Int. Pipeline Conf.*, ASME, Calgary, AB, Canada.

Arumugam, U., Gao, M., Krishnamurthy, R., Wang, R., Kania, R. (2018) Study of safe dig pressure level for rock dents in gas pipelines, *Int. Pipeline Conf.*, ASME, Calgary, AB, Canada.

Babbar, V., Clapham, L. (2009) Finite element modeling of magnetic flux leakage signals from mechanical damage containing corrosion pits, *Am. Inst. Phys. Conf. Proc.*, 1096, 453–458.

Baek, J., Kim, Y., Kim, W., Koo, J., Seok, C. (2012) Load bearing capacity of API X65 pipe with dent defect under internal pressure and in-plane bending, *Mater. Sci. Eng. A*, 540, 70–82.

Bao, J., Zhou, W. (2021) Influence of depth thresholds and interaction rules on the burst capacity evaluation of naturally corroded pipelines, *J. Pipeline Sci. Eng.*, 1, 148–165.

Bedairi, B., Cronin, D., Hosseini, A., Plumtree, A. (2012) Failure prediction for crack-in-corrosion defects in natural gas transmission pipelines, *Int. J. Press. Vessel. Pip.*, 96–97, 90–99.

Bernard, M., Fontanabona, J., Gaffard, V., Popineau, D., Wiet, P. (2013) New horizons for pipelines integrity assessment, *Offshore Technology Conference*, Houston, TX, USA.

Błachut, J., Iflefel, I.B. (2007) Collapse of pipes with plain or gouged dents by bending moment, *Int. J. Press. Vessel. Pip.*, 84, 560–571.

Błachut, J., Iflefel, I.B. (2008) Experimental and numerical investigation of plain and gouged dents in steel pipes subjected to pressure and moment loading, *J. Press. Vessel Technol.*, 130, 021203.

Bratton, J.P., Alexander, T., Bubenik, T.A., Finneran, S., Heggen, H.O. (2012) An approach for evaluating the integrity of plain dents reported by in-line inspection tools, *Int. Pipeline Conf.*, ASME, Calgary, AB, Canada.

Cai, J., Jiang, X., Lodewijks, G., Pei, Z., Wu, W. (2018) Residual ultimate strength of damaged seamless metallic pipelines with combined dent and metal loss, *Marine Struct.*, 61, 188–201.

Cheng, Y.F. (2013) *Stress Corrosion Cracking of Pipelines*, Wiley, New York, USA

Cheng, Y.F., Norsworthy, R. (2017) *Pipeline Coatings*, NACE, Houston, TX, USA

Choi, J.B., Goo, B.K., Kim, J.C., Kim, Y.J., Kim, W.S. (2003) Development of limit load solutions for corroded gas pipelines, *Int. J. Press. Vessel. Pip.*, 80, 121–128.

Coramik, M., Ege, Y. (2017) Discontinuity inspection in pipelines: a comparison review, *Measurement*, 111, 359–373.

Cosham, A., Hopkins, P. (2004) The assessment of corrosion in pipelines–guidance in the pipeline defect assessment manual (PDAM), *Pipeline Pigging Integrity Management Conf.*, Amsterdam, The Netherlands.

Cosham, A., Hopkins, P., MacDonald, K.A. (2007) Best practice for the assessment of defects in pipelines–corrosion, *Eng. Fail. Anal.*, 14, 1245–1265.

Cunha, S.B., Pasqualino, I.P., Pinheiro, B.C. (2009) Stress-life fatigue assessment of pipelines with plain dents, *Fatigue Frac. Eng. Mater.*, 32, 961–974.

Cunha, S.B., Pasqualino, I.P., Pinheiro, B.C. (2014) Pipeline plain dent fatigue: a comparison of assessment methodologies, *Int. Pipeline Conf.*, ASME, Calgary, AB, Canada.

Dawson, S.J., Russell, A., Patterson, A. (2006) Emerging techniques for enhanced assessment and analysis of dents, *Int. Pipeline Conf.*, ASME, Calgary, AB, Canada.

Dawson, J., Murray, I., Hedger, J. (2018) Techniques for the enhanced assessment of pipeline dents, *J. Pipeline Eng.*, 17, 127–135.

Det Norske Veritas (2010) *Corroded Pipelines—Recommended Practice*, RP F101, Norway.

Dubyk, Y., Seliverstova, I. (2019) Assessment of dents for gas pipelines, *Procedia Struct. Integrity*, 18, 622–629.

Freire, J., Paiva, V., Gonzáles, G., Vieira, R.D., Maneschy, J.E., Almeida, A., Ribeiro, A.S. (2019) Fatigue assessment and monitoring of a dented pipeline specimen, *Press. Vessel. Pip. Conf.*, ASME, San Antonio, TX, USA.

Gao, M., Krishnamurthy, R. (2015) Mechanical damage in pipelines: a review of the methods and improvements in characterization, evaluation, and mitigation, *Oil Gas Pipelines*, 289–326.

Gao, M., McNealy, R., Krishnamurthy, R., Colquhoun, I. (2008) Strain-based models for dent assessment: a review, *Int. Pipeline Conf.*, ASME, Calgary, AB, Canada.

Gao, M., Krishnamurthy, R., Tandon, S., Arumugam, U. (2013) Critical strain based ductile damage criterion and its application to mechanical damage in pipelines, *Int. Conf. Fracture*, Beijing, China.

Ghaednia, H., Silva, J., Kenno, S., Das, S., Wang, R., Kania, R. (2013) Pressure tests on 30-in. diameter X65 grade pipes with dent–crack defects, *J. Pipeline Eng.*, 12, 61–67.

Ghaednia, H., Das, S., Wang, R., Kania, R. (2014) Effect of dent depth on the burst pressure of NPS30 X70 pipes with dent-crack defect, *Int. Pipeline Conf.*, Calgary, AB, Canada.

Ghaednia, H., Das, S., Wang, R., Kania, R. (2015a) Effect of operating pressure and dent depth on burst strength of NPS30 linepipe with dent–crack defect, *J. Offshore Mech. Arctic Eng.*, 137, 031402.

Ghaednia, H., Das, S., Wang, R., Kania, R. (2015b) Safe burst strength of a pipeline with dent-crack defect: effect of crack depth and operating pressure, *Eng. Fail. Anal.*, 55, 288–299.

Gossard, J., Bratton, J., Kemp, D., Finneran, S., Polasik, S.J. (2016) Evaluating dents with metal loss using finite element analysis, *Int. Pipeline Conf.*, Calgary, AB, Canada.

Hafez, K.M. (2021) The role of a plain dent on the failure mode of a crude oil pipeline, *Eng. Fail. Anal.*, 122, 105291.

Han, C., Tan, S., Zhang, J., Zhang, C. (2018) Simulation investigation of dent behavior of steel pipe under external load, *Eng. Fail. Anal.*, 90, 341–354.

He, Z, Zhou, W. (2021) Fatigue reliability analysis of dented pipelines, *J. Pipeline Sci. Eng.*, 1, 290–297.

Heggen, H.O., Bratton, J., Kemp, D., Liu, J., Austin, J. (2014) Fitness for service of dents associated with metal loss due to corrosion, *Int. Pipeline Conf.*, Calgary, AB, Canada.

Hyde, T.H., Luo, R., Becker, A.A. (2007) Force–deflection analysis of offset indentations on pressurised pipes, *Int. J. Press. Vessel. Pip.*, 84, 368–377.

Iflefel, I.B., Moffat, D.G., Mistry, J. (2005) The interaction of pressure and bending on a dented pipe, *Int. J. Press. Vessel. Pip.*, 82, 761–769.

Kainat, M., Woo, J., Langer, D., Krausert, T., Cheng, J.R., Hassanien, S., Adeeb, S. (2019) Effects of loading sequences on remaining life of plain dents in buried liquid pipelines, *J. Pipeline Syst. Eng.*, 10, 4019001.

Kec, J., Cerny, I. (2017) Stress-strain assessment of dents in wall of high pressure gas pipeline, *Procedia Struct. Integrity*, 5, 340–346.

Lancaster, E.R., Palmer, S.C. (1996) Burst pressures of pipes containing dents and gouges, *J. Proc. Mech. Eng.*, 210, 19–27.

Leis, B.N. (2021) Evolution of metal-loss severity criteria: gaps and a path forward, *J. Pipeline Sci. Eng.*, 1, 51–62.

Li, C., Dang, S. (2017) Plastic damage analysis of oil and gas pipelines with unconstrained and constrained dents, *Eng. Fail. Anal.*, 77, 39–49.

Liu, J.H., Francis, A. (2004) Theoretical analysis of local indentation on pressured pipes, *Int. J. Press. Vessel. Pip.*, 81, 931–939.

Liu, Z.Y., Li, X.G., Cheng, Y.F. (2011) Effect of strain rate on cathodic reaction during stress corrosion cracking of X70 pipeline steel in a near-neutral pH solution, *J. Mater. Eng. Perf.*, 20, 1242–1246.

Liu, X., Zhang, H., Baodong, W., Mengying, X., Kai, T., Kai, W.U. (2017) Nonlinear finite element analysis of the limit state and pressure of dented X60 steel pipeline, *Mechanics*, 23, 814–819.

Lockey, A., Santamaria, W., Gonzalez, G. (2014) Modelling shallow dents using local regression methods and finite element analysis, *Int. Pipeline Conf.*, Calgary, AB, Canada.

Lukasiewicz, S.A., Czyz, J.A., Sun, C., Adeeb, S. (2006) Calculation of strains in dents based on high resolution in-line caliper survey, *Int. Pipeline Conf.*, ASME, Calgary, AB, Canada.

Luo, J., Zhang, Y., Li, L., Zhu, L., Wu, G. (2020) Fatigue failure analysis of dented pipeline and simulation calculation, *Eng. Fail. Anal.*, 113, 104572.

Ma, B., Shuai, J., Liu, D., Xu, K. (2013) Assessment on failure pressure of high strength pipeline with corrosion defects, *Eng. Fail. Anal.*, 32, 209–219.

MacDonald, K.A., Cosham, A. (2005) Best practice for the assessment of defects in pipelines–gouges and dents, *Eng. Fail. Anal.*, 12, 720–745.

MacDonald, K.A., Cosham, A., Alexander, C.R., Hopkins, P. (2007) Assessing mechanical damage in offshore pipelines – Two case studies, *Eng. Fail. Anal.*, 14, 1667–1679.

Naghipour, M., Ezzati, M., Elyasi, M. (2018) Analysis of high-strength pressurized pipes (API-5L-X80) with local gouge and dent defect, *Appl. Ocean Res.*, 78, 33–49.

Noronha, D.B., Martins, R.R., Jacob, B.P., de Souza, E. (2010) Procedures for the strain based assessment of pipeline dents, *Int. J. Press. Vessel. Pip.*, 87, 254–265.

Okodi, A. (2021) *Burst pressure of pipeline with longitudinal crack in dent defects using extended finite element method*, PhD thesis, University of Alberta, Edmonton, AB, Canada.

Okodi, A., Lin, M., Yoosef-Ghodsi, N., Kainat, M., Hassanien, S., Adeeb, S. (2020) Crack propagation and burst pressure of longitudinally cracked pipelines using extended finite element method, *Int. J. Press. Vessel. Pip.*, 184, 104115.

Okodi, A., Li, Y., Cheng, J., Kainat, M., Yoosef-Ghodsi, N., Adeeb, S. (2021) Effect of location of crack in dent on burst pressure of pipeline with combined dent and crack defects, *J. Pipeline Sci. Eng.*, 2, 252–263.

Okoloekwe, C., Aranas, N., Kainat, M., Langer, D., Hassanien, S., Cheng, J.J., Adeeb, S. (2018) Improvements to the ASME B31. 8 dent strain equations, *J. Press. Vessel Technol.*, 140, 041010.

Okoloekwe, C., Fowler, M., Virk, A., Yoosef-Ghodsi, N., Kainat, M. (2020) Reliability-based assessment of safe excavation pressure for dented pipelines, *Int. Pipeline Conf.*, ASME, Calgary, AB, Canada.

Oyane, M., Sato, T., Okimoto, K., Shima, S. (1980) Criteria for ductile fracture and their applications, *J. Mech. Work. Technol.*, 4, 65–81.

Pinheiro, B., Soares, C.G., Pasqualino, I. (2019) Generalized expressions for stress concentration factors of pipeline plain dents under cyclic internal pressure, *Int. J. Press. Vessel. Pip.*, 170, 82–91.

Pluvinage, G., Allouti, M., Schmitt, C., Capelle, J. (2011) Assessment of a gouge, a dent, or a dent plus a gouge, in a pipe using limit analysis or notch fracture mechanics, *J. Pipeline Eng.*, 10, 147–160.

Pournara, A.E., Papatheocharis, T., Karamanos, S.A., Perdikaris, P.C. (2019) Mechanical behavior of dented steel pipes subjected to bending and pressure loading, *J. Offshore Mech. Arctic Eng.*, 141, 011702.

Qin, G.J., Cheng, Y.F. (2021) A review on defect assessment of pipelines: principles, numerical solutions, and applications, *Int. J. Press. Vessel. Pip.*, 191, 104329.

Rafi, A.N.M. (2011) *Structural Behaviour of Dented Pipelines*, University of Windsor, Windsor, ON, Canada.

Rezaee, N., Sharifi, S.M.H., Rashed, G.R., Niknejad, A. (2018) Denting the oil pipelines by a rigid cylindrical indenter with conical nose by the numerical and experimental analyses, *Thin Wall. Struct.*, 124, 312–322.

Shahzamanian, M.M., Kainat, M., Ghodsi, N.Y., Adeeb, S. (2021) Systematic literature review of the application of extended finite element method in failure prediction of pipelines, *J. Pipeline Sci. Eng.*, 2, 241–251.

Shaik, M.R. (2015) Pipeline integrity assessment: methodology, *India Oil and Gas Pipeline Conference*, ASME, New Delhi, India.

Shirband, Z., Gosselin, A., Guest, S., Falcon, L. (2020) Pipeline plain dent fatigue assessment: shedding light on the API 579 level 2 fatigue assessment methodology, *Int. Pipeline Conf.*, ASME, Calgary, AB, Canada.

Shuai, J., Zhang, C.E., Chen, F., He, R. (2008) Prediction of failure pressure of corroded pipelines based on finite element analysis, *Int. Pipeline Conf.*, Calgary, AB, Canada.

Shuai, Y., Shuai, J., Xu, K. (2017) Probabilistic analysis of corroded pipelines based on a new failure pressure model, *Eng. Fail. Anal.*, 81, 216–233.

Shuai, Y., Shuai, J., Zhang, X. (2018) Experimental and numerical investigation of the strain response of a dented API 5L X52 pipeline subjected to continuously increasing internal pressure, *J. Nat. Gas Sci. Eng.*, 56, 81–92.

Shuai, Y., Wang, X., Shuai, J., Zhao, Y., Zhang, X., Tang, C. (2020) Mechanical behavior investigation on the formation of the plain dent of an API 5L L245 pipeline subjected to concentrated lateral load, *Eng. Fail. Anal.*, 108, 104189.

Sun, J.L., Cheng, Y.F., Woo, J., Kainat, M., Hassanien, S. (2021) Assessment of interaction between a dent and an adjacent corrosion feature on pipelines and the effect on pipeline failure pressure by finite element modeling, *J. Pipeline Sys. Eng. Pract.*, 12, 4021029.

Tang, X., Cheng, Y.F. (2011) Quantitative characterization by micro-electrochemical measurements of the synergism of hydrogen, stress and dissolution on near-neutral pH stress corrosion cracking of pipelines, *Corros. Sci.*, 53, 2927–2933.

Tee, K.F., Wordu, A.H. (2020) Burst strength analysis of pressurized steel pipelines with corrosion and gouge defects, *Eng. Fail. Anal.*, 108, 104347.

Tian, X., Zhang, H. (2017a) Failure criterion of buried pipelines with dent and scratch defects, *Eng. Fail. Anal.*, 80, 278–289.

Tian, X., Zhang, H. (2017b) Failure pressure of medium and high strength pipelines with scratched dent defects, *Eng. Fail. Anal.*, 78, 29–40.

Tian, X., Lu, M., Chen, Y. (2020) Numerical simulation method for failure pressure of buried dented pipeline, *J. Fail. Anal. Prev.*, 20, 944–949.

Tiku, S., Semiga, V., Dinovitzer, A., Vignal, G (2012) Full scale cyclic fatigue testing of dented pipelines and development of a validated dented pipe finite element model, *Int. Pipeline Conf.*, ASME, Calgary, AB, Canada.

Wang, R., Kania, R., Arumugam, U., Gao, M. (2012) A combined approach to characterization of dent with metal loss, *Int. Pipeline Conf.*, ASME, Calgary, AB, Canada.

Wang, Y.C., Xu, L.Y., Sun, J.L., Cheng, Y.F. (2021) Mechano-electrochemical interaction for pipeline corrosion: a review, *J. Pipeline Sci. Eng.*, 1, 1–18.

Wu, Y., Tang, N., Zhang, P. (2015) The comparison of dented pipeline displacement calculation methods, *Eng. Fail. Anal.*, 57, 562–573.

Wu, Y., Xiao, J., Zhang, P. (2016) The analysis of damage degree of oil and gas pipeline with type II plain dent, *Eng. Fail. Anal.*, 66, 212–222.

Wu, Y., Li, J., Li, L. (2019) Damage and springback analysis of two typical dented pipelines with different parameters, *J. Press. Vessel Technol.*, 141, 041701.

Xie, M., Tian, Z. (2018) A review on pipeline integrity management utilizing in-line inspection data, *Eng. Fail. Anal.*, 92, 222–239.

Xu, L., Cheng, Y.F. (2012a) Assessment of the complexity of stress/strain conditions of x100 steel pipeline and the effect on the steel corrosion and failure pressure prediction, *Int. Pipeline Conf.*, ASME, Calgary, AB, Canada.

Xu, L.Y., Cheng, Y.F. (2012b) An experimental investigation of corrosion of X100 pipeline steel under uniaxial elastic stress in a near-neutral pH solution, *Corros. Sci.*, 59, 103–109.

Xu, L.Y., Cheng, Y.F. (2013) Development of a finite element model for simulation and prediction of mechano-electrochemical effect of pipeline corrosion, *Corros. Sci.*, 73, 150–160.

Zarea, M., Hertz-Clemens, S., Batisse, R., Cardin, P. (2014) Experimental investigation on combined "dent and gouge" defects on vintage steel transmission pipelines, *Int. Pipeline Conf.*, ASME, Calgary, AB, Canada.

Zhang, Z.W., Ni, X., Cheng, Y.F. (2020a) Assessment by finite element modelling of the mechano-electrochemical interaction at double-ellipsoidal corrosion defect with varied inclinations on pipelines, *Constr. Build. Mater.*, 260, 120459.

Zhang, P., Huang, Y., Wu, Y., Mohamed, H.S. (2020b) Investigations on the re-rounding performance of dented-pipelines at the service and shutdown stages, *Eng. Fail. Anal.*, 116, 104746.

Zhang, Z.W., Wang, J.C., Zhang, J.H., Cheng, Y.F (2021a) Modeling of the mechano-electrochemical effect at corrosion defect with varied inclinations on oil/gas pipelines, *Petro. Sci.*, 18, 1520–1529.

Zhang, X., Lin, M., Okodi, A., Tan, L., Leung, J., Adeeb, S. (2021b) Numerical analysis of API 5LX42 and X52 vintage pipes with cracks in corrosion defects using extended finite element method, *J. Press. Vessel Technol.*, 143, 061302.

Zhao, J., Cheng, Y.F. (2022a) A new criterion based on strain determination for dent assessment of pipelines, *J. Pipeline Sci. Eng.*, 2, 18–28.

Zhao, J., Cheng, Y.F. (2022b) Assessment by finite element modeling of corrosion in dent on X52 steel pipelines, *Int. J. Press. Vessel Pip.*, 200, 104835.

Zhao, P., Shuai, J., Lv, Z., Xu, K. (2020) Strain response of API 5L X80 pipeline subjected to indentation, *Appl. Ocean Res.*, 94, 101991.

Zhao, P., Shuai, J., Sun, M., Lv, Z., Xu, K., Wang, Y. (2021) Burst pressure of thin-walled pipes with dent and gouge defects, *Thin Wall. Struct.*, 159, 107213.

Zhao, J., Lv, Y., Cheng, Y.F. (2022a) Standards and methods for dent assessment and failure prediction of pipelines: a critical review, *Petro. Sci.*, 19, 3029–3045.

Zhao, J., Lv, Y., Cheng, Y.F. (2022b) A new method for assessment of burst pressure capacity of corroded X80 steel pipelines containing a dent, *Int. J. Press. Vessel Pip.*, 199, 104742.

Zhu, X., Wang, R. (2019) Effect of residual stress or plastic deformation history on fatigue life simulation of pipeline dents, *Int. Pipeline Conf.*, Calgary, AB, Canada.

7

Assessment of Buckles on Pipelines and Buckling Failure Analysis

7.1 Introduction

Geometric anomalies on pipelines include not only dents but also buckles and ovalities. As stated, dents can occur either during pipeline construction or in-service, while buckles and wrinkles may result from manufacturing such as cold bending of pipe segments or from pipeline construction such as loss of stability during pipe laying. A buckle is a local geometric instability causing ovalization and flattening of the pipe, and possibly abrupt changes in local curvature [Yablonskikh et al., 2007]. There are two main types of global buckling of pipelines, i.e., lateral buckling and vertical buckling (also named upheaval buckling). When a pipeline is under operation at temperature and pressure higher than the ambient, it will tend to expand. If the pipe is not free to expand, an axial compressive force will be developed. The axial compressive loading can also be generated on the pipe due to pipe-soil interactions. When the force exerted by the pipe on the soil exceeds the vertical restraint against uplift movement created by the pipe's weight, its bending stiffness and the resistance of the soil cover, the pipe will move upward, and considerable vertical displacement may occur [Schaminee et al., 1990]. The buckling of pipelines in the lateral direction is most probable when the pipelines are laid on a flat area or an area typically in a desert and when it is not provided with sufficient lateral resistance [Chee et al., 2018]. When the pipelines are buried in a normal trench, the soil constraints in the lateral direction are sufficiently high, which remarkably reduces the chances of lateral buckling. Both types of buckling are a failure mode of pipelines and must be assessed in terms of the risk of causing pipeline failure.

Furthermore, compared with buckling occurring on a pipeline free of surface damages such as dents and corrosion defects, the presence of various defects or anomalies would remarkably increase the occurrence probability

Defect Assessment for Integrity Management of Pipelines, First Edition. Y. Frank Cheng.

of pipeline buckling under identical loading and soil conditions. Thus, in addition to buckling assessment on intact pipelines, attention is always paid to assessing and analyzing pipeline buckling at surface defects under given loading conditions, defining the buckling resistance, e.g., critical bending moment or critical compressive load, of the pipelines. In addition, a method to evaluate the burst capacity of pipelines, especially corroded pipelines, under axial compressive loading and bending moment is developed.

7.2 Buckling Failure Analysis of an X80 Steel Pipe Containing a Dent Under Bending Moment

Buckling can occur on pipelines under bending moment or axial compressive loading. When dents are present on the pipe body, local buckling would preferentially occur at the dents under geotechnically unstable conditions such as faults, landslides, and floods, which can induce a bending moment on the pipelines. The buckling phenomenon occurring on dented pipelines under bending moment has been widely studied [Błachut and Iflefel, 2007; Limam et al., 2010, 2012; Baek et al., 2012]. For example, numerical modeling might not always give accurate results about the buckling process under bending moment, which could not be consistent with the actual process. The numerical results underestimated the critical buckling moment of the dented pipelines. The critical buckling curvature of pipelines decreased with increased dent depth. Moreover, the plastic buckling behavior of the dented pipe was greatly affected by bending modes and dent geometry. To date, most reported work in the topic has aimed at small-diameter pipelines made of low-grade steels. The results are not applicable for high-grade and large-diameter pipelines. Meanwhile, little has been known about the effect of pipe steel properties on buckling behavior of the pipelines. A 3D nonlinear FE model is developed to investigate the local buckling behavior of an X80 high-grade steel pipe containing a dent under bending moment [Shuai et al., 2020a]. The effects of internal pressure, pipe outer diameter and wall thickness, and steel properties on the critical buckling moment were determined.

7.2.1 The Model Development for Numerical Analysis

7.2.1.1 The Model

The true stress–strain behavior of high-grade pipeline steels such as X80 steel can be described by R-O equation [Ramberg and Osgood, 1943]. It is a round-house-type of curve where a yielding plateau that usually appears in the stress–strain curves measured on low-grade steels is not observed [Liang and Cheng, 2013; Shuai et al., 2020a]. The stress increases continuously with the strain, even after the occurrence of necking.

Figure 7.1 shows the cross-sectional and global views of the developed FE model for a steel pipe containing a dent generated by an indenter and the support. The indenter, pipe segment, and support "saddle" are meshed separately. The meshing density of section L_2 and the pipe in the dented area was appropriately dense since the middle section was the focus of the modeling. The sections at both ends, i.e., L_1 and L_3, and the support grids were sparse appropriately, balancing the calculation accuracy and saved computational time. A surface-to-surface contact algorithm was used to simulate the non-linear contact between the indenter and the pipe, and the pipe and the support. The contact stiffness coefficient is 0.25.

Application of a load and boundary conditions for the modeling included the following steps. (1) An internal pressure was applied to the pipe, and an axial tension was applied on both ends of the pipe segment to simulate the influence of the end-cap conditions. (2) The displacement control method was used to apply a predetermined vertical downward displacement of 29.343 mm on the upper surface of the indenter. (3) The indenter was removed along the vertical direction of the pipe. (4) When the internal pressure was unloaded to 0 MPa, the axial tension at both ends of the pipe was also unloaded to 0 MPa. (5) The outer end nodes of bottom of the segments L_1 and L_3 were restricted in the vertical direction, and a vertical displacement of 1000 mm was applied to the lower surface nodes of the support cushion to separate them from the pipe. (6) The internal pressure was re-applied on the pipe and an axial tension was applied at both ends of the pipe. (7) Two symmetrical nodes were selected on the segments L_1 and L_2, and a vertical downward displacement of 600 mm was applied so that the dented pipe was subject to the action of an opening bending moment until buckling occurred.

The local buckling behavior of pipelines at a dent is complex, involving multiple nonlinearities of pipe and dent geometries, pipe steel and boundary conditions. Negative stiffness may occur in the load-displacement curve in solving the problems. To solve the highly nonlinear problem, there are mainly arc length method and nonlinear stability algorithm. The former is a dual objective control method, where the load and displacement increment are simultaneously used to control the step size in the iteration process. The load increment step varies with the stiffness matrix and the load control can be carried out automatically. Therefore, the solution process can overcome the extreme point and avoid the Snap-through and Snap-back phenomena in the single load control method [Jin and Li, 2010; Liu et al., 2010]. The characteristics of the arc length method include a common

(a)

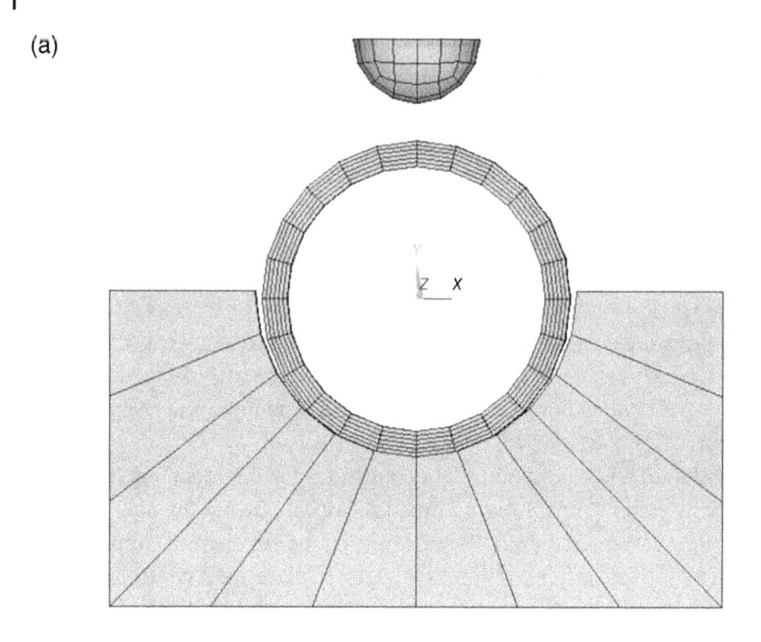

(b)

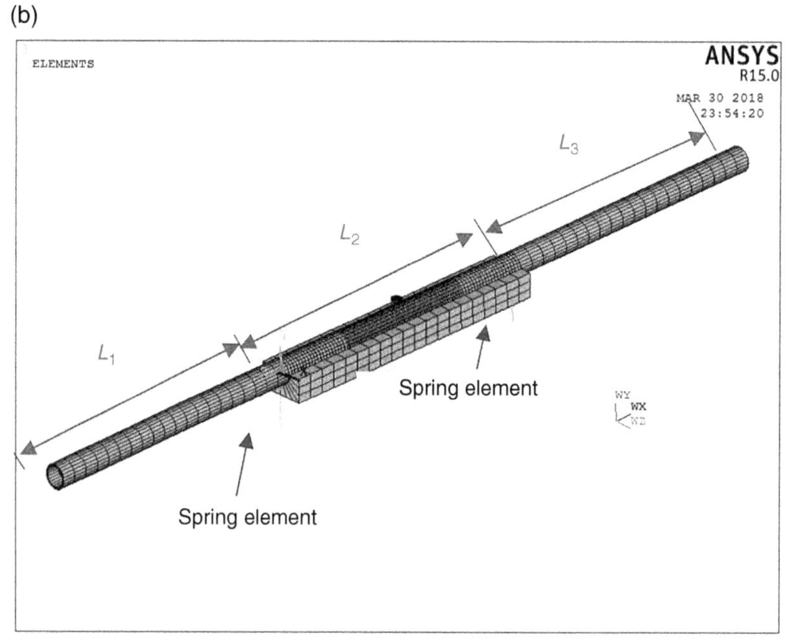

Figure 7.1 The developed FE model for a steel pipe containing a dent created by an indenter and support (a) cross-sectional view and (b) global view. *Source:* From Shuai et al. [2020a] / with permission from Elsevier.

applicability, easy convergence, and good stability, but the number of iterations is large, and the computational efficiency is not high, especially when solving nonlinear problems for complex structures. The nonlinear stability algorithm is better than the arc length algorithm in terms of computational efficiency [Liu et al., 2017]. The developed FE model is used to simulate the testing process and compare the calculation results of the two algorithms. The local buckling behavior of the pipeline is evaluated by the curve of bending moment as a function of the displacement at the dent vertex. The results show that the curves given by the two methods are almost identical [Shuai et al., 2020a]. The buckling instability of the test pipe occurs at a displacement of dent vertex of 774 mm, where the bending moment load is the so-called critical buckling load of the pipeline. Considering the reduced computational time, the nonlinear stability algorithm is chosen for the modeling.

7.2.2 Calculation of Bending Curvature of the Pipeline During Buckling

When a pipe buckles under a bending moment, a certain degree of bending usually occurs. The curvature of the pipe can be used to characterize the degree of bending during buckling of the pipe. Generally, the ultimate buckling strain of a pipe under bending moment is characterized by the average value of local buckling positions multiplied by the diameter along the pipe's axial direction [Suzuki et al., 2007; Oldfield et al., 2016]. Similarly, this method is used to calculate the curvature of the pipe during buckling, which is equal to the average curvature of nodes at the length of L_1 in Figure 7.1 multiplied by the pipe outer diameter (D) along the axial direction. As shown in Figure 7.2, the nodes in the central axis of the pipe marked in red line are those involved in the curvature calculation. Assume that the nodes in the selected area are marked as N_1, N_2, ..., N_n, and their curvatures are K_1, K_2, ..., K_n, respectively, the pipe curvature, K_{Buckling}, at the local buckling position is defined as:

$$K_{\mathrm{Buckling}} = \frac{K_1 + K_2 + K_3 + \dots + K_n}{n} \tag{7.1}$$

During bending, it is difficult to calculate the curvature of the pipe by directly fitting the curve function through the coordinates of each node. Since the deformation between any two adjacent nodes in the bending direction is small, the curvature of each node can be approximately solved by a curvature definition method [Shuai et al., 2020a].

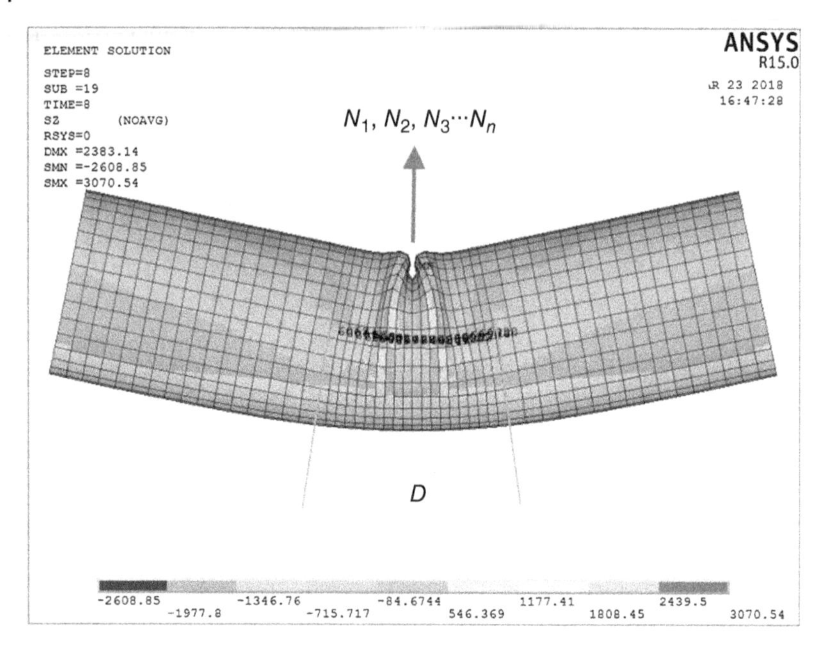

Figure 7.2 Schematic diagram illustrating the nodes that participate in curvature calculation. *Source:* From Shuai et al. [2020a] / with permission from Elsevier.

7.2.3 Effect of Operating Pressure on Critical Buckling Moment

The internal pressure affects the stiffness of a pipeline and has a direct impact on springback of an unconstrained dent on pipe body, specifically, the residual depth of the dent. Two cases are considered for the dent depth when the effect of internal pressure is investigated. (1) Under various internal pressures, the indenter is loaded with a fixed depth. Upon unloading of the indenter, different residual depths remain on the pipe body. (2) Under various internal pressures, the indenter is loaded to different depths on the pipe, but the residual depth is maintained identically upon unloading. Figure 7.3 shows the distribution of axial stress under various internal pressures when buckling occurs, as well as the buckling mode of the pipe for the first case. Obviously, the buckling mode of the dented pipe is greatly affected by the internal pressure. When the pressure is 0 or at small levels such as 1.538 MPa, the pipe presents a bending behavior, and buckling occurs toward the concave side of the dent center. With the increase in internal pressure, a folding of the pipe is observed, and buckling occurs on both sides of the dent. When the internal pressure is 10.68 MPa, the

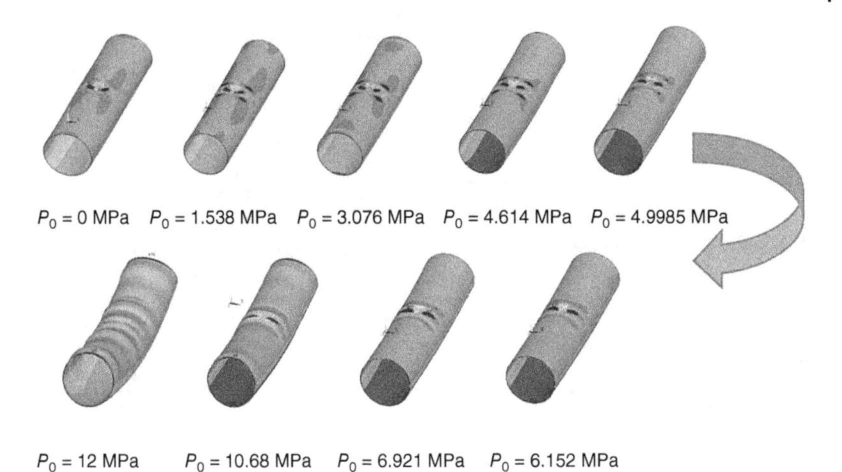

Figure 7.3 The buckling modes of the pipe under various internal pressures and an identical indentation depth. *Source:* From Shuai et al. [2020a] / with permission from Elsevier.

hoop stress of the pipe reaches the yield strength of X80 steel, and the pipe exhibits a wave-like multi-channel folded buckling.

Figure 7.4 shows the relationship between bending moment and curvature of the pipe under various internal pressures. When the internal pressure is not sufficient to induce a hoop stress up to the yield level, the curves have a similar shape without occurrence of buckling. A slight change of the pipe curvature causes a rapid rise of the bending moment. When the internal pressure reaches and exceeds 10.68 MPa, i.e., the hoop stress level up to yield strength of the steel, the bending moment-curvature curves show an obvious buckling plateau, indicating that the pipe has a strong buckling resistance.

Similarly, in the second case where the residual depth of the dent remains unchanged of about 6.1% of the pipe's outer diameter upon unloading, the buckling modes of the pipe and the bending moment-curvature curve are also like those obtained in the first case [Shuai et al., 2020a].

Figure 7.5 shows the influence of internal pressure on the critical buckling moment of the dented pipe under the two cases as described. It is obvious that the two curves have identical dependence on the internal pressure. The critical buckling moment increases first and then decreases with the internal pressure. This is attributed to the combined effect of pipeline stiffness and circumferential stress on the buckling resistance of the pipeline. When the internal pressure is low, the stiffness of the pipeline is low, and the buckling resistance is poor. With the increase in internal pressure, the

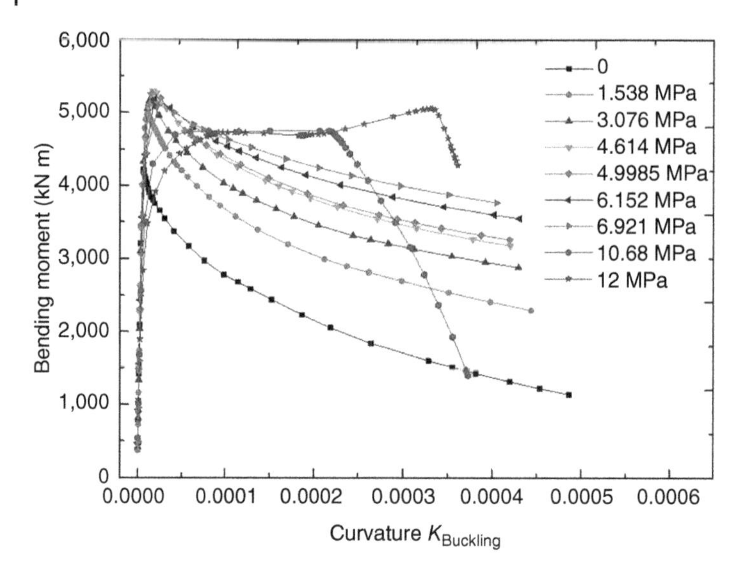

Figure 7.4 Relationship between bending moment and curvature of the pipe under various internal pressures. *Source:* From Shuai et al. [2020a] / with permission from Elsevier.

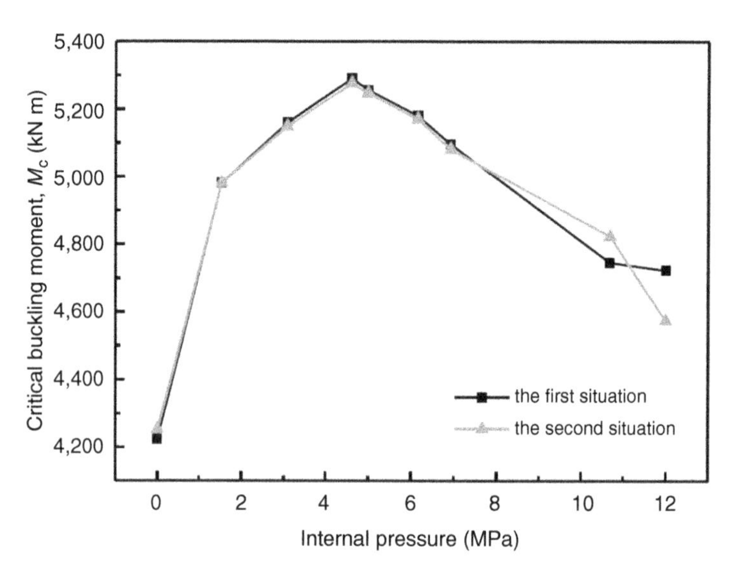

Figure 7.5 Influence of internal pressure on the critical buckling moment under the two cases as described. *Source:* From Shuai et al. [2020a] / with permission from Elsevier.

pipeline stiffness and critical buckling moment increase. When the internal pressure is up to a certain value, the circumferential stress of the pipeline increases with the increased internal pressure, causing a decrease of the critical buckling moment.

7.2.4 Effect of Pipe Dimension on Critical Buckling Moment

The pipe outer diameter and wall thickness affect the stiffness of a pipeline, and the hoop stress level of the pipeline under given internal pressures. Figure 7.6 shows the critical bending moment-curvature curves of the dented pipeline with an outer diameter of 720 mm and various wall thicknesses. It is seen that, as the pipe wall thickness increases, the critical buckling bending moment increases. This is due to the increased stiffness of the pipeline, and thus reduced hoop stress. In addition, before the buckling occurs, the slope of the moment-curvature curve increases with the pipe wall thickness, indicating that the pipeline has a strong ability to resist bending.

Figure 7.7 shows the relationship between the critical bending moment and the curvature of the dented pipeline with a pipe wall thickness of 10.3 mm and various outer diameters. An increased pipe outer diameter is associated with increased critical bending moment of buckling, and thus, the

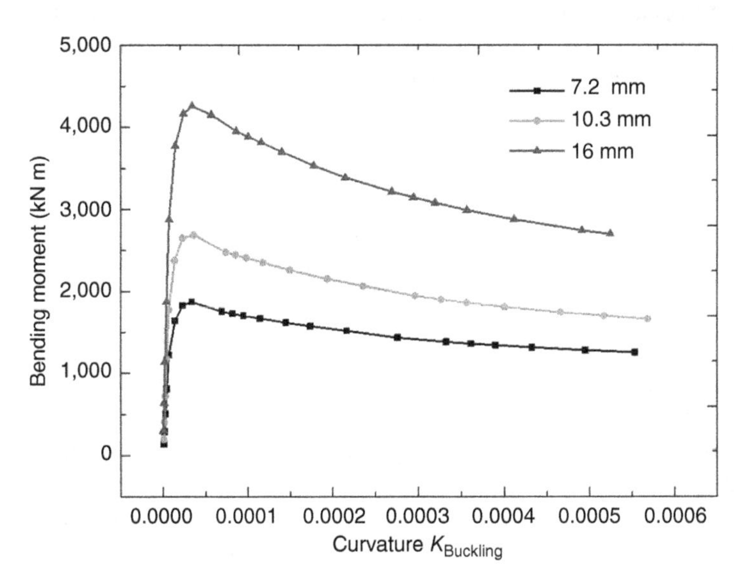

Figure 7.6 Effect of the pipe wall thickness on critical buckling bending moment of dented pipeline with an outer diameter of 720 mm. *Source:* From Shuai et al. [2020a] / with permission from Elsevier.

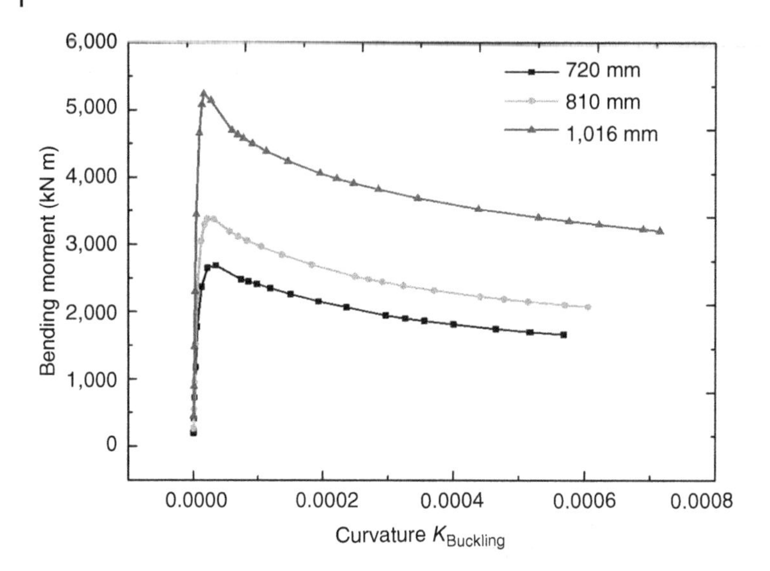

Figure 7.7 Effect of the pipe outer diameter on critical buckling bending moment of dented pipeline with a wall thickness of 10.3 mm. *Source:* From Shuai et al. [2020a] / with permission from Elsevier.

improved resistance of the pipeline to bending. The increase of pipe's outer diameter can not only improve the stiffness of the pipeline and increase its bending resistance but also increase the hoop stress, which decreases the bending resistance of the pipeline. However, the former plays a leading role in this process.

7.2.5 Effect of Steel Properties on Critical Buckling Moment

The local buckling of dented pipelines under a bending moment is an elastic–plastic behavior with a high degree of material nonlinearity. The influence of steel properties on critical buckling load should be considered. For example, it was found [Yang et al., 2013] that the critical buckling load of a cylindrical shell was positively related to the tangent modulus of the material, which was affected by yield strength of steels, stress hardening exponent, and the local equivalent stress at the buckling site on the pipeline.

Figure 7.8 shows the effect of the stress hardening exponent on critical buckling bending moment of a dented X80 steel pipe under various internal pressures. Generally, the critical buckling moment decreases with increased stress hardening exponent, but the effect is slight at low internal pressures such as 3 MPa. With the increase in internal pressure, the

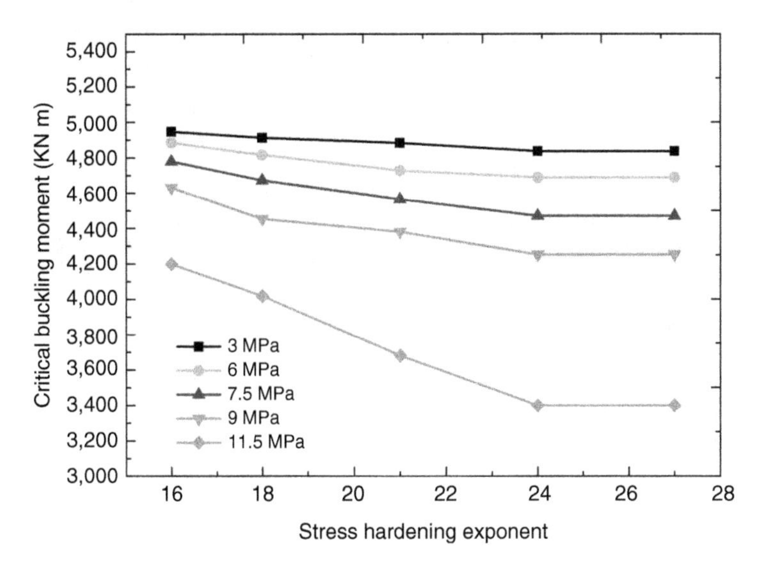

Figure 7.8 Effect of the strain hardening exponent on critical buckling moment of a dented X80 steel pipe under various internal pressures. *Source:* From Shuai et al. [2020a] / with permission from Elsevier.

critical buckling moment decreases rapidly with the stress hardening exponent. When the internal pressure is 11.5 MPa, the hoop stress exceeds the yield strength of the steel, and the critical buckling moment of the pipeline is greatly affected by the stress hardening exponent.

Figure 7.9 shows the effect of stress hardening exponent on critical buckling moment of pipelines made of steels with various yield strengths at the ratio of hoop stress to yield strength of 0.72, where the ratio represents the industry-acceptable operating condition. With decreased yield strength, the critical buckling moment of the pipelines decreases. At specific stress hardening exponents, a greater yield strength of steels would resist a greater hoop stress, improving the buckling resistance of the pipelines.

In summary, the local buckling behavior at a dent on high-strength steel pipelines under a bending moment is associated with a high degree of multiple nonlinearities. The nonlinear stability algorithm is appropriate to solve the buckling problem. Generally, the presence of dents on pipe body reduces the buckling resistance of the pipeline, especially for deep and large dents, which usually cause a significant decrease in the critical buckling moment. The applied internal pressure can improve the pipeline stiffness, enhancing the buckling resistance of the pipelines. However, a maximum value phenomenon is observed in the relationship between the critical buckling

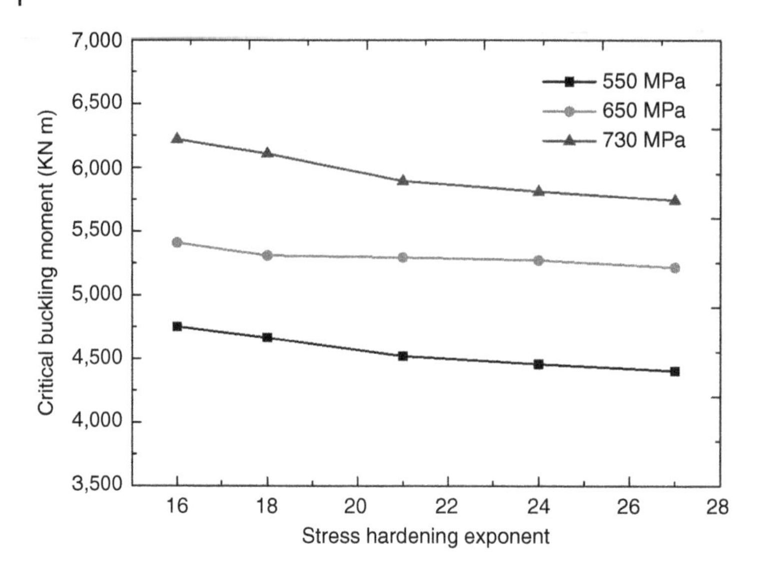

Figure 7.9 Effect of stress hardening exponent on critical buckling moment of pipelines made of steels with various yield strengths at the ratio of hoop stress to yield strength of 0.72, where the ratio represents the industry-acceptable operating condition. *Source:* From Shuai et al. [2020a] / with permission from Elsevier.

moment and internal pressure. The buckling mode of the pipelines is affected by the internal pressure. At low internal pressures, the pipeline buckles toward the concave center of the dent. With the increase in internal pressure, folds are observed on both sides of the dent when buckling occurs. When the circumferential stress exceeds the yield strength of pipe steels, a wavelike multi-fold buckling is caused on the pipelines. The critical buckling moment decreases with increased stress hardening exponent, but the effect is limited to low internal pressures. An increase of the yield strength can improve the buckling resistance of the pipeline. When the hoop stress exceeds the yield strength of the pipe steel, the effect of yield strength on the buckling resistance is limited.

7.3 Buckling Failure Analysis of a Corroded Pipe Under Axial Compressive Loading

For pipelines buried in geotechnically unstable regions, an axial compressive load is usually generated owing to pipe-soil interactions. As a result, local buckling potentially occurs, especially at the corrosion defects on the pipe surface [Suzuki et al., 2007; Höhler et al., 2016; Shuai et al., 2020]. The buckling of cylindrical thin-wall structures such as pipelines under an axial compressive load has been extensively investigated. According to Donnell's shallow shell theory and the assumption of a membrane pre-buckling state, a classical model for prediction of the elastic buckling load was proposed for the thin-wall cylinders [Ifayefunmi, 2016; Wang et al., 2019]:

$$F_c = \frac{2\pi E t^2}{\sqrt{3(1-\nu^2)}} \tag{7.2}$$

where F_c is critical buckling load, E is Young's modulus, ν is Poisson's ratio, and t is wall thickness of the thin-wall structure. For cylindrical structures with a thick wall thickness, they usually fail in the elastic–plastic range and the failure is governed by plastic collapse. The following empirical equation can be used to calculate the critical axial compressive load to cause yielding of the structure, which is defined as the reference buckling load, F_{ref} [Ifayefunmi, 2016; Wang et al., 2019]:

$$F_{ref} = \pi D t \sigma_y \tag{7.3}$$

where D is outer diameter of the cylindrical structure (e.g., pipe outer diameter).

The methods mentioned above are mainly used to determine the critical buckling load of pipelines free of corrosion and other types of defects subject to an axial compression. To date, works investigating the buckling of corroded pipelines have still been limited [Dewanbabee, 2009; Nazemi, 2009; Zhou et al., 2018a]. The synergism of corrosion defect, pipe geometry, and internal pressure on local buckling of pipelines has remained unclear. Moreover, no relevant work has been reported on buckling of corroded high-strength X80 steel pipelines. A new FE model is developed to determine the local buckling behavior of an X80 steel pipe containing a corrosion defect subject to axial compressive loading.

7.3.1 The Model Development and Numerical Algorithm

The local buckling behavior of an X80 steel pipe containing a corrosion defect under an axial compressive load was modeled by ANSYS, and the developed FE model is shown in Figure 7.10. The dimensions of the corrosion defect, including its depth (d), circumferential width (W), and longitudinal length (L), as well as the pipe outer diameter (D) and wall thickness (t), are shown in Figure 7.11. Corrosion is regarded as metal loss, where the material properties do not change. Due to stress and strain gradients at

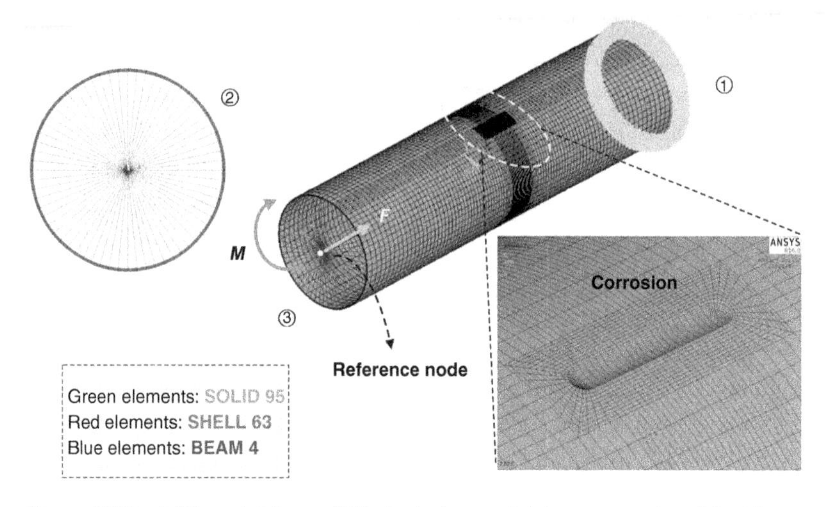

Figure 7.10 A FE model for an X80 steel pipe containing a corrosion defect for local bulking analysis. *Source:* From Shuai et al. [2020b] / with permission from Elsevier.

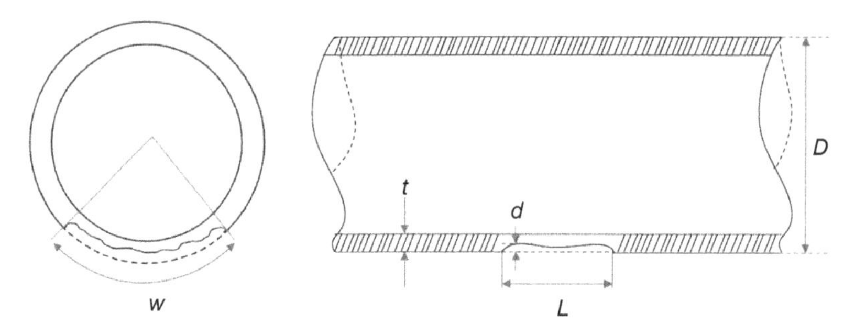

Figure 7.11 The dimension of the corrosion defect on the steel pipe. *Source:* From Shuai et al. [2020b] / with permission from Elsevier.

the defect, the mesh in the corrosion zone was refined. To save computational; time, the grid generation of the pipe body which is away from the corrosion defect was relatively sparse. To apply the axial compressive load, a reference node was arranged at the cross-sectional center of the pipe at the left end (i.e., section ② in Figure 7.10) [Mondal and Dhar, 2019]. Application of the axial compressive load was implemented through coupling of multiple nodes. A surface-to-surface contact algorithm with a friction coefficient of 0.3 was used to simulate the nonlinear contact between the shell elements and the solid elements. The total numbers of the elements and nodes of the FE model were 29,830 and 165,640, respectively.

The testing included two steps [Dewanbabee, 2009]. An internal pressure was applied on the pipe first, and full constraints were imposed on the nodes of the cross-section at the right side of the pipe (i.e., section ① in Figure 7.10). An axial tensile stress was applied on the left end of the pipe to obtain an equivalent action of the end cap. A continuously increasing axial compressive load was applied on the reference point until buckling happened. The element sizes at the defect area and the pipe body were determined by a mess sensitivity analysis. The nonlinear stability algorithm was selected for solving the nonlinear buckling problem of the corroded pipe subject to axial compression.

7.3.2 Effects of Depth, Length, and Width of the Corrosion Defect on Critical Buckling Load

To determine the effect of corrosion depth on buckling of the pipe under an axial compressive load, FE modeling with various d/t ratios (where d is corrosion defect depth and t is pipe wall thickness) is conducted. Generally, when the d/t ratio exceeds 0.6, the pipe should be repaired. Thus, the d/t ratio ranging from 0.15 to 0.6 is considered. Other parameters as inputs of the modeling include pipe outer diameter of 813 mm, pipe wall thickness of 12 mm, operating pressure of 0.5 of yield pressure, and X80 steel as the material. The relationship between the critical buckling load F_c and d/t (t is fixed) with various corrosion lengths and widths is shown in Figure 7.12. It is seen that F_c decreases with increased corrosion depth (d). Therefore, the defect depth is an important parameter in assessing the bulking behavior of corroded pipelines. When the corrosion defect depth increases, the local stiffness of the steel reduces. As a result, the pipeline becomes more prone to under the axial compression load.

Figure 7.13 shows the buckling mode of the pipe containing a corrosion defect with various depths. It is seen that, with the increasing ratio d/t, the height of the buckling wave increases. The bulge at the bulking location becomes more obvious. Thus, the pipeline buckling tends to occur as the corrosion defect depth increases.

Figures 7.14 and 7.15 show the relationships between the critical buckling load and the dimensionless corrosion defect length, $\frac{L}{\sqrt{Dt}}$ (with the ratio d/t of 0.5) and the dimensionless corrosion defect width, $\frac{w}{\pi D}$, respectively. It is seen that, when the corrosion length $\frac{L}{\sqrt{Dt}} \leq 4.86$ (i.e., the critical length of corrosion defect that is defined as a short defect in ASME B31G), the

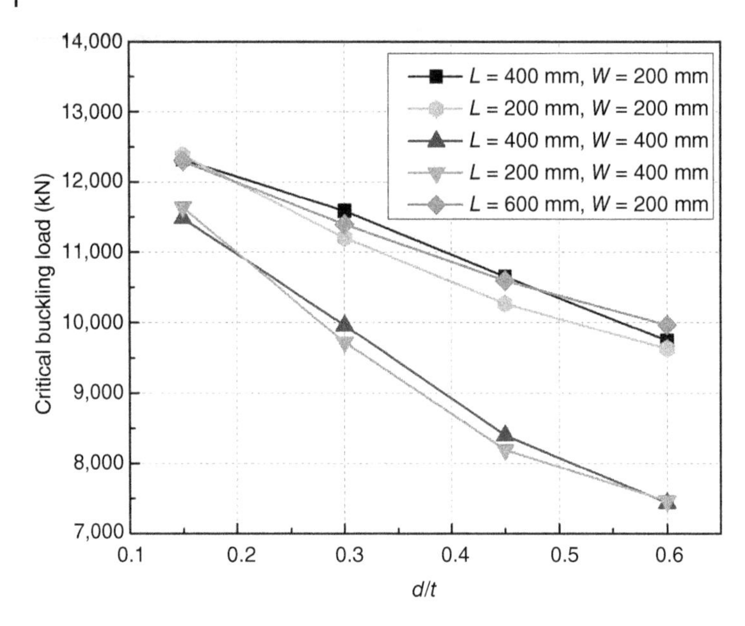

Figure 7.12 Relationship between the critical buckling load and *d/t* ratio with various corrosion defect lengths and widths. *Source:* From Shuai et al. [2020b] / with permission from Elsevier.

critical buckling load decreases first and then increases with increased corrosion length. When $\frac{L}{\sqrt{Dt}} > 4.86$ (i.e., long corrosion defects), the critical buckling load is approximately independent of the defect length. Thus, the long and short corrosion defects affect the pipeline buckling in diverse ways. Further, the F_c decreases rapidly with increased $\frac{w}{\pi D}$ when the corrosion width $\frac{w}{\pi D} < 0.5$. After that, the defect width does not affect F_c. Generally, as the defect width increases, there is a larger area on the pipe suffering from reduced wall thickness. As a result, the pipe stiffness decreases. Under the axial compression load, the pipe is more likely to experience eccentric instability and buckling. When the defect is sufficiently wide, i.e., $\frac{w}{\pi D} > 0.5$, although the local stiffness of the pipe decreases, the eccentricity effect tends to reduce as well. As a result, the influence of the defect width on critical buckling load becomes not important.

In summary, the modeling results regarding the effect of corrosion defect dimension (i.e., depth, length, and width) on the critical buckling load of corroded pipelines show that the susceptibility of the corroded pipe to buckling failure under an axial compressive loading is mainly affected by the

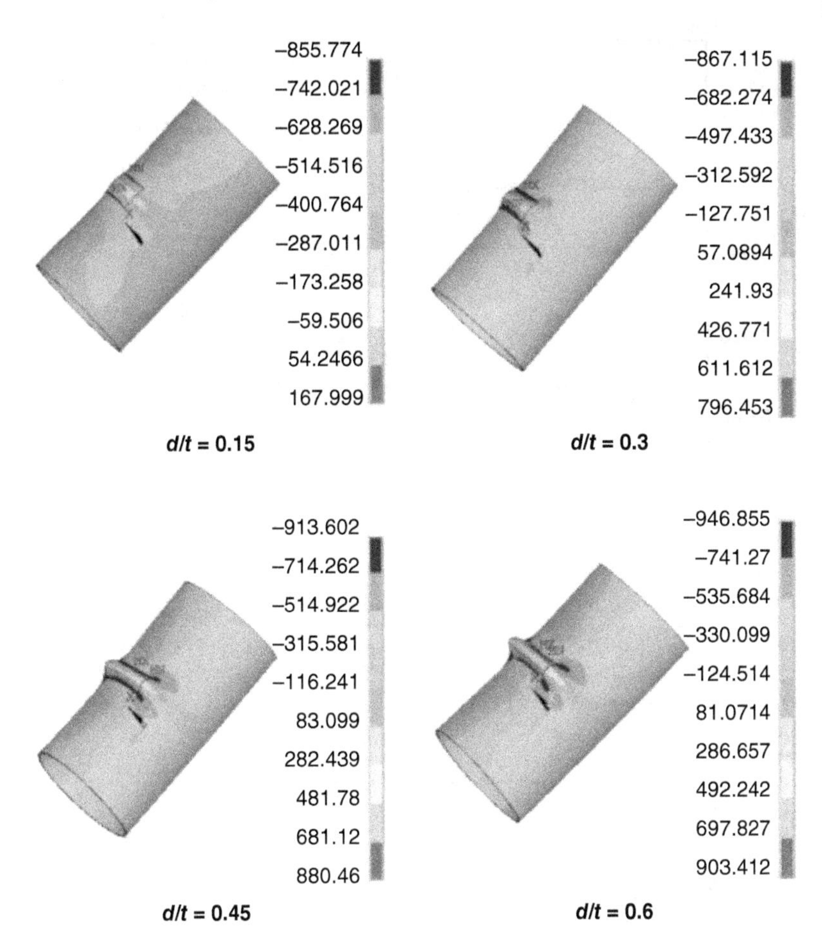

Figure 7.13 Buckling modes of the pipe containing a corrosion defect with various depths while the pipe wall thickness is fixed. *Source:* From Shuai et al. [2020b] / with permission from Elsevier.

corrosion depth and width, while the corrosion length is the least important. When the pipelines experience an axial compression, the corrosion depth should be paid primary attention for buckling analysis. The corrosion width should also be included in buckling assessment although it is often ignored owing to a limited impact of corrosion width on ultimate failure pressure of the pipeline. After the corrosion defect geometry is characterized by ILI data analysis, the deepest and the widest corrosion defects should be focused to perform geohazard risk assessment or axial stress monitoring. Moreover, the corrosion defects with a small depth but a great width should not be ignored.

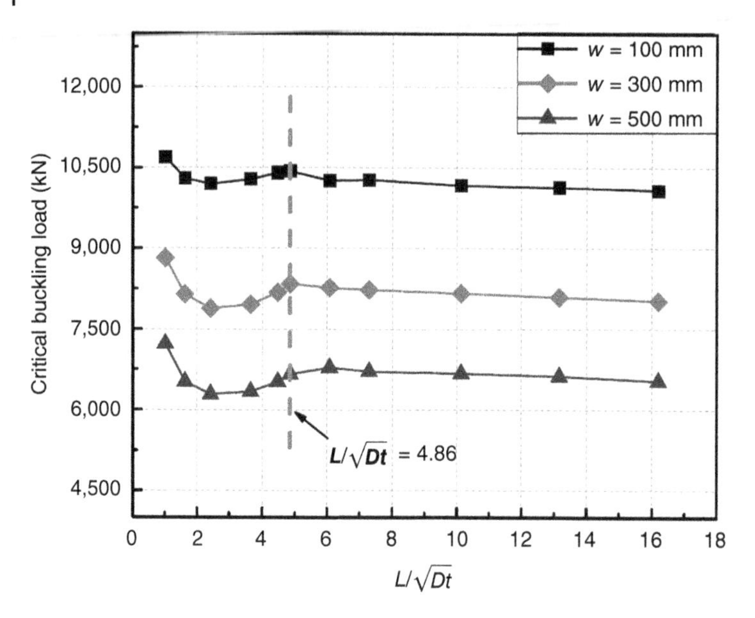

Figure 7.14 Critical buckling load of the corroded pipe as a function of the dimensionless corrosion defect length $\frac{L}{\sqrt{Dt}}$ at $d/t = 0.5$. *Source:* From Shuai et al. [2020b] / with permission from Elsevier.

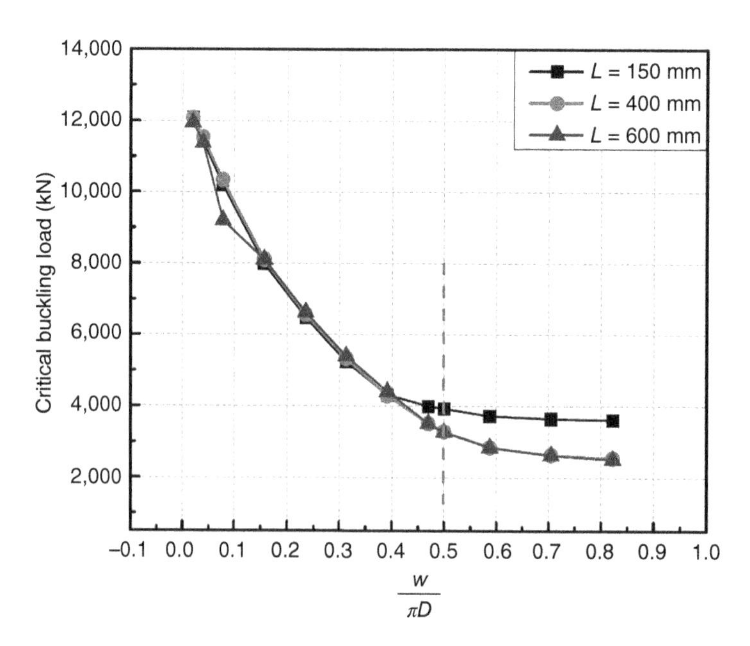

Figure 7.15 Effect of the corrosion defect width $\frac{w}{\pi D}$ on critical buckling load of the pipe. *Source:* From Shuai et al. [2020b] / with permission from Elsevier.

7.3.3 Effect of Pipe Dimension on Critical Buckling Load

Pipeline dimension including the outer diameter and wall thickness affects the stiffness and stress state of the pipe and cannot be ignored in buckling analysis. Figure 7.16 shows the relationship between the critical buckling load F_c and pipe outer diameter D, with a fixed pipe wall thickness of 18.4 mm. It is shown that, as D increases, the value of F_c increases. Generally, an increased outer diameter improves stiffness of the pipe, enhancing its compression resistance. However, the increased diameter also elevates hoop stress and von Mises equivalent stress, reducing the capacity of the pipe to resist buckling. However, the former plays a dominant role in buckling response.

Figure 7.17 shows the effect of pipe wall thickness (t) on critical buckling load F_c of the pipe with a fixed outer diameter of 1016 mm. It is seen that F_c increases linearly with t. This is attributed to greater stiffness as the pipe wall thickness increases. As a result, the pipeline is more resistant to local deformation. Therefore, when pipelines are buried in regions with potential geohazards, the designed wall thickness of the pipeline can be appropriately increased to improve the resistance to buckling under axial loading caused by ground movement.

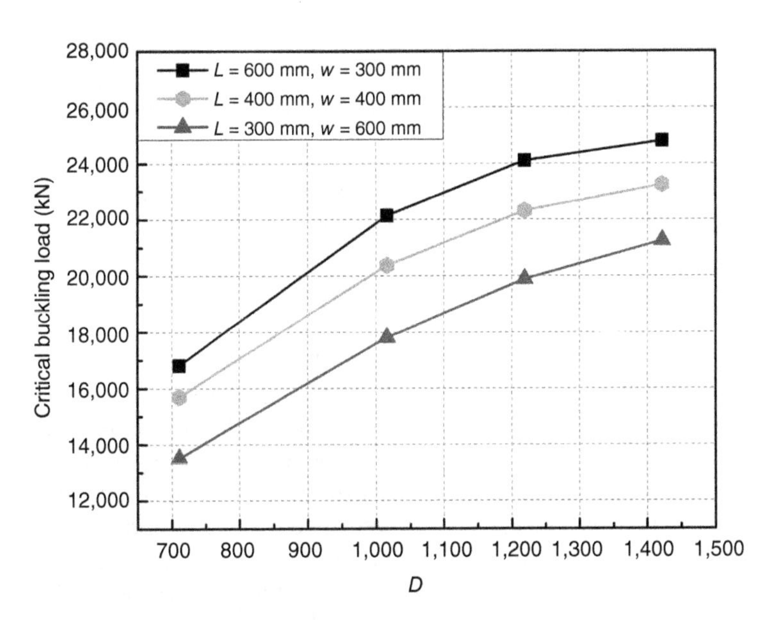

Figure 7.16 Relationship between the critical buckling load and the pipe outer diameter, D, with a fixed pipe wall thickness of 18.4 mm. *Source:* From Shuai et al. [2020b] / with permission from Elsevier.

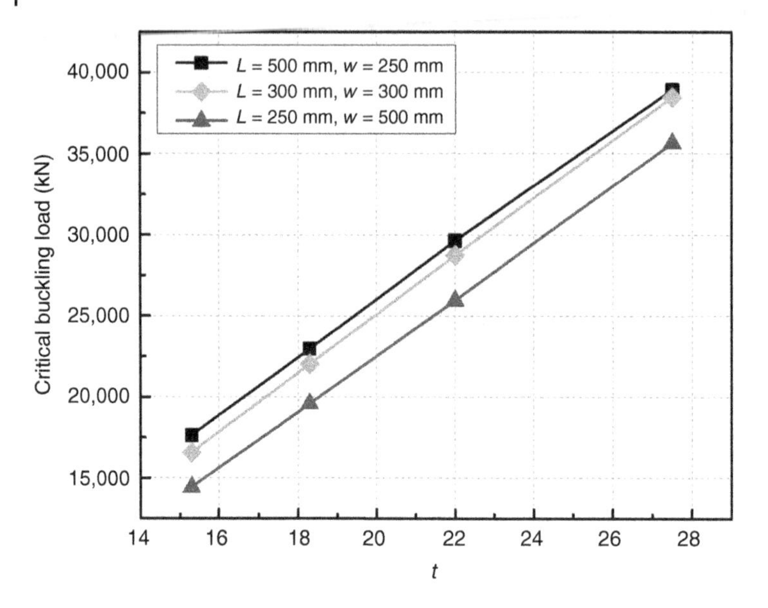

Figure 7.17 Critical buckling load of the steel pipe as a function of the pipe wall thickness with a fixed pipe outer diameter of 1016 mm. *Source:* From Shuai et al. [2020b] / with permission from Elsevier.

7.3.4 Effect of Operating Pressure on Critical Buckling Load

The operating pressure is the primary source of stress applied on a pipeline, directly affecting its buckling resistance. To investigate the effect of operating pressure (P) on buckling of a corroded pipe, the FE modeling with various P/P_y ratios (where P_y is the critical internal pressure when pipe steel yields) are conducted. It is noted that the internal pressure does not fail the pipeline at the corrosion defect before an axial compression load is applied. Figure 7.18 shows the relationship between the critical buckling load F_c and the dimensionless internal pressure P/P_y at a corrosion defect with various lengths and widths (the corrosion defect depth is fixed at $d/t = 0.4$). It is shown that F_c decreases with increased operating pressure. Generally, as the internal pressure increases, the equivalent stress at the corrosion defect increases, causing a decreased tangent modulus and reduced critical buckling load. Moreover, at specific operating pressures, the critical buckling load is greater as the corrosion defect width decreases (while the corrosion defect length is fixed) and the corrosion defect length increases (while the corrosion width unchanges). Thus, the effect of operating

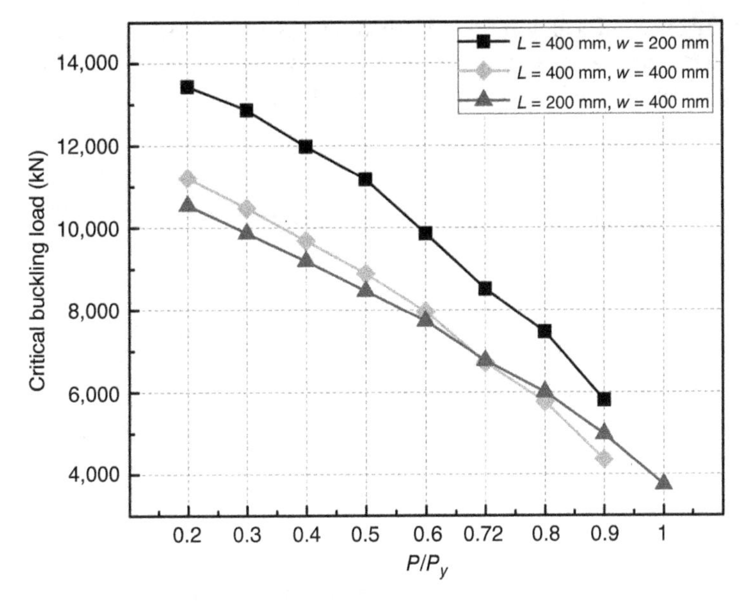

Figure 7.18 Relationship between the critical buckling load and the dimensionless internal pressure P/P_y at a corrosion defect with various lengths and widths (the corrosion defect depth is fixed at $d/t = 0.4$). *Source:* From Shuai et al. [2020b] / with permission from Elsevier.

pressure on critical buckling load of the corroded pipeline is also related to the dimension of the corrosion defect.

7.4 Buckling Resistance of Corroded Pipelines under Bending Moment

Corrosion is one of the primary reasons causing pipeline failures in North America. In general, corrosion reduces the pressure-bearing capacity of pipelines. Standards and models, as summarized by Qin and Cheng [2021], have been proposed for failure prediction of corroded pipelines. Long-distance transmission pipelines usually pass through a wide variety of geographical regions with varied and complex geotechnical conditions. In addition to the hoop stress induced by internal pressure, the pipelines potentially experience external bending loads due to geohazards such as earthquakes, faults, and landslides, resulting in local buckling failures [Suzuki et al., 2007; Liu et al., 2017; Shuai et al., 2020a, 2020b, 2020c].

Buckling of cylindrical thin-walled structures has been investigated under a bending moment. The classical model to predict the critical elastic bending moment to cause pipeline buckling is [Iflefel et al., 2005]:

$$M_c = D^2 t \sigma_y \tag{7.4}$$

where M_c is critical buckling moment. The model predicted the critical buckling moment of an intact pipeline free of corrosion or another type of defect without internal pressure but subject to a bending moment. Kanninen et al. [1978] proposed a Net Section Collapse (NSC) criterion to predict the ultimate moment of pipelines containing cracks. Kim and Park [2003] found that this criterion gave conservative results in calculations of the ultimate bending moment of corroded pipelines. Miyazaki et al. [1999, 2002] found that the failure mode of a corroded pipeline under bending moment was related to the defect geometry, and the failure criterion based on ultimate tensile strength was more accurate than the NSC criterion. Later, Kim et al. [2006] compared several ultimate moment calculation methods and noted that the accuracy of a criterion based on material flow stress was better than the tensile strength criterion. Through four-point bending tests, it was determined [Oh et al., 2009] that the failure mode of pipelines was greatly affected by corrosion length and internal pressure. When the pipeline was subject to an opening bending moment (i.e., the corrosion defect was under tension), an increased internal pressure would reduce the ultimate bending moment. However, when the pipeline was subject to a closing bending moment (i.e., the corrosion defect was under compression), the increased internal pressure would increase the ultimate bending capacity of the pipeline. Furthermore, Shim et al. [2004] found that, when the depth of a corrosion defect increased, the internal pressure significantly influenced the ultimate moment-bearing capacity of the pipeline. When the corrosion defect was shallow, the influence was not obvious. However, Gao et al. [2016] thought that the influence of corrosion width on the ultimate moment-bearing capacity of pipelines was greater than corrosion depth and length. Chen et al. [2014] studied the bending capacity of corroded pipelines and developed a model for ultimate bending moment calculations. However, the work was based on assumption that the pipeline did not experience local buckling, thus generating conservative prediction results.

While efforts have been made to determine the bending resistance of corroded pipelines, controversial results exist in terms of the effects of corrosion dimension, internal pressure, and other relevant factors on the bending-bearing capacity of the pipelines. In Section 7.2, a nonlinear FE model was developed to study buckling of an X80 steel pipe at a dent under bending moment, where the essential role of dent in buckling failure of the pipe was determined. To date, the role of corrosion defect in pipeline buckling under a bending moment has remained unknown.

7.4.1 The Model Development

The buckling of a corroded steel pipe subject to a bending moment through four-point bending was modeled, as shown in Figure 7.19, where a corrosion defect is present on the external surface of the pipe. For thin-walled structures such as pipelines, the buckling is usually accompanied with the generation of local wrinkles, resulting in a large deformation. Thus, proper solid elements during modeling such as a 20-node hexahedral solid element named C3D20R [Shuai et al., 2021] should be selected to simulate the pipe with a strong deformation ability. Owing to a large stress gradient from intact pipe segment to the corrosion defect, a dense mesh was defined around the corrosion region. To save computational time, the grid density of the pipe segment away from the corrosion defect was sparse. To avoid collapse failure at loading points, the material in segments L_1 and L_2 was assigned as elastic, with an elastic modulus slightly greater than the elastic modulus of segment L_p. The segment L_p was long enough to ensure that the

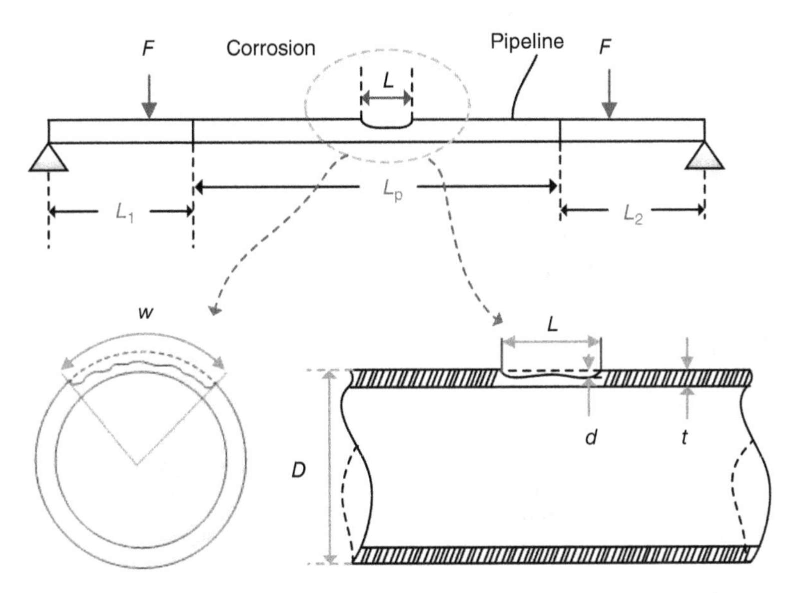

Figure 7.19 Schematic diagram of the four-point bending test on a steel pipe containing a corrosion defect with defined dimension. *Source:* From Shuai et al. [2021] / with permission from Elsevier.

corrosion defect would not be affected by the stress state at both ends of the pipe. The assignments of L_p, L_1, and L_2, as defined by [Shuai et al., 2021], made a convergence solution to be obtained. The FE model developed to study buckling behavior of the corroded pipe is shown in Figure 7.20.

The bending test on the corroded pipe included two steps [Kim et al., 2006]. First, an internal pressure was applied on the pipe. The circumferential and radial directions of the nodes at both ends of the pipe were constrained. The middle node at the pipe bottom was constrained at the axial direction to prevent rigid body displacement. An effective axial tensile force was applied on the end nodes of L_1 and L_2 segments to simulate the effect of welded end caps. Second, a bending moment was applied on the pipe. The two nodes at both ends of the pipe were constrained in the vertical direction so that the pipe could bear the bending moment. A displacement control method was used to accelerate convergence of the bending moment application. Two symmetrical nodes (i.e., points A and B at pipe segments L_1 and L_2, respectively, in Figure 7.20) were selected as loading positions. A vertical, downward displacement was applied simultaneously at the two nodes until a buckling failure occurred.

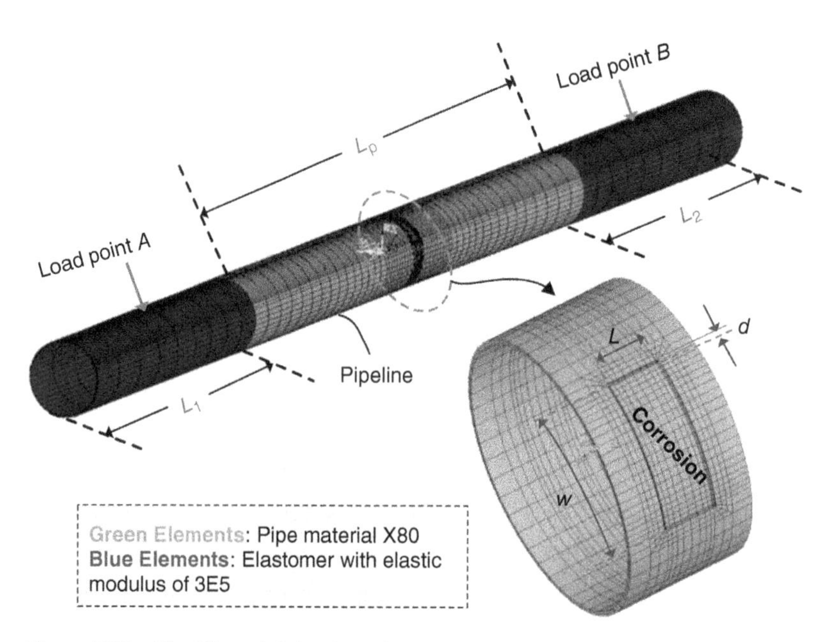

Figure 7.20 The FE model developed to study the buckling behavior of a steel pipe containing a corrosion defect. *Source:* From Shuai et al. [2021] / with permission from Elsevier.

A grid sensitivity study was conducted to obtain an exact convergence solution. By multiple trial calculations, the mesh density was continuously increased. When the critical buckling moment did not change, as compared with previous calculation results, the mesh was assumed to converge. The least number of meshes in all convergence solutions was used as the optimal mesh size. The element size around the corrosion defect area was 5 mm × 5 mm, and the grid size for the pipe away from the corrosion area was 21.40 mm × 16.63 mm. The nonlinear stability algorithm was used to solve the nonlinear problem of pipeline buckling at the corrosion defect.

7.4.2 Effect of Corrosion Defect Dimension

Generally, corrosion features on pipelines include pits, grooved corrosion, and uniform corrosion, as schematically shown in Figure 7.21. A semi-spherical shape is used for modeling a corrosion pit, a rectangular shape with a uniform thinning of pipe wall thickness for uniform corrosion, and a semi-ellipsoid shape for grooved corrosion. FE models are developed for the three types of corrosion defect to investigate the effect of corrosion shape on critical buckling moment, which is usually written as M_c/M_o, where M_o is critical elastic buckling moment. Figure 7.22 shows the M_c/M_o

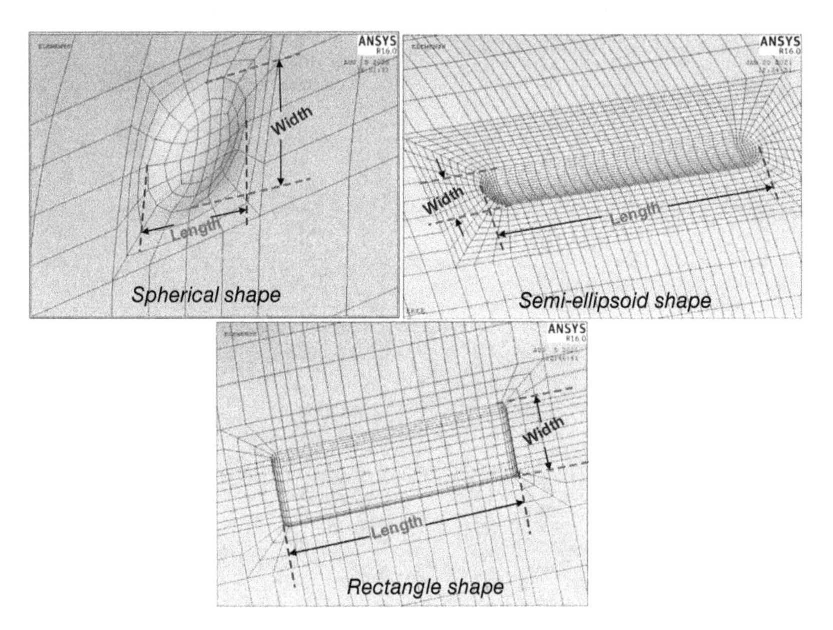

Figure 7.21 Three typical types of corrosion features present on pipelines for modeling. *Source:* From Shuai et al. [2021] / with permission from Elsevier.

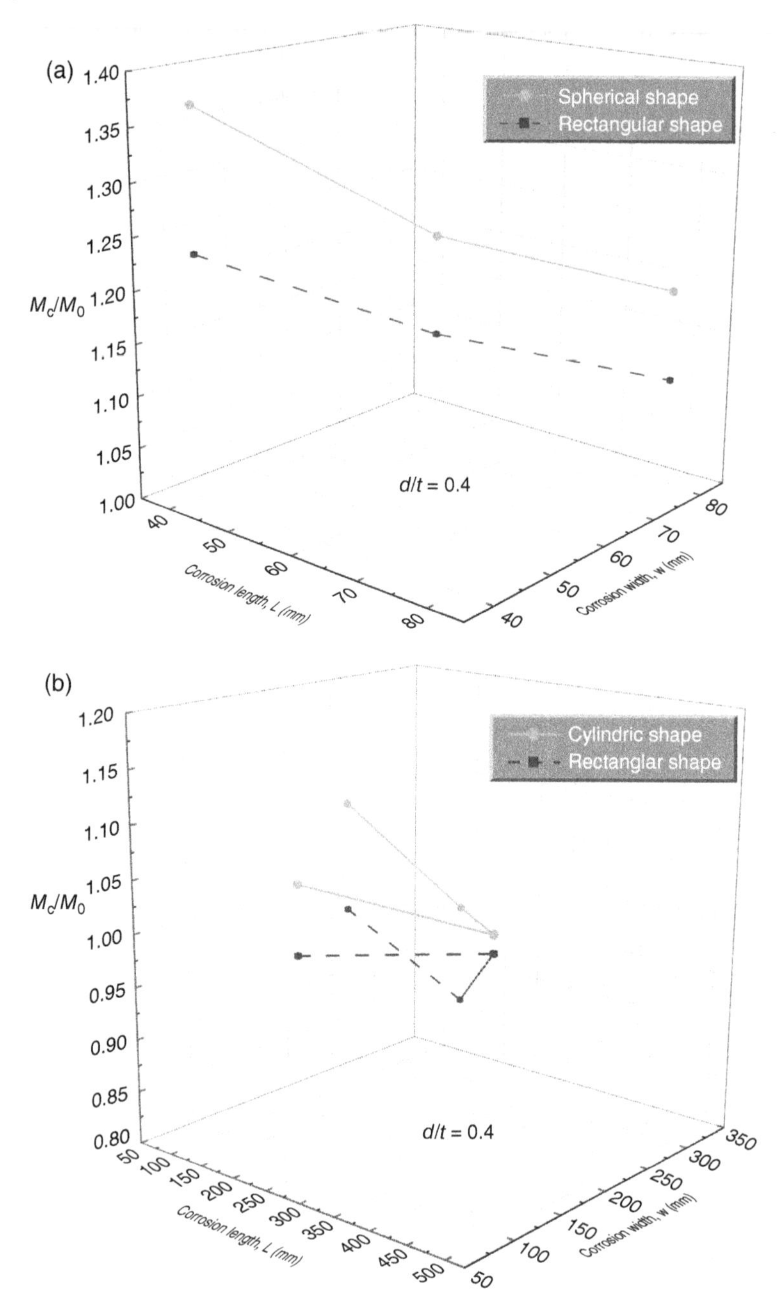

Figure 7.22 Critical buckling moment, i.e., M_c/M_0, of a pipe containing a corrosion feature as a function of the corrosion length and width, where the ratio of corrosion depth, d, to pipe wall thickness, t, is 0.4. The corrosion feature is modeled by (a) spherical and rectangular and (b) cylindrical and rectangular shapes, respectively. *Source:* From Shuai et al. [2021] / with permission from Elsevier.

of an X80 steel pipe containing a corrosion feature as a function of the corrosion length and width, where the ratio of corrosion depth, d, to pipe wall thickness, t, is 0.4. The corrosion feature is modeled by spherical, rectangular, and cylindrical shapes, respectively. Generally, the critical buckling moment of the rectangular corrosion model is smaller than those of the spherical and cylindrical corrosion models with identical corrosion dimensions. Thus, investigations will focus on the buckling behavior of the steel pipe containing a rectangular-shaped corrosion feature, which presents the most aggressive corrosion shape to reduce the pipe resistance to buckling failure.

To determine the effect of corrosion dimension, including corrosion depth, length and width, on buckling of the steel pipe under a bending moment, modeling is conducted with various d/t, $L/\sqrt{Dt}$, and $w/\pi D$ values, which are defined as normalized corrosion depth, length and width, respectively. Figure 7.23 shows the critical buckling moment, M_c/M_o, of the steel pipe as a function of the corrosion depth, length, and width, as defined above, respectively. It is seen that, with increased corrosion depth, the critical buckling moment decreases. Therefore, the corrosion depth must be considered in buckling assessment on corroded pipelines. It is more likely for the pipes to fail by buckling at a deeper corrosion defect under a bending moment. The effect of corrosion length on the critical buckling moment depends on the specific lengths. For $\frac{L}{\sqrt{Dt}} > \sqrt{20}$, the M_c/M_o ratio decreases first due to an increased equivalent stress and then increases with increased $L/\sqrt{Dt}$. When $\frac{L}{\sqrt{Dt}} > \sqrt{20}$, the effect of the corrosion length on M_c/M_o is negligible. The effect of corrosion width also depends on the specific width values. The M_c/M_o decreases with increased $w/\pi D$. For $\frac{w}{\pi D} > 0.5$, a further increase of the $w/\pi D$ does not affect M_c/M_o. Generally, the wider the corrosion defect, the larger the thinning area of pipe-wall in the corrosion region. As a result, the stiffness of the pipe decreases. Under a bending moment, the pipe tends to suffer from an eccentric instability and then buckling. However, when the corrosion width $\frac{w}{\pi D} > 0.5$, the local stiffness of the pipe decreases, while the eccentricity becomes weak. There is little influence of the corrosion width on critical buckling load. Additionally, there is a greater effect of corrosion width on the critical buckling moment than corrosion length.

In summary, the corrosion dimension remarkably affects the critical buckling resistance of corroded pipelines under bending moment. For pipelines located in geotechnically active regions where a bending moment is generated, the geometrical feature of corrosion defects should be identified and

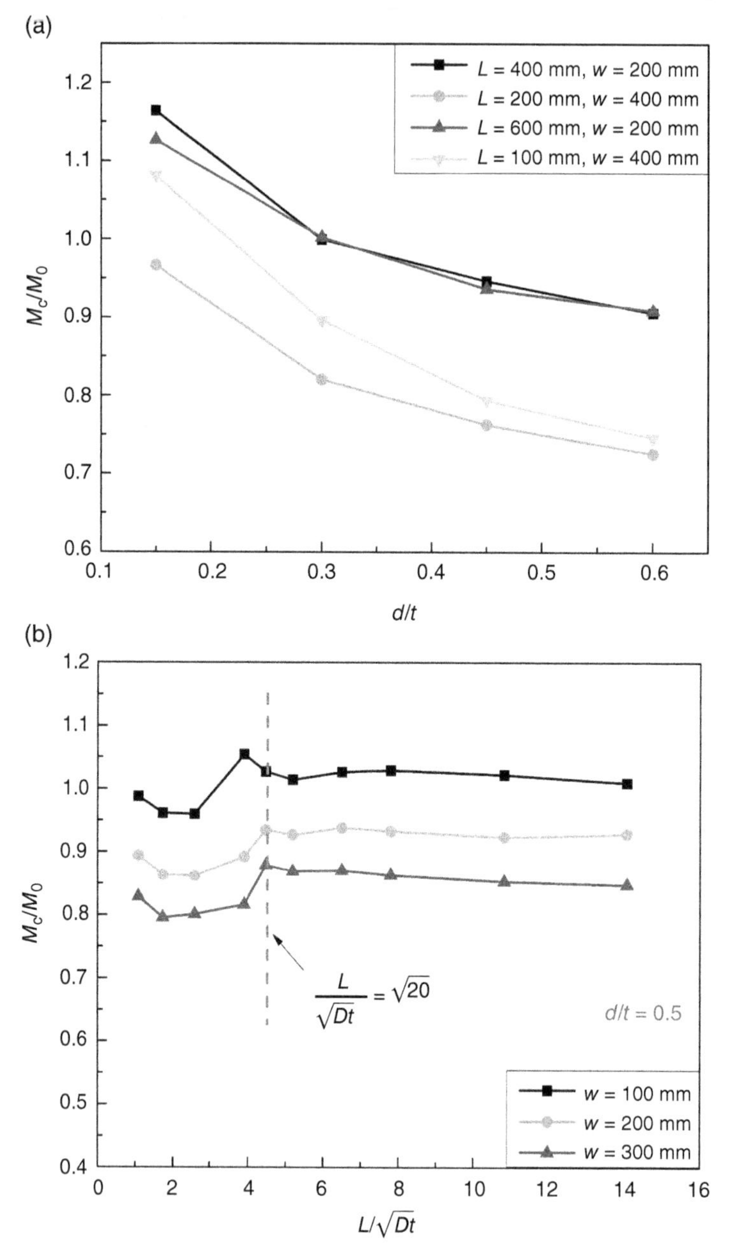

Figure 7.23 Critical buckling moment, i.e., M_c/M_o, of a pipe containing a corrosion feature as a function of (a) corrosion depth, d/t; (b) length, $L/\sqrt{Dt}$; and (c) width, $w/\pi D$. *Source:* From Shuai et al. [2021] / with permission from Elsevier.

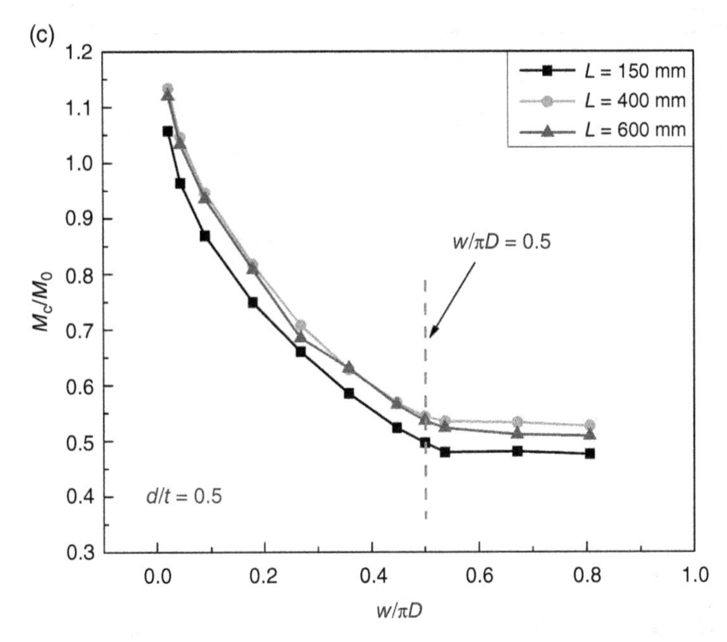

Figure 7.23 (Continued)

determined by ILI data analysis. Particularly, deep corrosion defects should be paid much attention to since they can decrease the resistance of the pipelines to buckling. Moreover, the influence of corrosion width should not be ignored.

7.4.3 Effect of Pipe Dimension

Pipe outer diameter and wall thickness are important parameters for safety design with an improved buckling resistance. The relationship between the critical buckling moment, M_c, and pipe outer diameter, D, with a fixed wall thickness (t) of 18.4 mm of an X80 steel pipe is shown in Figure 7.24, where the pipe outer diameters, i.e., 711, 1016, 1219, and 1422 mm, are selected from typical X80 steel pipelines [Zhang et al., 2019]. It is seen that, as the outer diameter increases, M_c increases continuously. Generally, an increased pipe diameter possesses dual roles in affecting pipeline integrity. It can improve the pipe stiffness, enhancing bending resistance of the pipeline. At the same time, an increased pipe diameter increases the hoop stress and von Mises equivalent stress, reducing capacity of the pipeline to resist buckling. The former usually plays a dominant role in the buckling process.

Figure 7.25 shows the critical buckling moment, M_c, of the X80 steel pipe with various wall thicknesses while the outer diameter is fixed at 1016 mm,

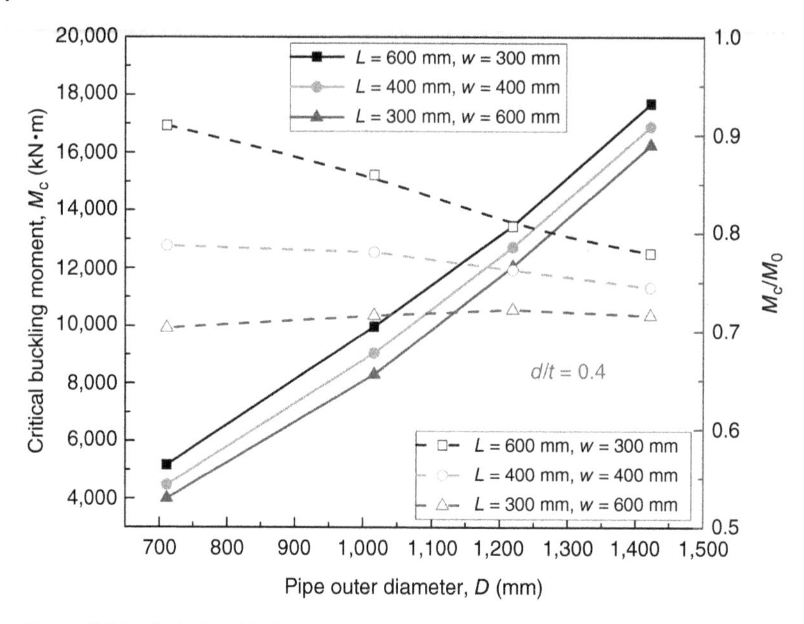

Figure 7.24 Relationship between the critical buckling moment and the pipe outer diameter with a fixed pipe wall thickness of 18.4 mm. *Source:* From Shuai et al. [2021] / with permission from Elsevier.

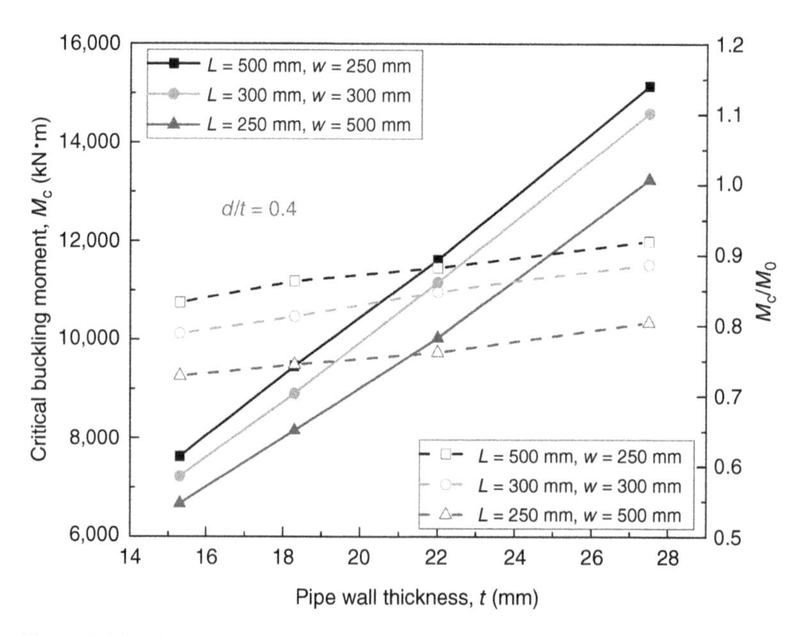

Figure 7.25 Critical buckling moment of the X80 steel pipe with various wall thicknesses while the outer diameter is fixed at 1016 mm. *Source:* From Shuai et al. [2021] / with permission from Elsevier.

where four design wall thicknesses (i.e., 15.3, 18.3, 22, and 27.5 mm) of pipelines are selected for modeling. The critical buckling moment increases with the increased pipe wall thickness. This is attributed to the increased stiffness and reduced hoop stress for the pipe as the pipe wall thickness increases. Thus, for pipelines buried in geohazard regions where a bending effect is usually applied, an increased pipe wall thickness allowance can improve the resistance of the pipelines to buckling.

7.4.4 Effect of Internal Pressure

To determine the effect of internal pressure on buckling behavior of the X80 steel pipe subject to a bending moment, numerical modeling is conducted with various P/P_y ratios. Figure 7.26 shows the relationship between the critical buckling moment, M_c/M_o, and normalized internal pressure. It is seen that the M_c/M_o increases with increased internal pressure. This result is controversial to the general finding that an increased internal pressure reduces the bending resistance of pipelines due to increased equivalent stress [Liu et al., 2017; Shuai et al., 2020a].

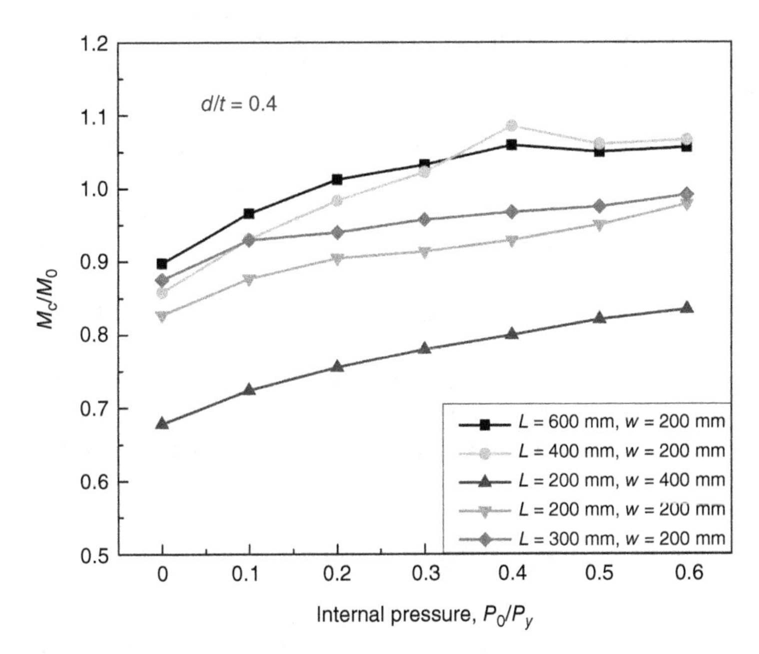

Figure 7.26 Relationship between the critical buckling moment, M_c/M_0, and normalized internal pressure, P/P_y. *Source:* From Shuai et al. [2021] / with permission from Elsevier.

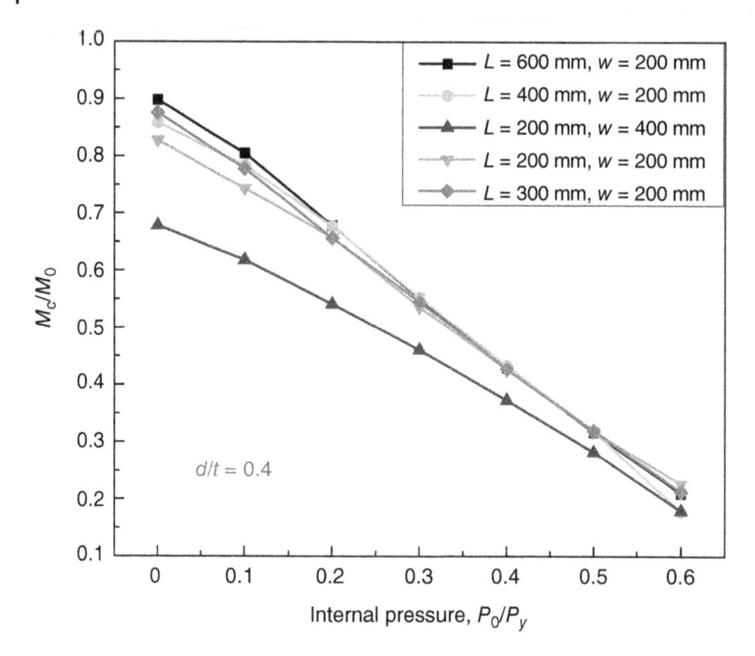

Figure 7.27 Critical buckling moment as a function of internal pressure when both ends of the pipe are not restrained so that an axial stress is not applied. *Source:* From Shuai et al. [2021] / with permission from Elsevier.

It is noted that the modeled pipe segment is restrained at both ends under an axial tensile stress. However, for buried pipelines, there are no caps at both ends. Thus, there is no axial stress caused by the end-caps. To study the critical buckling moment of a buried pipeline at corrosion defect, the axial stress is removed from the FE model, and the results are shown in Figure 7.27. Obviously, the M_c/M_o decreases with increased internal pressure when the pipe end is not under an axial tensile stress. A comparison of Figures 7.26 and 7.27 can determine the influence of axial stress caused by pipe end-caps on the bending resistance of corroded pipelines. This is important especially when the sensitivity of pipeline buckling to internal pressure is discussed.

7.4.5 Implications on Pipeline Integrity Management

The shape of corrosion defects affects the buckling resistance of pipelines. Therefore, the actual geometrical shape of corrosion defects should be accurately defined, providing a sound base to assess the buckling behavior of the corroded pipelines. In pipeline integrity management program, the defect geometry including the defect shape is usually determined by ILI data analysis. The susceptibility of a corroded pipeline to buckling failure under a bending moment is mainly dependent on corrosion depth and width, while the corrosion length is not a dominant factor. When a pipeline is under bending due to geohazards, the corrosion depth should be paid primary attention for buckling analysis. Moreover, the corrosion width, which is often ignored in today's burst assessment, should be included in buckling failure evaluation. Upon determination of the corrosion defect geometry, the deepest and the widest corrosion defect should be assessed to carry out a geological hazard risk assessment and/or stress monitoring. The corrosion defects with a small depth but a great width should not be ignored. Additionally, internal pressure plays a great role in bending resistance of corroded pipelines, especially for the buried pipelines with no end caps. The results provide important technical recommendations for improved integrity management of pipelines, especially for those experiencing significant ground movement.

Nowadays, most standards, such as DNV RP-F101 [Det Norske Veritas, 2004] and DNV RP-F105 [Det Norske Veritas, 2008], developed for evaluation of pipeline buckling have been focused on intact pipes, where the effect of defects is rarely considered. The results confirm that corrosion defects can significantly impact the buckling behavior of pipelines. Therefore, improvements should be made to existing standards for accurate and reliable assessment of pipeline buckling.

7.5 Prediction of Burst Capacity of Corroded Pipelines under a Combined Bending Moment and Axial Compressive Load

7.5.1 Introduction

Pipelines containing corrosion defects are associated with reduced burst capacity. Most of the commonly used standards and codes for prediction of burst pressure on corroded pipelines, as reviewed in Chapters 2 and 3, generally consider internal pressure only. Buried pipelines are usually under complex stress conditions, such as a bending moment or axial loading caused by factors such as soil subsidence, landslides, and frost heaving of soils. Therefore, the codes and models used today may obtain conservative results or overestimations.

A pressurized pipeline may experience buckling or burst failure under bending moment and axial loading, particularly occurring at corrosion defects on the pipe body. The buckling behavior of pipelines containing a corrosion defect, or a dent was studied under either axial compression or bending moment in the previous sections, where the critical buckling load or moment was defined by developed FE models while considering the effects of defect geometry, internal pressure, and pipe dimension. An axial compressive force or a bending moment that is applied on pipelines affects the burst capacity [Lasebikan and Akisanya, 2014; Taylor et al., 2015]. Zhao et al. [2019] determined that an additional axial force applied on a corroded pipeline could decrease allowable operating pressure. A model was proposed for detemination of the burst capacity of corroded pipelines under an axial stress. However, the model was based on the assumption that the cross-section of the pipe was under the same axial stress, and thus was not applicable for non-uniform axial stress caused by bending moment. Zhou et al. [2018b] investigated the effect of a longitudinal strain caused by bending load on the burst capacity of a corroded pipeline by full-scale burst testing, where the longitudinal strain at the defect reduced the burst pressure. A strain-based burst pressure formula was developed for pipelines containing a corrosion defect. However, the model was limited for the corrosion defects with a depth $d/t \leq 0.4$, a length $L/t > 1$ and $L/\sqrt{Dt} \leq 1.3$, and a width $\frac{w}{t} > 1$ and $w/\sqrt{Dt} \leq 1.3$. Thus, it was applicable for small defects only. Liu et al. [2009] and Chauhan and Swankie [2010] proposed interaction diagrams to evaluate the allowable operating pressure of corroded pipelines under either a bending load or an axial force, while it does not apply for pipelines that are subject to simultaneous bending moment and axial force. The widely used standard DNV-RP-F101 [Det Norske Veritas, 2017] for burst prediction of corroded pipelines considers the effect of external loads by modifying the failure pressure model through a compressive longitudinal stress factor. However, the method is mainly applicable for the condition that the corrosion defect is located at the compressive side of the bending. When corrosion is on other circumferential positions of the pipe, it is difficult to calculate the longitudinal stress from the bending moment and axial force, limiting its application scope.

Obviously, most works have focused on determination of the buckling resistance of corroded pipelines under either an axial compression force or bending moment. There have been limited results obtained under a combined axial force and bending moment which usually co-apply on buried pipelines. More important, it is critical to develop a reliable method to address the problem.

7.5.2 Numerical Model and Pipe Failure Criterion

The burst process of a corroded pipeline under the combined action of internal pressure, axial compressive load, and bending moment was modeled by FE analysis. Generally, corrosion defects on a pipeline are most likely irregular. For modeling convenience, the corrosion defect was modeled as a semi-ellipsoid shape for grooved corrosion, a spherical shape for corrosion pit, and a rectangle shape for uniform corrosion, as schematically shown in Figure 7.21. These are also suggested in ASME B31G [American Society of Mechanical Engineering, 2012], DNV RP F-101 [Det Norske Veritas, 2017], and many other standards and publications [Zhou et al., 2018a; Mondal and Dhar, 2019; Shuai et al., 2021]. It is expected that a high stress gradient is associated with the defect, whereas the stress distribution away from the defect is relatively uniform. Therefore, the mesh of the corroded area was refined to obtain accurate modeling, and a relatively sparse grid density meshed at the region away from the corrosion.

The developed FE model is like that shown in Figure 7.10. The right-end cross-section ① of the pipe was fully constrained, and the left end was free so that a combined bending moment and axial force could be applied. The details of the method for modeling are described in [Shuai et al., 2022]. Since the burst pressure of the corroded pipe was targeted, the loading path of the model was the external loads first and then internal pressure. The bending moment M and the axial compressive force F_{comp} were simultaneously applied. Subsequently, a monotonically increasing internal pressure was applied on the interior of the pipe until the pipe bursts. The arc-length method was used to solve the FE modeling. The pipeline failure criterion, i.e., the ultimate tensile stress failure criterion [Zhu, 2021], was used. According to this criterion burst failure occurs when the von Mises equivalent stress throughout the entire ligament thickness at the local corrosion area is up to the ultimate tensile strength of the material. An excessive axial compressive load and a bending moment applied on the pipe may cause local buckling at the corrosion defect. However, the assessment focuses on burst failure dominated by the circumferential stress. Thus, the applied axial force F_{comp} and bending moment M were controlled at low levels. It is assumed that the axial compressive force F_{comp} and the bending moment M are less than $0.5F_c$ and $0.5M_c$, respectively, where F_c is the critical compressive force of buckling for an intact pipe, and M_c is the critical bending moment of buckling for an intact pipe. This can make the longitudinal stress smaller than the circumferential stress. Under actual conditions such as landslides, ground subsidence and pipelines in suspension, the external load applied on the pipelines is usually small. The local buckling due to longitudinal compression was not considered.

7.5.3 Effect of Loading Conditions on Burst Failure of a Corroded Pipe

To study the effect of combined bending moment and axial compressive force on burst failure of corroded pipelines, four loading conditions were considered for FE modeling on an X52 steel pipe with the following conditions: pipe outer diameter 914.4 mm, pipe wall thickness 9.525 mm, and a corrosion defect with a relative depth (d/t) 0.3, length 160 mm and width 40 mm. The four loading conditions included a dimensionless axial compression force $F/F_c = -0.2$ ($F_c = \pi Dt\sigma_y$), a dimensionless axial tensile force $F/F_c = 0.2$, a dimensionless closing bending moment (i.e., a bending moment causing an axial compression at the corrosion defect) $M/M_c = 0.25$ ($M_c = D^2 t\sigma_y$), and a dimensionless opening bending moment (i.e., a bending moment causing an axial tension at the corrosion defect) $M/M_c = 0.25$. Figure 7.28 shows the effects of axial tensile and compressive loads on burst

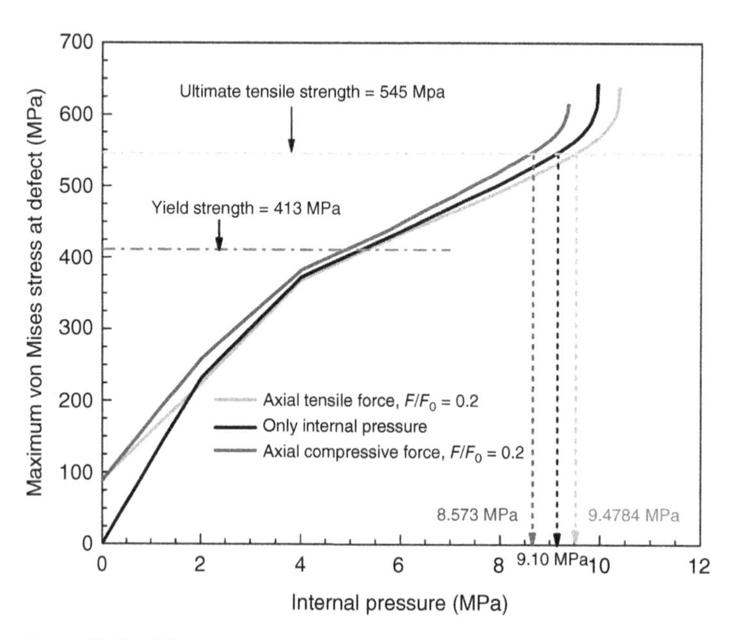

Figure 7.28 Effects of axial tensile and compressive loads on burst failure of the corroded X52 steel pipe, where the yield strength of 413 MPa and ultimate tensile stress of 545 MPa of the steel are labeled by two horizontal dotted lines. *Source:* From Shuai et al. [2022] / with permission from Elsevier.

failure of the corroded X52 steel pipe, where the yield strength of 413 MPa and ultimate tensile stress of 545 MPa of the steel are labeled by two horizontal dotted lines. It is seen that the maximum von Mises stress at the corrosion defect increases with increased internal pressure for both the tensile and compressive loads. The von Mises stresses are always greater under the axial compression than that under axial tension at specific internal pressures. In other words, an axial compressive load makes the local stress reach the yield and tensile strengths at a lower internal pressure. When the stress reaches the ultimate tensile stress (i.e., the burst failure occurs), the axial compressive load results in a reduction of the burst pressure, whereas the axial tension improve the burst resistance of the pipe slightly when compared with the pipe under the internal pressure only.

Figure 7.29 shows the effects of closing and opening bending moments on the maximum von Mises stress at the corrosion defect on X52 steel pipe. The maximum stress at the defect under a closing bending moment is always greater than the stress under an opening bending moment at specific internal pressures. Thus, the closing bending moment makes the corrosion defect reach the yield and ultimate tensile strengths at a lower internal pressure. At

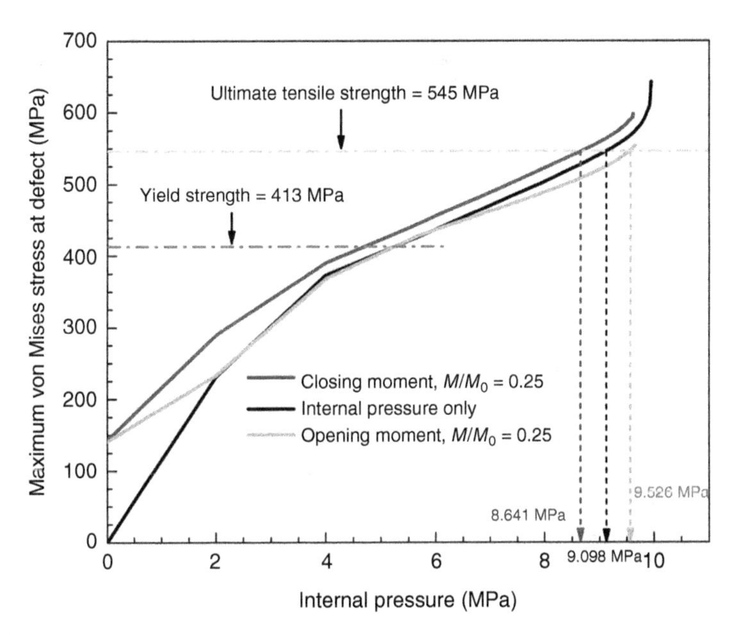

Figure 7.29 Effects of closing and opening bending moments on the maximum von Mises stress at the corrosion defect on the corroded X52 steel pipe. *Source:* From Shuai et al. [2022] / with permission from Elsevier.

the burst moment, the closing bending moment can reduce the burst pressure, whereas the opening bending moment increases the burst pressure when compared with the pipe under internal pressure only.

In summary, the axial compressive force and the closing bending moment that generates a compressive stress at the corrosion defect can reduce the burst capacity of the corroded pipe. The axial tensile load and the opening bending moment that generates a tensile stress at the defect will improve the burst pressure of the pipe slightly.

7.5.4 Parametric Sensitivity Study

7.5.4.1 Bending Moment

For convenience of modeling, the axial force and the bending moment are normalized by $F_c = \pi D t \sigma_y$ and $M_c = D^2 t \sigma_y$, respectively. To investigate the effect of bending moment on burst capacity of the corroded pipe, two types of load combinations, i.e., an axially compressive force and a closing bending moment, and an axial tensile force and a closing bending moment, are modeled. The corrosion defect is located on the compression side under the bending load. Figure 7.30 shows the relationship between the burst pressure and

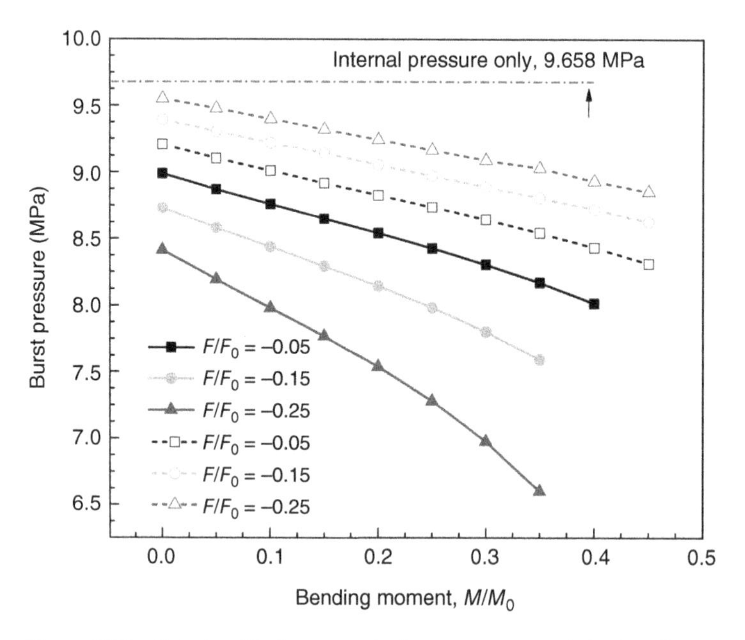

Figure 7.30 Relationship between the burst pressure and the bending moment, M/M_c, under various axial forces F/F_c. *Source:* From Shuai et al. [2022] / with permission from Elsevier.

the bending moment, M/M_c, under various axial forces F/F_c. Generally, the burst pressures of the pipe under various load combinations are smaller than 9.658 MPa, i.e., the burst pressure of the pipe is subject to internal pressure only. Apparently, the applied bending load combinations facilitate burst occurrence on the pipe. Moreover, the burst pressure decreases with increased bending moment M/M_c. Furthermore, at specific bending moments, the burst pressure of the pipe under an axial compression is smaller than the burst pressure under an axial tension, indicating that axial compression will facilitate the burst failure.

7.5.4.2 Axial Compressive Load

As modeled, the axial compressive load poses a greater threat to corroded pipelines for burst to occur than an axial tensile load. Figure 7.31 shows the burst pressures of a corroded X52 steel pipe as a function of the axial compressive force under bending moments of $M/M_c = 0.05$, 0.15, and 0.30, respectively. It is seen that the burst pressure decreases with increased axial compressive force.

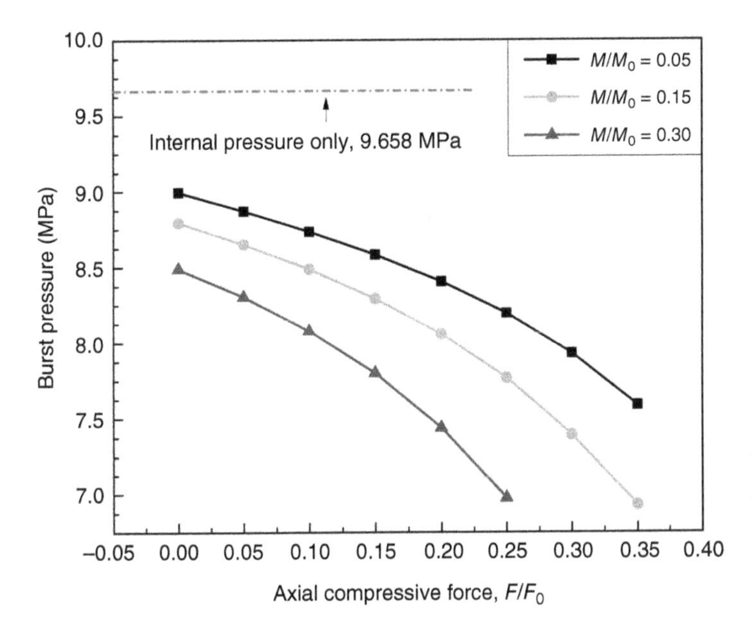

Figure 7.31 Relationship between burst pressure of the corroded X52 steel pipe and the axial compressive force under various bending moments. *Source:* From Shuai et al. [2022] / with permission from Elsevier.

7.5.4.3 Corrosion Depth, Length, and Width

The FE modeling is conducted to investigate the influences of corrosion defect depth, length, and width on burst pressure of the steel pipe under combined axial compressive force and closing bending moment, as shown in Figure 7.32. It is seen that the burst pressure decreases linearly with the increased defect depth. In addition, as the corrosion length increases, the burst pressure decreases more rapidly with the increase of the defect depth.

For the effect of corrosion length on burst pressure under the combined load and bending moment, a critical corrosion length that affects the burst capacity is determined as $\frac{L}{\sqrt{Dt}} = \sqrt{20}$. When $\frac{L}{\sqrt{Dt}} < \sqrt{20}$, the burst pressure decreases with increased defect length. After the corrosion length $\frac{L}{\sqrt{Dt}}$ exceeds $\sqrt{20}$, the burst pressure is almost independent of the defect length. Moreover, the effect of corrosion length on burst pressure depends on the defect depth. For shallow corrosion defects, the effect of corrosion length

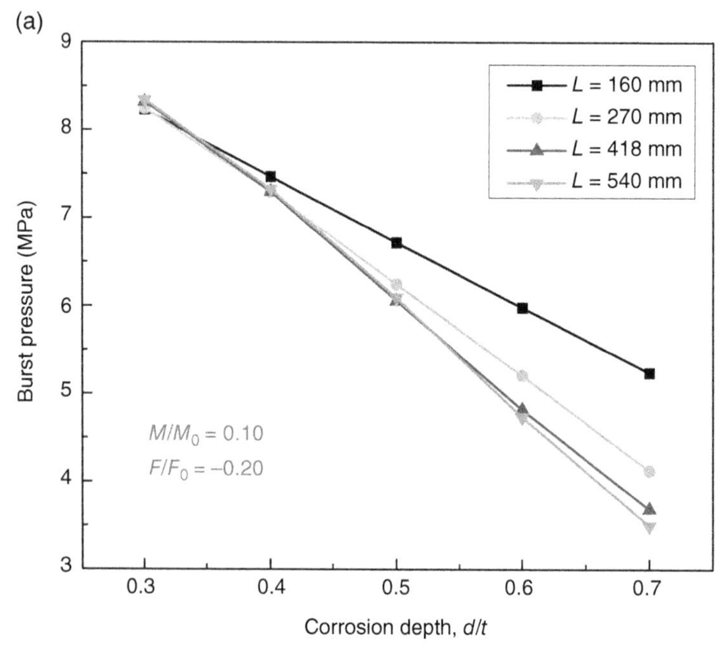

Figure 7.32 Effects of (a) corrosion depth, (b) length and (c) width on burst pressure of the pipe under a combination of axial compressive force and a bending moment. *Source:* From Shuai et al. [2022] / with permission from Elsevier.

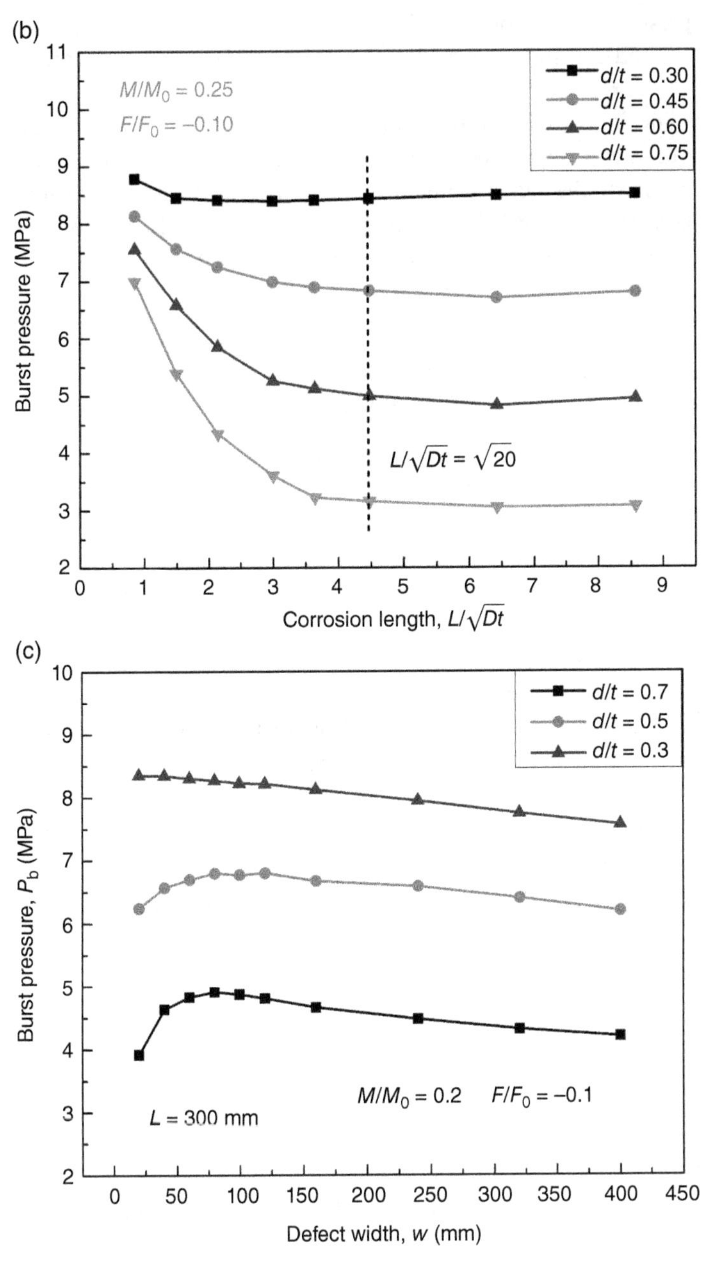

Figure 7.32 (Continued)

on burst pressure is not obvious. Thus, both corrosion depth and length should be considered simultaneously for burst assessment of the corroded pipeline.

The influence of corrosion width on burst pressure depends on corrosion depth. For deep corrosion defects, the burst pressure increases initially, and then decreases with the defect width. For shallow defects, the failure pressure decreases approximately linearly with the defect width. Generally, the effect of corrosion width on the pipe's burst capacity is not remarkable under the combined load condition.

7.5.5 A New Burst Model for Corroded Pipelines Under Combined Axial Compressive Load and Bending Moment

A new model for assessment and determination of burst capacity of pipelines containing a corrosion defect under a combined axial compressive force and bending moment is developed. An upper limit pressure, $P_{\max}$, can be determined as the ultimate burst pressure of an intact pipeline, as what Zhu and Leis [2007] proposed based on the von Mise yield strength criterion:

$$P_{\max} = \frac{4\sigma_u}{\left(\sqrt{3}\right)^{n+1}} \frac{t}{D} \tag{7.5}$$

The strain-hardening exponent, n, can be obtained from standard uniaxial tensile testing or empirical calculation by [Zhu and Leis, 2005, 2007]:

$$n = 0.239\left(\frac{\sigma_u}{\sigma_y} - 1\right)^{0.596} \tag{7.6}$$

The lower limit of failure pressure, $P_{\min}$, can be obtained from a corroded pipeline:

$$P_{\min} = \frac{4\sigma_u}{\left(\sqrt{3}\right)^{n+1}} \frac{t-d}{D} \tag{7.7}$$

It is then obtained:

$$\frac{P_{\min}}{P_{\max}} = \frac{t-d}{t} \tag{7.8}$$

The burst pressure P_b at a corrosion defect on the pipeline is between the lower limit pressure and the upper limit pressure:

$$P_b = P_{\min} + (P_{\max} - P_{\min}) g\left(\frac{d}{t}, \frac{L}{\sqrt{Dt}}, \frac{w}{\pi D}\right) \tag{7.9}$$

where $\left(\frac{d}{t}, \frac{L}{\sqrt{Dt}}, \frac{w}{\pi D}\right)$ is a function representing the geometric factor of the corrosion defect, where d/t, $L/\sqrt{Dt}$, and $w/\pi D$ are defect depth, length and width normalized by the pipe size. Eq. (7.9) can be rewritten to obtain:

$$P_b = P(0)\left\{1 - \frac{d}{t}\left[1 - g\left(\frac{d}{t}, \frac{L}{\sqrt{Dt}}, \frac{w}{\pi D}\right)\right]\right\} \tag{7.10}$$

When the pipeline is under external loads, a load factor function, f, is introduced to modify the failure pressure:

$$P_b = P(0)f\left(\frac{F}{F_c}, \frac{M}{M_c}, \theta\right)\left\{1 - \frac{d}{t}\left[1 - g\left(\frac{d}{t}, \frac{L}{\sqrt{Dt}}, \frac{w}{\pi D}\right)\right]\right\} \tag{7.11}$$

where $f\left(\frac{F}{F_c}, \frac{M}{M_c}, \theta_b\right)$ is the external load factor function, $\frac{F}{F_c}$ and $\frac{M}{M_c}$ are normalized axial compressive force and normalized bending moment, respectively, and θ_b is orientation of the bending load. The values of the functions g and f range from 0 to 1.

According to the FE model, the failure pressures of corroded pipelines under various conditions are determined, while considering the effects of pipe dimension, corrosion defect geometry (depth, length, and width), and external loading conditions (axial compressive force and bending moment). A quantitative model for prediction of burst pressure at a corrosion defect on pipelines under a combined axial compressive force and bending moment is obtained:

$$P_b = \frac{4\sigma_u}{(\sqrt{3})^{n+1}} \cdot \frac{t}{D} \cdot \left(1 - 0.24\frac{F}{F_c}\right)^2 \cdot \left(1 - 0.13\frac{M}{M_c}\right)^{3.61\cos\theta_b - 1}$$
$$\cdot \left\{1 - \frac{d}{t}\left(1 - \left(0.12 \cdot \left(1 - \left(\frac{w}{\pi D}\right)^2\right)^4 + 0.88\exp\left(\frac{-1.16L}{\sqrt{Dt}}\right)\right) \cdot \left(1 - \frac{d}{t}\right)^{-0.08}\right)\right\} \tag{7.12}$$

It is noted that an excessive axial compression force or bending moment may cause local buckling of a pipeline. The axial force and bending moment applied on the pipe in the developed model are in the ranges of $0 \le M < 0.45M_c$ and $0 \le F < 0.45F_c$, avoiding local buckling before the burst occurs. It is thus suggested that $0 \le M < 0.45M_c$ and $0 \le F < 0.45F_c$ are the application limits of the model. In addition, the corrosion depth is recommended at $0.10 \le d/t \le 0.75$.

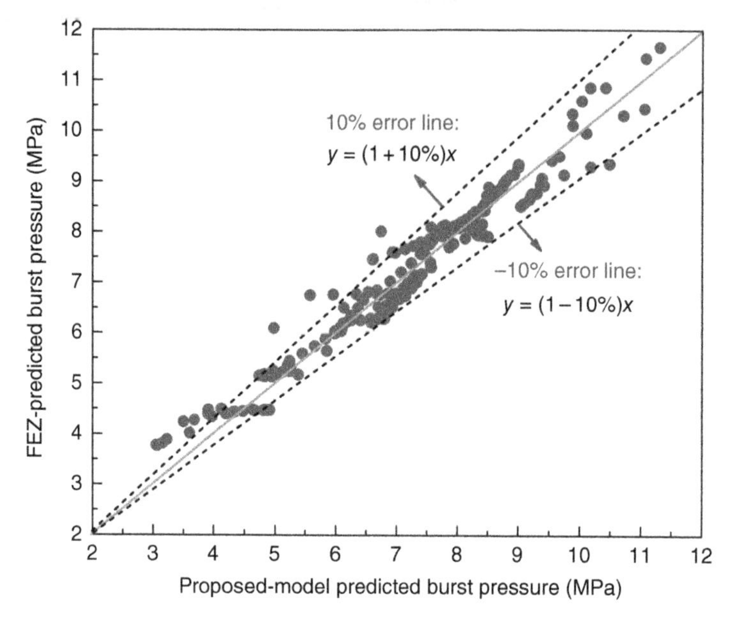

Figure 7.33 Comparison of burst pressures obtained from the developed model and FE predictions. *Source:* From Shuai et al. [2022] / with permission from Elsevier.

Figure 7.33 shows the comparison of burst pressures of various case analysis results obtained from the developed burst model in Eq. (7.12) and FE predictions. It is seen that the two groups of failure pressure are linearly related around $y = x$, with an average relative error of 4.52% only. Thus, the new model is in good agreement with the FE analysis method.

The burst pressures of corroded pipelines under the combined axial compressive load and bending moment predicted by the developed model are also compared with 22 burst tests, as well as other industry codes, and the results are shown in Figure 7.34, where the two red dashed line-s represent the critical dividing lines with an error of $\pm 15\%$. It is seen that, most of the data points in Figure 7.34(1)–(4) are located below the blue line $y = x$, indicating that the burst failure pressures predicted by ASME B31G, Modified B31G and CSA Z662 are smaller than the test data, making the predictions conservative. The prediction of Shell 92 model is also conservative, with only six points above the blue line, as seen in Figure 7.34(4). Generally, the DNV, PCORRC, CUP and the developed model show a better prediction accuracy and reliability. The error for prediction of the burst pressure in most cases is controlled within 15%.

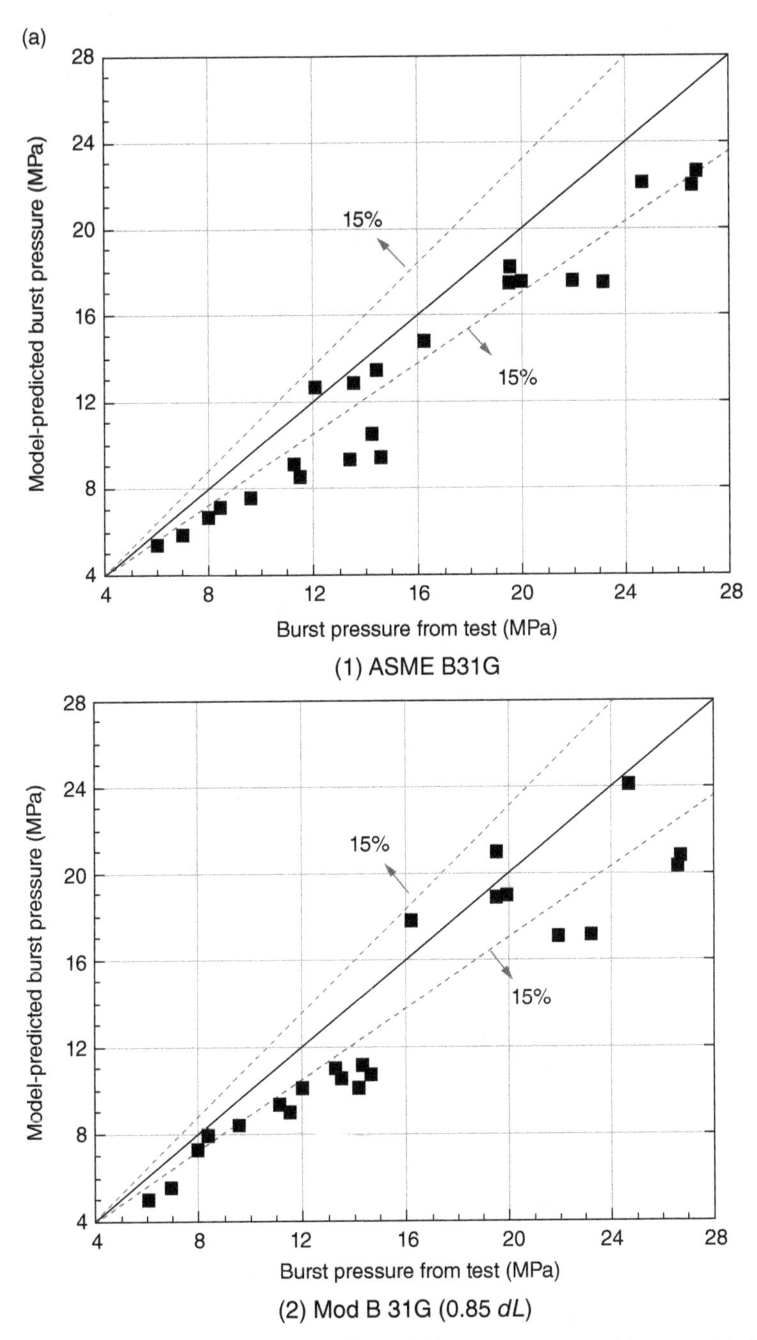

Figure 7.34 Performance comparison of failure pressure prediction models for corroded pipelines (a) ASME B31G, Mod B31G, CSA Z662 and Shell 92 models, (b) DNV RP-F101, PCORRC, CUP and the proposed model in this work. *Source:* From Shuai et al. [2022] / with permission from Elsevier.

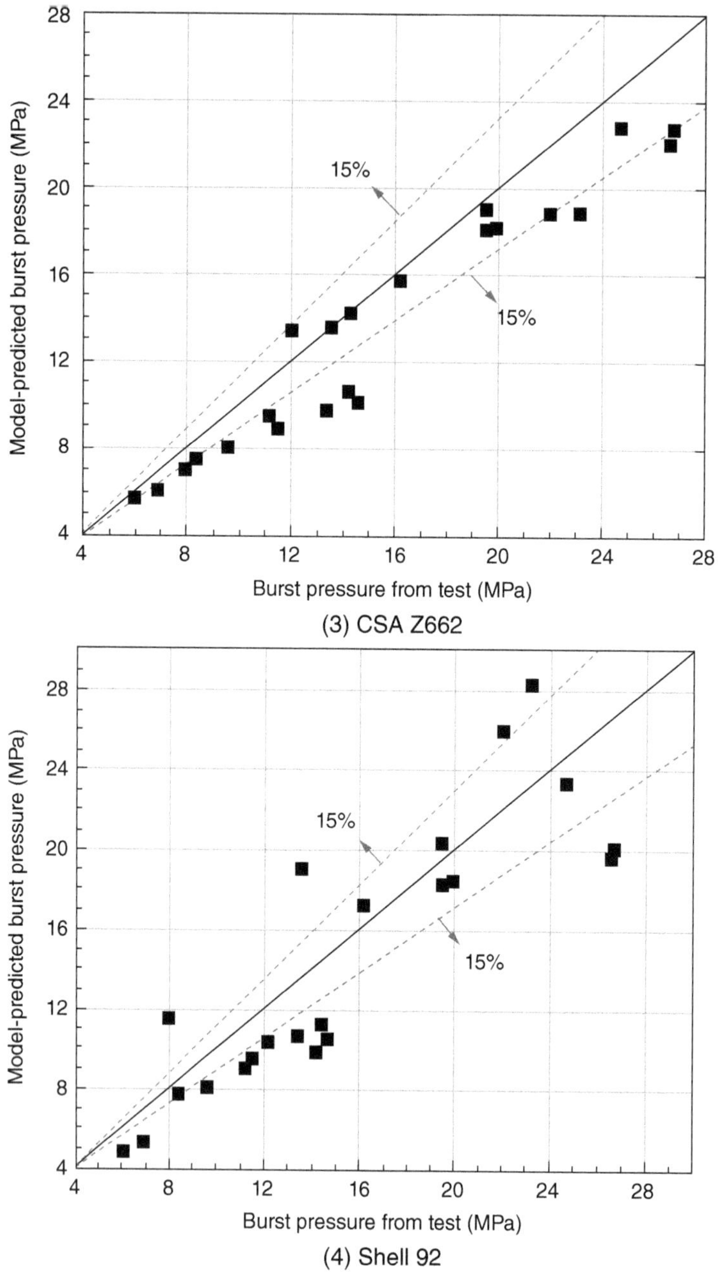

Figure 7.34 (Continued)

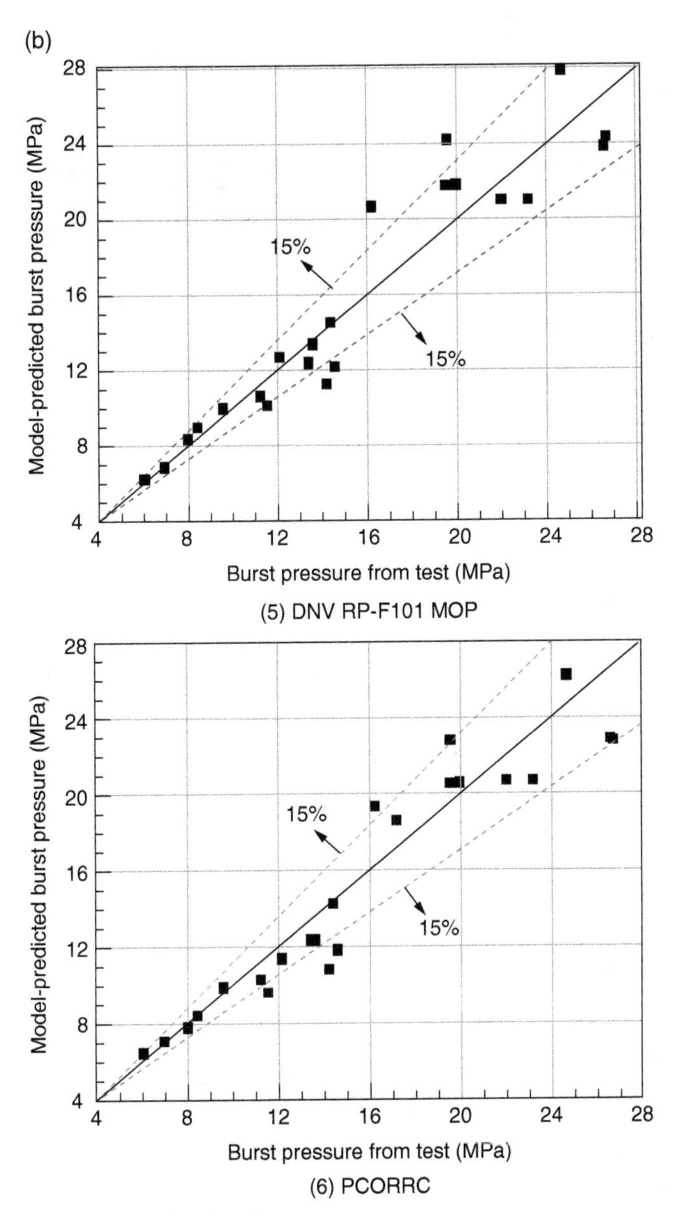

Figure 7.34 (Continued)

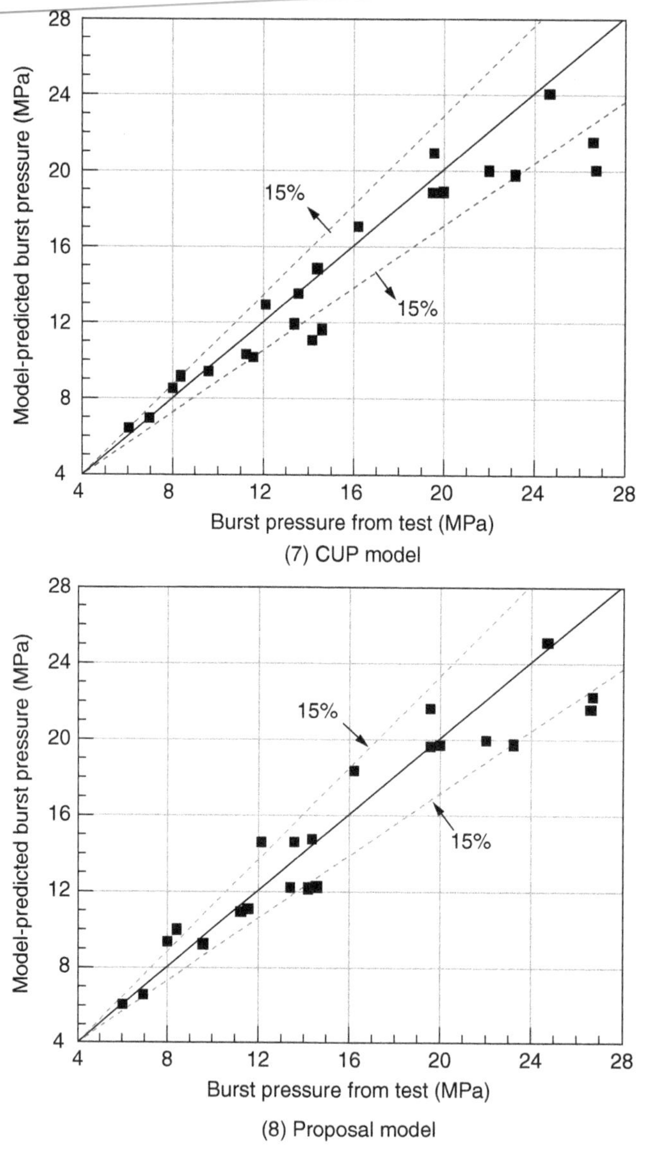

Figure 7.34 (Continued)

References

American Society of Mechanical Engineering (2012) *Manual for Determining the Remaining Strength of Corroded Pipelines*, ASME B31G, Washington DC, USA.

Baek, J.H., Kim, Y.P., Kim, W.S., Koo, J.M., Seok, C.S. (2012) Load bearing capacity of API X65 pipe with dent defect under internal pressure and in-plane bending, *Mater. Sci. Eng.*, 540, 70–82.

Błachut, J., Iflefel, I.B. (2007) Collapse of pipes with plain or gouged dents by bending moment, *Int. J. Press. Vessel. Pip.*, 84, 560–571.

Chauhan, V., Swankie, T. (2010) *Guidance for Assessing the Remaining Strength of Corroded Pipelines*, report no. 9492, PHMSA, Department of Transportation, Washington DC, USA.

Chee, J., Walker, A., White, D. (2018) Controlling lateral buckling of subsea pipeline with sinusoidal shape pre-deformation, *Ocean Eng.*, 151, 170–190.

Chen, Y.F., Zhang, H., Zhang, J., Liu, X.B., Li, X., Zhou, J. (2014) Residual bending capacity for pipelines with corrosion defects, *J. Loss. Prevent. Proc.*, 32, 70–77.

Det Norske Veritas (2004) *Recommended Practice of Corroded Pipelines*, DNV RP-F101, Norway.

Det Norske Veritas (2008) *Recommended Practice of Spanning Pipelines*, DNV RP-F105, Norway.

Det Norske Veritas (2017) *Recommended Practice of Corroded Pipelines*, DNV RP-F101, Norway.

Dewanbabee, H. (2009) Behavior of Corroded X 46 Steel Pipe under Internal Pressure and Axial Load, University of Windsor, Windsor, ON, Canada.

Gao, J., Li, X., Zhou, J. (2016) Bending failure of the corroded pipeline subjected to initial internal pressure and axial compressive force, *Ocean Eng.*, 34, 74–82.

Ifayefunmi, O. (2016) Buckling behavior of axially compressed cylindrical shells: comparison of theoretical and experimental data, *Thin Wall. Struct.*, 98, 558–564.

Iflefel, I.B., Moffat, D.G., Mistry, J. (2005) The interaction of pressure and bending on a dented pipe, *Int. J. Press. Vessel Pip.*, 82, 761–769.

Jin, L., Li, H.J. (2010) Nonlinear response analysis of buried pipeline crossing thrust fault, *Disaster Prev. Mitigation Eng.*, 30, 130–134.

Kanninen, M.F., Broek, D., Hahn, G.T. (1978) Toward an elastic fracture mechanics predictive capability for reactor piping, *Nuclear Eng. Des.*, 48, 117–134.

Kim, J.W., Park, C.Y. (2003) Effect of length of thinning area on the failure behavior of carbon steel pipe containing a defect of wall thinning, *Nuclear Eng. Des.*, 220, 274–284.

Kim, Y.J., Oh, C.K., Park, C.Y., Hasegawa, K. (2006) Net-section limit load approach for failure strength estimates of pipes with local wall thinning, *Int. J. Press. Vessel Pip.*, 83, 546–555.

Lasebikan, B.A., Akisanya, A.R. (2014) Burst pressure of super duplex stainless steel pipes subject to combined axial tension, internal pressure and elevated temperature, *Int. J. Press. Vessels Pip.*, 119: 62–68.

Liang, G.C., Cheng, Y.F. (2013) Strain aging of X100 steel in service and the enhanced susceptibility of pipelines to stress corrosion cracking, *J. Mater. Eng. Perf.*, 22: 3778–3782.

Limam, A., Lee, L.H., Corona, E. (2010) Inelastic wrinkling and collapse of tubes under combined bending and internal pressure, *Int. J. Mech. Sci.*, 52, 637–647.

Limam, A., Lee, L.H., Kyriakides, S. (2012) On the collapse of dented tubes under combined bending and internal pressure, *Int. J. Mech. Sci.*, 55, 1–12.

Liu, J., Chauhan, V., Ng, P., Wheat, S., Hughes, C. (2009) *Remaining Strength of Corroded Pipe under Secondary (Biaxial) Loading*, report no. R9068, PHMSA, Department of Transportation, Washington, DC, USA.

Liu, P.F., Zheng, J.Y., Zheng, B.J., Shi, P. (2010) Failure analysis of natural gas buried X65 steel pipeline under deflection load using finite element method, *Mater. Des.*, 31, 1384–1391.

Liu, X., Zhang, H., Wu, K., Xia, M.Y., Chen, Y.F., Li, M. (2017) Buckling failure mode analysis of buried X80 steel gas pipeline under reverse fault displacement, *Eng. Fail. Anal.*, 77, 50–64.

Miyazaki, K., Kanno, S., Ishiwata, M. (1999) Fracture behavior of carbon steel pipe with local wall thinning subjected to bending load, *Nuclear Eng. Des.*, 191, 195–204.

Miyazaki, K., Kanno, S., Ishiwata, M. (2002) Fracture and general yield for carbon steel pipes with local wall thinning, *Nuclear Eng. Des.*, 211, 61–68.

Mondal, B.C., Dhar, A.S. (2019) Burst pressure of corroded pipelines considering combined axial forces and bending moments, *Eng. Struct.*, 186, 43–51.

Nazemi, N. (2009) Behavior of X60 Line Pipe under Combined Axial and Transverse Loads with Internal Pressure, University of Windsor, Windsor, ON, Canada.

Oh, C.K., Kim, Y.J., Park, C.Y. (2009) Effects of local wall thinning on net-section limit loads for pipes under combined pressure and bending, *Nuclear Eng. Des.*, 239, 261–273.

Oldfield, T., Turner, T., Young, A. (2016) Influence of cold formed bending on strain-based design buckling limits, *Proc. 11th Int. Pipeline Conf.*, Calgary, AB, Canada.

Qin, G.J., Cheng, Y.F. (2021) A review on defect assessment of pipelines: principles, numerical solutions, and applications, *Int. J. Press. Vessel Pip.*, 191, 104329.

Ramberg, W., Osgood, W.R. (1943) Description of Stress-Strain Curves by Three Parameters, Technical Report Archive & Image Library, USA.

Schaminee, P.E.L., Zorn, N.F., Schotman, G.J.M. (1990) Soil response for pipeline upheaval buckling analyses: full-scale laboratory tests and modelling, *Offshore Technology Conference*, Paper Number: OTC-6486-MS, Houston, TX, USA.

Shim, D.J., Choi, J.B., Kim, Y.J. (2004) Failure strength assessment of pipes with local wall thinning under combined loading based on finite element analyses, *Int. J. Press. Vessel Pip.*, 126, 179–183.

Shuai, Y., Zhou, D.C., Wang, X.H., Yin, H.G., Zhu, S.D., Li, J., Cheng, Y.F. (2020a) Local buckling failure analysis of high strength pipelines containing a plain dent under bending moment, *J. Nat. Gas Sci. Eng.*, 77, 103266.

Shuai, Y., Wang, X.H., Cheng, Y.F. (2020b) Modeling of local buckling of corroded X80 gas pipeline under axial compression loading, *J. Nat. Gas Sci. Eng.*, 81, 103472.

Shuai, Y., Wang, X.H., Shuai, J. (2020c) Mechanical behavior investigation on the formation of the plain dent of an API 5L L245 pipeline subjected to concentrated lateral load, *Eng. Fail. Anal.*, 108, 104189.

Shuai, Y., Wang, X.H., Cheng, Y.F. (2021) Buckling resistance of an X80 steel pipeline at corrosion defect under bending moment, *J. Nat. Gas Sci. Eng.*, 93, 104016.

Shuai, Y., Zhang, X., Feng, C., Han, J.Y., Cheng, Y.F. (2022) A novel model for prediction of burst capacity of corroded pipelines subjected to combined loads of bending moment and axial compression, *Int. J. Press. Vessel. Pip.*, 196, 104621.

Suzuki, N., Kondo, J., Shimamura, J. (2007) Strain capacity of X80 high-strain line pipes, *Proc. 26th. Int. Conf. Offshore Mech. Arctic. Eng.*, San Diego, CA, USA.

Taylor, N., Clubb, G., Atkins, I.M. (2015) The effect of bending and axial compression on pipeline burst capacity, *Proc. Offshore Europe Conf.*, SPE-175464-MS, Aberdeen, UK.

Wang, H.K., Yang, Y., Xu, J.X., Han, M.X. (2019) Effect of pitting defects on the buckling strength of thick-wall cylinder under axial compression, *Construct. Build. Mater.*, 224, 226–241.

Yablonskikh, I.L., Dawson, S.J., Venkatanarayanan, R. (2007) Assessment and Analysis of Pipeline Buckles, Pigging Products and Services Association, Ipswich, UK.

Yang, L., Chen, Z., Cao, G., Yu, C., Guo, W. (2013) An analytical formula for elastic-plastic instability of large oil storage tanks, *Int. J. Press. Vessel. Pip.*, 101, 72–80.

Zhang, S.Z., Cheng, Y.F., Feng, X.D., Wei, L.J., Wang, R.J., Duo, Y.Q., Han, Y.X. (2019) Performance characteristics and technical challenges of X80 pipeline steel, *J. Oil. Gas. Storage Trans.*, 38, 481–495.

Zhao, P.C., Shuai, J., Tang, Y., Xu, K. (2019) Impact of axial stress on ultimate internal pressure of corroded pipelines, China, *Safety Sci. J.*, 29, 70–75.

Zhou, H.G., Wang, Y.Y., Stephens, M. (2018a) Tensile and compressive strain capacity in the presence of corrosion anomalies, *Proc. 12th Int. Pipeline Conf.*, Calgary, AB, Canada.

Zhou, H.G., Wang, Y.Y., Stephens, M. (2018b) Burst pressure of pipelines with corrosion anomalies under high longitudinal strains, *Proc. 12th Int. Pipeline Conf.*, Calgary, AB, Canada.

Zhu, X.K. (2021) A comparative study of burst failure models for assessing remaining strength of corroded pipelines, *J. Pipeline Sci. Eng.*, 1, 36–50.

Zhu, X.K., Leis, B.N. (2005) Influence of yield-to-tensile strength ratio on failure assessment of corroded pipelines, *J. Press. Vessel Technol.*, 127, 436–442.

Zhu, X.K., Leis, B.N. (2007) Theoretical and numerical predictions of burst pressure of pipelines, *J. Press. Vessel Technol.*, 129, 644–652.

Index

Note: Page numbers in *italics* refer to figures and those in **bold** refer to tables.

Defect Assessment for Integrity Management of Pipelines, First Edition. Y. Frank Cheng.

Printed in the USA
CPSIA information can be obtained
at www.ICGtesting.com
JSHW012056200224
57544JS00001B/25